Studies in History and Philosophy of Science

Volume 67

Studies in History and Philosophy of Science is a peer-reviewed book series, dedicated to the history of science and historically informed philosophy of science. The series publishes original scholarship in various related areas, including new directions in epistemology and the history of knowledge within global and colonial contexts. It includes monographs, edited collections, and translations of primary sources in the English language. These cover a broad temporal spectrum, from antiquity to modernity, and all regions of the world.

Michal Hubálek

Dissolving Naturalism and Historicism Into Each Other

Springer

Michal Hubálek
Philosophical Faculty
University of Hradec Králové
Hradec Králové, Czech Republic

ISSN 1871-7381 ISSN 2215-1958 (electronic)
Studies in History and Philosophy of Science
ISBN 978-3-032-07534-5 ISBN 978-3-032-07535-2 (eBook)
https://doi.org/10.1007/978-3-032-07535-2

This work was supported by Philosophical Faculty, University of Hradec Králové.

This Springer imprint is published by the registered company Springer Nature Switzerland AG
The registered company address is: Gewerbestrasse 11, 6330 Cham, Switzerland

The original version of the book has been revised. A correction to this book can be found at https://doi.org/10.1007/978-3-032-07535-2_8

To Jindřich Hubálek, my past
To Lenka, Iveta, and Jana Hubálková,
my present
*To Helena and Jindřich Hubálek*ová,*
my future

Preface

This book is a reworked version of my Ph.D. dissertation *Naturalism Without Adjectives: Science, History, and Philosophy*, which I defended in the fall of 2024. Back then, I was also considering making "Against First Philosophy" the first part of the title, mainly to avoid discouraging readers who do not consider themselves naturalists or who even think that naturalism is a flawed (meta-)philosophical position, but my drive to rehabilitate the term eventually proved stronger. The rehabilitation of natural*ism* is the main starting point of both my dissertation and this book.

This book, however, goes further in the attempt to make explicit that, properly understood, naturalism stands in a relationship of mutual containment with isms often deemed in conflict with it, especially then with historicism. Still, I sympathize with those who regard ismism and big labels for philosophical positions only as a necessary evil, for the actual pragmatic and intellectual significance lies in the specific assumptions, arguments, and exemplifications. I thus maintain that this book is simply for anyone interested in contemporary philosophy and its relationship with science, history, and broader culture.

Hradec Králové, Czech Republic — Michal Hubálek

You must say something new and yet nothing but what is old.
You must indeed say only what is old – but *all the same* something new!
Different "interpretations" must correspond to different applications.
The poet too must always be asking himself: "is what I am writing really true then?"
which does not necessarily mean: "is this how it happens in reality?"
It's true you must assemble old material. But for a *building*.
—Ludwig Wittgenstein

To Historian
You who celebrate bygones,
Who have explored the outward, the surfaces of the races,
  the life that has exhibited itself,
Who have treated of man as the creature of politics,
  aggregates, rulers and priests,
I, habitan of the Alleghanies, treating of him as he is in
  himself in his own rights,
Pressing the pulse of the life that has seldom exhibited itself,
  (the great pride of man in himself,)
Chanter of Personality, outlining what is yet to be,
I project the history of the future.
—Walt Whitman

Acknowledgments

The acknowledgments to the many people who made this work possible have already been given, so I wish only to once again thank my wife, Lenka Hubálková—without her unfailing patience and understanding, this book would never have come into existence.

Some text previously appeared in the following published works:

Hubálek, Michal. 2021. A brief (hi)story of just-so stories in evolutionary science. *Philosophy of the Social Sciences* 51 (5):447–468.

Hubálek, Michal. 2025. On two underappreciated motifs of Quine's naturalism; Or, Quine on reality and naturalistic philosophy. *Organon F* 32 (2):224–259.

Hubálek, Michal. forthcoming. Understanding naturalism with Quine from within science and history. *Journal for the History of Analytic Philosophy*.

My work on this book was made possible thanks to the Philosophical Faculty at the University of Hradec Králové and the Institute of Philosophy at the Czech Academy of Sciences.

Competing Interests The author has no competing interests to declare that are relevant to the content of this manuscript.

Acknowledgments

Contents

Chapter 1
In Medias Res: Philosophy in Natural*ism*

Abstract In this opening chapter, I primarily set the stage by outlining the overall aims and central arguments of the book. Although naturalism remains one of the most popular isms among philosophers, its contemporary interpretations are far from uniform. I deliberately take as the minimal delamination of naturalism in philosophy Quine's dictum that philosophy is continuous with science. The book as a whole may be then read as a development and defense of this dictum. Secondarily, this chapter also sketches the prospects and purposes of a philosophy understood in naturalistic terms. Following the insights of Wilfrid Sellars as well as Eric Schliesser, I propose that properly understood naturalistic philosophy should have synoptic and synthetizing aims. Finally, I briefly elaborate on how this particular understanding of philosophy's aims brings us close to pragmatism à la Richard Rorty and Robert Brandom—an approach that also holds the potential to bridge analytic and continental traditions of doing philosophy.

Keywords Naturalism · Meta-philosophy · Pragmatism · Quine · Sellars

> The main fault of ismism is that it generates straw men whose isms could never have been embraced by flesh and blood.
> —W. V. O. Quine[1]

> Cracker-barrel pragmatists like me always ask, as William James did, "what difference to practice is that funny little difference in theory supposed to make?"
> —Richard Rorty[2]

What does natural*ism* typically mean for doing philosophy? Within the last four decades, a significant number of thinkers have come to appreciate that while

[1] This statement is taken from a note Quine wrote to Hilary Putnam. Unfortunately, the date is not known. The transcription of the statement is provided by Sander Verhaegh (2018, 155). The statement is, of course, double-edged; as Verhaegh himself convincingly argues, Quine needed to exploit natural*ism* to unify and communicate his views (see ibid., 151ff).

[2] (Rorty 2021, 190)

M. Hubálek, *Dissolving Naturalism and Historicism Into Each Other*, Studies in History and Philosophy of Science 67,
https://doi.org/10.1007/978-3-032-07535-2_1

naturalism is one of the most cherished *meta-philosophical* positions, it is also one of the vaguest isms.[3] Such a paradox is resolved by several strategies in literature; from (re)defining naturalism (see, e.g., Roth 2020; Price 2011; De Caro and Macarthur 2004); to denying its usefulness (see, e.g., van Fraassen 1996); to classifying its already worked-out forms (see, e.g., Bryant 2020; Rosenberg 1996). David Papineau deals with the delineation of naturalism in the "Stanford Encyclopedia of Philosophy" and bluntly opens the entry as follows:

> This entry will not aim to pin down any more informative definition of "naturalism". It would be fruitless to try to adjudicate some official way of understanding the term. Different contemporary philosophers interpret "naturalism" differently. This disagreement about usage is no accident. For better or worse, "naturalism" is widely viewed as a positive term in philosophical circles – only a minority of philosophers nowadays are happy to announce themselves as "non-naturalists". This inevitably leads to a divergence in understanding the requirements of "naturalism". *Those philosophers with relatively weak naturalist commitments are inclined to understand "naturalism" in an unrestrictive way, in order not to disqualify themselves as "naturalists", while those who uphold stronger naturalist doctrines are happy to set the bar for "naturalism" higher.* (Papineau 2021, emphasis mine)

However, most of the past[4] and present naturalists would agree, at least by my lights, that naturalism qua metaphilosophical position concerns the relationship of philosophy to (natural) *science*, and subsequently to the whole of culture. I contend, moreover, that the majority of past and present naturalists would agree, some of course

---

[3] Moreover, as Martin Kusch always reminds us: "All big isms are spectrum concepts" (see his attempt to portray the spectrum of relativ*ism* in Kusch 2019). Also, I feel obligated, on the one hand, to stress that naturalism might be cherished predominantly within the analytic philosophy, i.e., Western and/or Anglo-American tradition of philosophy. But, on the other hand, I feel that this claim is not very informative in our contemporary globalized world. The analytic tradition of philosophy is now a quite common way of doing philosophy within many geographically non-western countries of what is known as Continental Europe. Thus, I would agree with those who, from a recent perspective, see the difference between analytic and continental tradition in philosophy "merely" as a difference in writing styles (cf., e.g., Rorty 1982, 200ff; and Norris 1989). Be that as it may, I do not offer here any account of analytic philosophy, and when I use the term "analytic philosophy" in this book, I have in mind only this minimalistic distinction. I am content to simply accept that naturalism might be more popular within the analytic tradition since, historically, this tradition has always been close to something one might label as *scientific* writing style. On that premise, naturalism can be seen as a position emerging from this particular style, or so I assume in this chapter, and in the rest of the book, too.

[4] Canonically, e.g., Ernest Nagel, Roy Wood Sellars, John Dewey. For further historically informative elaboration see Randall Jr. (1944). As I read Randall, he argues for a unitary understanding of naturalism primarily *from an epistemological perspective*, which is also my ultimate aim here in this book. He writes: "Whatever the label, however, the major fact stands out: the "new" or "contemporary" naturalism […] stands in fundamental opposition not only to all forms of supernaturalism, but also to all types of the reductionist thinking which up to this generation often arrogated to itself the adjective "naturalistic," and still is suggested by it to the popular mind. Second only to the unanimity with which ["new" or "contemporary" naturalists] reject supernaturalism and acclaim scientific procedures is their agreement that the richness and variety of natural phenomena and human experience cannot be explained away and "reduced" to something else. *The world is not really "nothing but" something other than it appears to be: it is what it is, in all its manifold variety, with all its distinctive kinds of activity*," (Randall Jr. 1944, 361, emphasis mine).

after a fashion, with W. V. O. Quine's now (in)famously known dictum that *philosophy is continuous with science*. Mario De Caro and David Macarthur, as I interpret them, even prophesied that the fate of analytic philosophy depends on a particular reading of this dictum (see De Caro and Macarthur 2004, 6–9). Most of the time, the dictum is isolated from Quine's essay "Natural Kinds:"

> *I see philosophy not as an a priori propaedeutic or groundwork for science, but as continuous with science.* I see philosophy and science as in the same boat – a boat which, to revert to Neurath's figure as I so often do, we can rebuild only at sea while staying afloat in it. There is no external vantage point, no first philosophy. *All scientific findings, all scientific conjectures that are at present plausible, are therefore in my view as welcome for use in philosophy as elsewhere.* (Quine 1969a, 126–127, emphasis mine)

Compare this passage with what Quine much later writes in "Naturalism; Or, Living Within One's Means:"

> *The naturalization of epistemology […] is both a limitation and liberation.* The old quest for a foundation for natural science, firmer than science itself, is abandoned: that much is the limitation. Liberation is free access to the resources of natural science, without fear of circularity. The naturalistic epistemologist settles for what he can learn about the strategy, logic, and mechanics by which our elaborate theory of the physical world is in fact projected, or might be, or should be, from just that amorphous neural intake. *Is this sort of thing still philosophy? Naturalism brings a salutary blurring of such boundaries. Naturalistic philosophy is continuous with natural science. It undertakes to clarify, organize, and simplify the broadest and most basic concepts, and to analyze scientific method and evidence within the framework of science itself. The boundary between naturalistic philosophy and the rest of science is just a vague matter of degree.* (Quine 1995, 256–257, emphasis mine).

Quine sees naturalism fundamentally as an epistemological/methodological position, emphasizing, inter alia, that philosophy does not have a special method, nor special grounding status towards science. Quine in effect thinks that naturalism does not need an ahistorical, extra-scientific demarcation criterion of science itself. As I further demonstrate in more detail in Chap. 2, this amounts to maintaining that philosophy and science are both in the same epistemic/evolutionary/historical enterprise of producing and justifying beliefs/knowledge. In these respects, naturalism is conceptually intertwined with science, or better said, with the *currently* recognized scientific practices and paradigms. Paul A. Roth, when defending the Quinean understanding of naturalism in a similar context, aptly notes: "But just how vague is the notion of naturalism? No more vague, I suggest, than our ability to catalog the methods of science," (Roth 2020, 121; see also Roth 1999).

Some fear, and others welcome, that such an epistemological position leads to a disdain for philosophy; to saying that philosophy as such is *useless*. That philosophy is not needed *anymore*. I do not share this outlook. Even if one accepts that philosophy and science possess the same catalog of methods (at a given time), and even if one accepts that philosophy and science very often (or always) inquire, in principle, into the same subject matters, it is still possible to see philosophy as a fruitful discipline interested in different *aims* than science, or alternatively, as a discipline asking different kinds of questions about the same subject matters.[5]

Nor do I share the outlook that sees this methodological position as a slippery slope to some kind of commonly identified *scientism*. There is more science than natural science, period. And I think that other traditional disciplines, or other language games if you will, for example fiction writing, can have a cognitive component, too. Literature and the arts "bring enlightenment" and "teach us," write in the same vein Quine together with Joseph Ullian (1978, 4). Lastly, I do not want to plead for some "anything goes" form of relativ*ism*, or in fact skeptic*ism*, which insist that without the demarcation criterion of science, one cannot satisfactorily arbitrate between rational/irrational belief, or between better/worse theories and explanations.

So far, my procedure commits me to elaborate on at least three issues that I consider to be complementary: (a) how to construe the aims of naturalistic philosophy, i.e., philosophy continuous with science; (b) how to overcome the worry or challenge that, without a firm, ahistorical demarcation criterion of science, nothing of a philosophical substance can be said against pseudo-scientific and anti-scientific views/theories; so alternatively (b*) how to contrast science, or simply "fact," with fiction and/or fabulation, absent a firm, ahistorical demarcation criterion of science; (c) how to understand the claim that the arts (e.g., fiction writing) have a cognitive component.

Throughout the book, I repeatedly, whether directly or indirectly, come back to address these three issues. In this opening chapter, however, I want to say right away a little more about the issue (a), since that will also serve me as a shortcut to expose my overall goals and points of departure in this book. However, in no way do I want to give the impression that philosophy would be better off having only one particular aim or identifying somehow its historically invariant set of aims. I am a pluralist. The fact that, after the naturalistic turn, some philosophers went experimental and some focused instead on conceptual engineering is, as far as I am concerned, one of the many positive side-effects of liberalizing epistemology from apriorism.

Pluralism means here the recognition that philosophy is better off striving to have multiple aims and employing and experimenting with multiple styles and methods to achieve these aims. And the reason for this is pragmatic. Unless the practices of philosophy and other relevant sciences themselves reveal the need for a unity of any kind, there are no reasons to enforce it, I believe. As long as one keeps in mind that every aim marks a different set of values (priorities, commitments, etc.), and that every style and method has its explicatory and explanatory pros and cons (of course, some of these pros and cons might be at a given time only tacit) no one seems to be worse off. By the same token, such pluralism does not prevent us

[5] Another Quinean naturalist who hints at a similar interpretation is, in my view, Jaroslav Peregrin. Peregrin wholeheartedly embraces Quine's continuity dictum, and then sensibly points out: "Note that this says that both science and philosophy are parts of one and the same endeavor of revealing the nature of our (natural) world, yet not very much about a division of labor among them," (Peregrin 2023, 54).

from favoring one aim/style/method over another; it just reminds us, to give one example, that comparing methods without specifying our explanatory aims is futile. Naturally, different aims may conflict, requiring careful, case-by-case adjudication. My claim here is only that making our aims explicit is half the battle.

1.1 Prospects and Purposes of a Naturalistic Philosophy

I myself see the value of contemporary, and of any genuinely useful, philosophy in what one might call *reflective mindedness*. As also Wilfrid Sellars, one of the well-known card-carrying naturalists, succinctly put it: "It is [...] the "eye on the whole" which distinguishes the philosophical enterprise," (Sellars 2007, 371). Or more forcefully and distinctively put:

> *The aim of philosophy, abstractly formulated, is to understand how things in the broadest possible sense of the term hang together in the broadest possible sense of the term.* Under "things in the broadest possible sense" I include such radically different items as not only "cabbages and kings", but numbers and duties, possibilities and finger snaps, aesthetic experience and death. To achieve success in philosophy would be, to use a contemporary turn of phrase, to "know one's way around" with respect to all these things, not in that unreflective way in which the centipede of the story knew its way around before it faced the question, "how do I walk?," but in that reflective way which means that no intellectual holds are barred. (Sellars 2007, 369, emphasis mine)

What is worth paying attention to is that Sellars not only points towards the reflective aim of philosophy, but also towards the *pragmatic* dimension of this reflectivity. Pragmatic at least in the sense that the "eye on the whole" does not seek just to glimpse a generalized version of our world and its state of affairs; it instead seeks, within the various horizons of our world, to make sense of and hang together the given state of affairs *in a preferred way*.

Put simply, the interest and investment in generalization is not an end in itself, it is the background for further reasoning, for achieving further (explanatory and other) purposes. Willem A. deVries reads Sellars on this point very similarly: "Thus, philosophy is a reflectively conducted higher-order inquiry that is continuous with but distinguishable from any of the special disciplines, and the understanding it aims at must have *practical force*, guiding our activities, both theoretical and practical," (deVries 2021; cf., deVries 2005, 7–15).

True, every specialist (e.g., historian) can be reflective in the sense sketched above; he/she can be obviously philosophically-minded as a specialist. "There is little to distinguish [...] the philosopher of history from the persistently reflective historian," (Sellars 2007, 371). However, to put it crudely, specialists do not need to stay constantly reflective-minded to achieve (at least some of) their aims, which is *not* the case for philosophers:

> A philosopher could scarcely be said to have his eye on the whole in the relevant sense, unless he has reflected on the nature of philosophical thinking. It is this reflection on the place of philosophy itself, in the scheme of things which is the distinctive trait of the phi-

> losopher as contrasted with the reflective specialist; and in the absence of this critical reflection on the philosophical enterprise, one is at best but a potential philosopher. (ibid.)

I read Sellars as saying, among other things, that the fundamental part of doing philosophy is to query, implicitly or explicitly, how philosophy can contribute to the development of our culture in the broadest possible sense of the term. In the same spirit, Eric Schliesser recently attempted to reintroduce the term *synthetic philosophy,* coined initially by Herbert Spencer at the end of the nineteenth century.

> *By "synthetic philosophy" I mean a style of philosophy that brings together insights, knowledge, and arguments from the special sciences with the aim to offer a coherent account of complex systems and connect these to a wider culture or other philosophical projects (or both).* Synthetic philosophy may, in turn, generate new research in the special sciences, a new science connected to the framework adopted in the synthetic philosophy, or new projects in philosophy. (Schliesser 2019, 1–2, emphasis mine, cf., Schliesser 2024)

Schliesser's attempt initiated further discussion which, inter alia, instructively situates the concept of synthetic philosophy within today's philosophical milieu (see Dutilh Novaes 2023 for a cogent take and further references). Expectably, one of the discussed issues is the conceptual relation of synthetic philosophy to particular analytic aims of philosophy; and even to analytic philosophy as such. On the very same issue, Sellars himself (rightly, in my view) remarks:

> It has often been said in recent years that the aim of the philosopher is not to discover new truths, but to "analyze" what we already know. *But while the term "analysis" was helpful in its implication that philosophy as such makes no substantive contribution to what we know, and is concerned in some way to improve the manner in which we know it, it is most misleading by its contrast to "synthesis".* For by virtue of this contrast, these statements suggest that philosophy is ever more myopic, tracing parts within parts, losing each in turn from sight as new parts come into view. *One is tempted, therefore, to contrast the analytic conception of philosophy as myopia with the synoptic vision of true philosophy. And it must be admitted that if the contrast between "analysis" and "synthesis" were the operative connotation in the metaphor, then a purely analytic philosophy would be a contradiction in terms.* (Sellars 2007, 371–372, emphasis mine)

What Schliesser's conception of synthetic philosophy underscores, and what I see as a congenial link to Sellars' views, is the insight that the stimulus for generality always comes from, therefore in fundamental respects presupposes, "a *general* theory [...] that is thin and flexible enough to be applied in different special sciences, but rich enough that, when applied, it allows for connection to be developed among them," (Schliesser 2019, 7–8; cf., deVries 2005, 13–15ff). Put another way, the chosen general theory offers "the synthetizing glue" (ibid., 5). Schliesser's favored synthetizing theory is Darwinism, and in passing he also mentions game theory, information theory, or Bayesian statistics.

However, he holds that his concern is purely methodological, meaning that synthetic philosophy can, as a "theory-driven program," coexist with various isms and philosophical traditions (Schliesser 2020). Schliesser is furthermore very clear that by "special sciences" he means de facto any form of organizing information: "All that is required is that there is a body of knowledge that can be taught, shared, and criticized (in neighboring fields) by others," (ibid.).

Taking cues from both Sellars and Schliesser, one can scratch the transcendentalist itch by accepting the synoptic aims of philosophy, while staying sensitive to the contingent, historical conditions in which science and philosophizing take place by acknowledging one's pragmatic theoretical and explanatory choices. As Schliesser (2019, 2020) also realizes, such an approach preserves, in very specific contexts, the image of philosophy as providing science with substantive insights.

Nevertheless, this never happens with-out but always with-in the body of knowledge generated by the special sciences. So, we get philosophy as a synoptic and synthetizing enterprise that is continuous with science, but that is not a science in any technical sense of that term (it does not have a distinct method or subject matter). And, I maintain that, in this view, the substantive insights philosophy can yield as an intellectual effort and institutionalized discipline will consist of epistemological rather than metaphysical/ontological considerations and consequences.

If we give up the specialized conception of philosophy, we are left with something akin to what Richard Rorty at one point sarcastically called a "bland" conception of philosophy (he in fact explicitly meant the Sellarsian conception of philosophy I put forward above, see Rorty 1982, 226). One of the advantages of this bland type of philosophy is a shift of philosophers' attention from arguing for the best unifying method/style to appreciating subtler differences that really matter in practice and to accepting that, within philosophy as such, there must be a kind of *unity in diversity*, or so I read Rorty at this point:

> Since both *post-positivistic analytic philosophy* and *post-phenomenological Continental philosophy* abjure the Reichenbachian notion of a permanent neutral conceptual matrix within which all thought and language must fit, they should join in abdicating the role of sage, of final authority on the meaningfulness or the rationality of assertions or actions. Having joined in this renunciation, they can agree to differ on whether finding interesting new problems about which to argue, or telling sweeping historical stories, is the more profitable residual enterprise. (Rorty 1982, 226–227, emphasis mine)

In line with Rorty, this book also aims to further elaborate on these two residual parts of the philosophical enterprise simultaneously by thematizing *the epistemological nature of producing historical stories*. So, in particular, I argue, as well as collect and present several views, for considering our past (history) and our attempts to explain it (historiography) as an "interesting" and in some respects also a "new problem" for us philosophers to argue about.

In addition, I later show (most explicitly in Chap. 3) that this foreshadowed naturalistic approach to philosophy is fully reconcilable with Quine's understanding of naturalism, and its subsequent impacts on the whole (empirical) inquiry. That, I hope, will illuminate once again the much-debated relationship between naturalistic and historicist discourse. And I also think that to fruitfully mediate this relationship is a pivotal drive animating the pragmatist approach to inquiry and to philosophy.

As Robert Brandom succinctly puts it: "Pragmatism ought to be seen as comprising complementary vocabularies generated by the perspectives of naturalism and historicism, of common purposes and novel purposes, rather than as restricting itself to one or the other," (Brandom 2000, 171).[6] Consider now, against the background of this section, the following observations written by Quine:

> *The philosopher* [...] *focuses his scrutiny on the conceptual scheme itself. Here is the task of making things explicit that had been tacit, and precise that had been vague*; of uncovering and resolving the paradoxes, smoothing out the kinks, lopping off the vestigial growths, *clearing* the ontological slums.
>
> *It is understandable, then, that the philosopher should seek a transcendental vantage point, outside the world that imprisons natural scientist and mathematician.* He would make himself independent of the conceptual scheme which it is his task to study and revise. "Give me που στω {a place to stand}," Archimedes said, "and I will move the world." However, there is no such cosmic exile. *The philosopher cannot study and revise the fundamental conceptual scheme of science and common sense, without having meanwhile some conceptual scheme, whether the same or another no less in need of philosophical scrutiny, in which to work.* The philosopher is in the position rather, as Neurath says, "of a mariner who must rebuild his ship on the open sea [...]," (Quine 1944 in Verhaegh 2018, 176, long emphasis mine).

One of the main reasons for my exegesis of Quine in this thesis is that while he is generally considered to be an intellectual father of all the current versions of naturalism, no serious attention is being paid to his peculiar post-positivist conception of science.[7] For some philosophers, it has become a custom to claim allegiance to Quine (1969b, 82), who tries to make epistemology "a chapter of psychology and hence of natural science," and simultaneously, completely overlook Quine (1961, 20) who blurs "the supposed boundary between speculative metaphysics and natural science."

The apparent vagueness of natural*ism* in philosophy is indeed not an accident; it stems simply from the fact that different philosophers prefer different conceptions of science, which often means that they have different views on what counts as a proper science. Ipso facto, they have different concepts of, e.g., "explanation," "evidence," "rationality," or even "metaphysics," "nature," and "history". *In this regard, Quine's continuity dictum must always be supplemented by a particular concept of science in order to have an actual practical force (for doing philosophy).*

Thus, my exposition of Quine's concept of science can also be read as an attempt to square naturalism, and broadly conceived pragmatism, in philosophy with post-positivist, i.e., *historicized*, conceptions of science. I specifically aim to analyze the

---

[6] "The reductive naturalist must be reminded that she is leaving out of her story an absolutely crucial practical capacity that vocabularies give us: the capacity to frame genuinely novel purposes, *and so in a real sense to remake ourselves*. The uncritical historicist must be sprung from the dilemma of flabby relativism, on the one hand, and self-satisfied parochialism, on the other, *by the reminder that there are purposes that transcend vocabularies and permit us to make comparative assessments. The theological fanatics should not be permitted to claim theoretical progress over traditional natural science until and unless that progress can be certified technologically as well.* The question is: can they on the basis of their theories both keep the machines running and continue to make the sort of progress at securing common practical ends that would have convinced Aristotle of our greater prowess, and ought to convince contemporary scientists that their successors had indeed made corresponding progress?," (ibid., emphasis mine).

[7] Exceptions are in this matter, e.g., the works on Quine by Paul A. Roth, and also Verhaegh (2018).

relationship of philosophy to selected post-positivist conceptions of science; post-Popperian/post-Hempelian/post-Kuhnian. I take it that, in these accounts, science is understood as a community-based practice without any strict, i.e., ahistorical demarcation criterion.

References

Brandom, Robert. 2000. Vocabularies of pragmatism: Synthesizing naturalism and historicism. In *Rorty and his critics*, ed. Robert Brandom, 156–183. Malden: Blackwell.

Bryant, Amanda. 2020. Naturalisms. *Think* 19 (56): 35–50.

De Caro, Mario, and David Macarthur, eds. 2004. *Naturalism in question*. Cambridge, MA: Harvard University Press.

deVries, Willem A. 2005. *Wilfrid Sellars*. Chesham, Bucks: Acumen Publishing and Montreal & Kingston: McGill-Queen's University Press.

deVries, Willem A. 2021. Wilfrid Sellars. In *The Stanford encyclopedia of philosophy*, ed. Edward N. Zalta. https://plato.stanford.edu/archives/fall2021/entries/sellars/.

Dutilh Novaes, Catarina. 2023. *A plea for synthetic philosophy*. Daily Nous. https://dailynous.com/2023/05/30/a-plea-for-synthetic-philosophy-guest-post/.

van Fraassen, Bastiaan. 1996. Science, materialism, and false consciousness. In *Warrant in contemporary epistemology*, ed. Jonathan L. Kvanvig, 149–182. Latham, Md.: Rowman & Littlefield.

Kusch, Martin, ed. 2019. *The Routledge handbook of philosophy of relativism*. Abington: Routledge.

Norris, Christopher. 1989. Philosophy as *not* just a "kind of writing": Derrida and the claim of reason. In *Redrawing the lines: Analytic philosophy, deconstruction, and literary theory*, ed. Reed W. Dasenbrock, 189–203. Minneapolis: University of Minnesota Press.

Papineau, David. 2021. Naturalism. In *The Stanford encyclopedia of philosophy*, ed. Edward N. Zalta. https://plato.stanford.edu/archives/spr2021/entries/naturalism/.

Peregrin, Jaroslav. 2023. *Normative species: How naturalized inferentialism explains us*. New York: Routledge.

Price, Huw. 2011. *Naturalism without mirrors*. New York: Oxford University Press.

Quine, Willard Van Orman. 1961. Two dogmas of empiricism. In *From a logical point of view: Logico-philosophical essays*, 20–46. New York: Harper and Row, Publishers. (Originally published 1951).

Quine, Willard Van Orman. 1969a. Natural kinds. In *Ontological relativity*, 114–138. New York: Columbia University Press.

Quine, Willard Van Orman. 1969b. Epistemology naturalized. In *Ontological relativity*, 69–90. New York: Columbia University Press.

Quine, Willard Van Orman. 1995. Naturalism; or, living within one's means. *Dialectica* 49 (2–4): 251–263.

Quine, Willard Van Orman, and Joseph S. Ullian. 1978. *The web of belief*. 2nd ed. New York: Random. (Originally published 1970).

Randall, J. H., Jr. 1944. The nature of naturalism. In *Naturalism and the human spirit*, ed. Yervant Hovhannes Krikorian, 354–382. New York: Columbia University Press.

Rorty, Richard. 1982. *Consequences of pragmatism: Essays, 1972–1980*. Minneapolis, MN: University of Minnesota Press.

Rorty, Richard. 2021. *Pragmatism as anti-authoritarianism*. Cambridge, MA: Harvard University Press.

Rosenberg, Alexadner. 1996. A field guide to recent species of naturalism. *The British Journal for the Philosophy of Science* 47 (1): 1–29.

Roth, Paul A. 1999. The epistemology of "Epistemology Naturalized". *Dialectica* 53 (2): 87–110.

Roth, Paul A. 2020. *The philosophical structure of historical explanation*. Evanston: Northwestern University Press.

Schliesser, Eric. 2019. Synthetic philosophy. *Biology & Philosophy* 34 (2): 1–9.

Schliesser, Eric. 2020. *On Kitcher (vs Schliesser) on synthetic philosophy*. Digressions & Impressions. https://digressionsnimpressions.typepad.com/digressionsimpressions/2020/07/on-kitcher-vs-schliesser-on-synthetic-philosophy.html.

Schliesser, Eric. 2024. *Synthetic philosophy, redux*. Digressions & Impressions. https://digressionsimpressions.substack.com/p/synthetic-philosophy-redux.

Sellars, Wilfrid. 2007. Philosophy and the scientific image of man. In *In the space of reasons: Selected essays of Wilfrid Sellars*, ed. Robert Brandom and Kevin Scharp, 369–408. Cambridge, MA: Harvard University Press. (Originally published 1962).

Verhaegh, Sander. 2018. *Working from within: The nature and development of Quine's naturalism*. Oxford: Oxford University Press.

Chapter 2
Quine and Naturalism Historicized

Abstract In this chapter, I present Quine's fragmented and often forgotten views on writing history. However, this needs to be done in conjunction with an examination of his notion of science as our "total world-picture." Quine elegantly avoids the task of specifying a demarcation criterion of science. The result is to position science as a Wittgensteinian language game that gradually expanded from its purpose of predicting our experiences and became socio-historically institutionalized set of practices. Thus, this chapter has two aims; first, I analyze Quine's idiosyncratic, post-positivist concept of science with a particular focus on his inclusion of sciences that are traditionally labeled as soft or social. Then, second, I indicate how Quine deals with historical inquiry. When taken together, this chapter can also be read as a case study analysis of Quine's take on all non-experimental modes of scientific inquiry, and, more broadly, his special account of naturalism and pragmatism.

Keywords Quine · Science · History · Post-positivism · Epistemology naturalized

> Philosophers long made a mummy of science. When they finally unwrapped the cadaver and saw the remnants of an historical process of becoming and discovering, they created for themselves a crisis of rationality. That happened around 1960.
>
> —Ian Hacking[1]

> I philosophize from the vantage point only of our own provincial conceptual scheme and scientific epoch, true; but I know no better.
>
> —W. V. O. Quine[2]

[1] (Hacking 1983, 1).

[2] (Quine 1969a, 25).

M. Hubálek, *Dissolving Naturalism and Historicism Into Each Other*, Studies in History and Philosophy of Science 67,
https://doi.org/10.1007/978-3-032-07535-2_2

2.1 Introduction: Quine In-Between Isms

Quine is undoubtedly one of the most influential philosophers of the second half of the twentieth century. Within the analytic tradition of philosophy, Quine significantly contributed to the demise of positivism and is often *retrospectively* characterized as one of the first post-positivist or post-analytic philosophers. Some even suggested that Quine is a post-modern philosopher (e.g., Patterson 1996; Ciupe 2007). On the contrary, John H. Zammito claims (and he is certainly not the only one claiming this) that, with respect to a variety of issues, Quine "remains an unregenerate positivist," (Zammito 2004, 43, see also 47; cf., Hacking 1983, 41ff, for an instructive analysis of "positivism;" then, e.g., Putnam 1992; informative are also the early ambiguous receptions of Quine's most systematic book *Word and Object* published in 1960 as documented by Verhaegh 2018, 151–155).[3]

In any case, Quine's analyses of meaning, inscrutability of reference, indeterminacy of translation, and evidential undetermination of theories have had a decisive influence on diverse theorizers and philosophers; I would like to point out people such as Nelson Goodman, Thomas Kuhn, Donald Davidson, and Richard Rorty. And more importantly for my aims here in this book, Quine is (in)famous for his endorsement of naturalism. However, as in the case of every original and radical thinker, Quine's philosophical positions and the tenets of his naturalism require a subtle and sophisticated treatment. Roger F. Gibson suggested that: "There is a key to unlocking a correct interpretation of Quine which many of his critics and commentators have overlooked. That key is Quine's commitment to naturalism," (Gibson 1992, 17; cf. also Hylton 2007).

Although I do agree with Gibson in general outlines, I suggest, and argue in the following sections, that the interpretative key is rather his conception of science. This is simply because Quine's commitment to naturalism is a scientific commitment (however "scientific" in the sense that must be further specified).[4] I propose that Quine co-opts the concept of science for his own explicatory aims, and reconfigures not only its relation to epistemology but to rationality as such. Naturalism is then simply a historical label for this reconfiguration. Expanding on this, I highlight that this reconfiguration entails both recognizing that science/rationality is historically dynamic and recognizing that science/rationality derives its epistemic authority in part from its own history. I thus purposely analyze Quine's views on history

[3] I strongly disagree with Zammito. What is, however, true is that Quine often writes as a positivist; he uses the central concepts similarly. But I argue that he does it only after he radically re-assesses the positivist starting points. I am therefore convinced that the study of Quine's philosophical system is an indirect study of the transition from positivism to post-positivism which happened around 1960.

[4] I thus cannot, for example, wholly agree with Gibson's conclusion that "Quine's epistemological perspective (empiricism) presupposes his ontological perspective (physicalism); and the key to interpreting Quine is the recognition that the findings of empiricism do not repudiate the ontology in which those findings are articulated," (Gibson 1992, 27; cf., especially Section 2.2).

(meaning the natural/cultural past), as well as on historical theorizing itself (meaning history as a scientific discipline).

This interpretative approach to Quinean naturalism could, I hope, foster a better understanding of science/natural*ism*, and its relation to contemporary philosophy and/or our intellectual culture. Also, I see this approach as a contribution to the ongoing attempts to (philosophically, historically, sociologically, etc.) illuminate the development of post-positivist accounts of science. Roth recently stated that "some attention must be paid to how the notion of science has itself evolved post-positivism in order to appreciate what one endorses if one declares for naturalism," (Roth 2020, 118). I consider myself doing precisely that when exposing that Quine's notion of science is strikingly post-positivist.

Also, I must note that my aims are *primarily* conceptual, not chronological or contextual. I try to fully exploit the fact that I analyze Quine's thoughts with the help of his later works, taking stock of his views and also the fact that Quine is a stubbornly consistent thinker, trying to answer the same set of questions throughout his whole professional career. So, instead of providing arguments for what Quine *really* thinks, likewise *where* and *when*, I propose a particular reading and understanding of the Quinean naturalism. Put another way, although I engage with Quine's work across his entire career, I regard his later writings as more authoritative for interpreting his naturalistic legacy.

2.2 What, for Quine, Is Science?

It has been more than 50 years since Quine famously called for a movement to naturalize epistemology.[5] As is also well known, Quine construes epistemology unorthodoxly as the study of the foundations of science (Quine 1969b, 69). From a global perspective, his epistemological project was a successful one. It has inspired and instructed scholars and scientists to make their own "naturalistic turns," and natural*ism* has become one of the leading isms for contemporary philosophers. But what, for Quine, is science? This question becomes particularly pressing insomuch as Quine asserts that naturalized epistemology just is "science self-applied" (Quine 1975a, 293).

For various reasons, Quine's epistemological naturalism is often understood only as a defense of natural-science-oriented models of explanation, i.e., of direct *observations* and *experimental* inquiries. Some thus understand Quine's epistemological project as a scientistic doctrine that is hostile to other non-experimental inquires. The case of historiography is then very illustrative in this sense. Intellectual historian and philosopher of history, Frank Ankersmit, for example, confidently proclaims that Quine "never spoke about the writing of history and regarded the

[5] Count from the first publication of the essay "Epistemology Naturalized" (Quine 1969b). But even generally viewed, Quine has actively labeled his position as "naturalistic" since the late 1960s.

discipline with distrust, if not outright contempt," (Ankersmit 2013, 571). This proclamation is ill-judged, not only for its absoluteness ("never spoke"), but also for its unwarranted (and, for that matter, undocumented) accusations.

In the next section, I start with parsing Quine's reconfiguration of epistemology/science. This creates both the necessary context for showing how Quine deals with historiography, and also indirectly reacts to many confusions and simplifications in interpreting Quine that ultimately portray his naturalism as a kind of methodological or disciplinary exclusivism/hierarchism, or even cultural imperialism (see, e.g., Haack 1993; Zammito 2004, 46ff; Rorty 2021, 181–184). I also argue that Quine's concept of science shapes what he understands as naturalism and, for example, pragmatism, not the other way around. Then, I finally indicate how Quine deals with historical research. As I am going to demonstrate here, Quine employs a very broad notion of science that includes all the so-called soft and social sciences. And historiography is, more than once, explicitly cited for inclusion. In fact, Quine exploits the example of history as a discipline to illustrate how one can assign scientific status to a non-experimental form of knowledge.

2.2.1 *Quine and Science as Total World-Picture*

I will argue that the uneasiness some scholars and scientists feel when looking at Quine's picture of science springs from interpretations fusing two central parts of his naturalizing project in epistemology, which fundamentally aims at eschewing any a priori and necessary knowledge; "The naturalistic epistemologist dismisses [the] dream of prior sense-datum language, arguing that the positing of physical things is itself our indispensable tool for organizing and remembering what is otherwise, in James' words, a 'blooming, buzzing confusion,'" Quine (1995b, 225). Quine thus aims at developing and defending an a posteriori/empirical epistemological account.

Firstly, there is the much-debated attempt to explain *how* our knowledge of the external world is even possible; how humans as natural and social beings acquire the knowledge of their world. To put it in Quine's own colorful language:

> How we, physical denizens of the physical world, can have projected our scientific theory of that whole world from our meager contacts with it: from the mere impacts of rays and particles on our surfaces and a few odds and ends such as the strain of walking uphill. (Quine 1995a, 16)

Quine then suggests, *from his own historically embedded perspective*, that physics, neuroscience, psychology, and sciences alike, are going to provide us with useful insights into the conditions and possibilities of our knowledge acquisition. According to Quine, this inquiry would then constitute a new type of epistemological perspective, i.e., science that could finally abandon the requirement to be somehow deduced from prior axioms of preferred "first philosophy" (see Quine 1981, 72ff; for cogent analysis, see Roth 1999, 2008).

However, this explanation of our theory-building cannot in itself sufficiently demarcate what science as such and/or the scientific method(s) are, nor what science should be. For this effort simply presupposes specific subject matter (natural and social beings), and then chooses individual disciplines/sciences (i.e., the body of knowledge that *at the given time* seems to be epistemically suitable) to approach it.

In a certain sense, Quine also had to specify more generally what science is, otherwise his naturalizing project would remain incomplete, if not unintelligible.[6] Put another way, it would be too ambiguous to claim that epistemology must surrender to natural sciences like psychology or neuroscience, without at least hinting at what makes, e.g., psychology a (natural) science. Therefore, the second constituting part of Quine's naturalizing project is the attempt to outline—from within the very same epistemological perspective/science—what counts as science itself (cf., Quine 1995b, 251).

To be sure, these two parts of Quine's naturalization are meant to be complementary, but they are not identical: the first part explains how it is possible to have knowledge (about the external world), based solely on our sensory input; and the second deals with how it is possible to *justify* our knowledge, relying solely on our sensory input. The second part has two crucial impacts on Quine's conception of science; it creates a *normative force* for the new epistemology, and simply by targeting the justification of knowledge, Quine, qua empiricist, faces what is called the problem of induction (the problem of grounding the validity of inductive inferences that has troubled philosophers at least since Hume). However, qua naturalist, Quine cannot bypass the problem of induction, as his account of scientific knowledge must also be thoroughly empirical (cf., Quine 1969b, 74–79).

Thus, Quine cannot but accept a form of fallibilism and falsificationism. "The Humean predicament is the human predicament," quips Quine (ibid., 72). The problem of induction is an unavoidable consequence of conceiving our *experience* as the only evidential source of our knowledge, and science is our best tool for guiding inductive practices. Even the scientific knowledge is then, strictly speaking, only provisional. So, what is then *science* for Quine?

In stark contrast to many specialists theorizing science in the twentieth century, Quine does not strive to find a clear-cut conceptual distinction between *scientific* and *non-scientific* modes of inquiry (1995b, 252). Historically speaking, this means that he was not trying to distinguish science from philosophy and/or metaphysics, or even from common sense. Already in 1957, Quine argued that there is no *qualitative* difference between common-sense reasoning and science. The scientist and the layman both begin with related standards of evidential reasoning, but science surpasses common sense by developing higher standards of carefulness, patience and systematicity (see Quine 1957, 5–6). In *World and Object*—one of his most systematic statements—first published in 1960, Quine writes: "Scientific neologism is itself just linguistic evolution gone self-conscious, as science is self-conscious

[6] As I further insist, I am convinced that the same holds even for Quine's very specific relationship to other traditionally recognized *isms*, such as empiricism, scientism, physicalism, and pragmatism. See also Verhaegh (2018) for a contemporary and cogent account.

common sense," (Quine 2013, 3). Put another way, scientists develop theoretical systems of (evidential) reasoning.

For Quine, science is the expression and extension of the dispositions inherent to human rationality that has its natural history and, ipso facto, natural underpinnings. To anchor our scientific cognition in nature, Quine in his texts very often sketches a minimalistic evolutionary explanation via the concept of *prediction*. I reconstruct it here from Quine's *Quiddities*:

> Prediction is rooted in a general tendency among higher vertebrates to expect that similar experiences will have sequels similar to each other [...]. In our expectation that subjectively similar events will have similar sequels, we and other animals are often deceived [...]. What is uncanny, however, is how overwhelmingly much more often our expectations are fulfilled than disappointed. Our standard of similarity, for all its subjectivity, is remarkably attuned to the course of nature [...]. For all its subjectivity, in short, it is remarkably objective. In the light of Darwin's theory of natural selection we can see why this might be. Veridical expectation has survival value in the wild [...]. A man sees the dawn's early light and, in the light of past experience, he expects birds to sing. Or, blessed with language, he sees the dawn and says "The birds will sing." This is prediction. In the fullness of time, he verbalizes his habitual expectation and says "When the sun comes up the birds sing." This, which I call an observation categorical, is the first step in scientific theory. It is a hypothesis that generates predictions and can be refuted by failure of prediction. (Quine 1987, 159–161)

This passage reveals that, for Quine, certain innate capacities must be in place for establishing an intersubjective "standard of similarity" in given causal contexts, and this standard is also necessary for developing and learning a language (e.g., Quine 1969c). For this reason, "observation sentences serve in both ways – as vehicles of scientific evidence and as entering wedge into language," (Quine 1992, 5). Once language is in place, one can not only verbalize their observations and expectations, but also connect two or more observations (observation sentences) into hypotheses (ideally observation categoricals).

All in all, Quine holds that scientific hypotheses (sentences) imply, *directly or indirectly*, hypothetical observation categoricals that carry the following structure: *If A, then B* (When the sun comes up the birds sing). At the same time, these categoricals determine (in our case very elementarily) the observational/experimental consequences for their own "testing" as scientific hypotheses. These categoricals determine expectations, i.e., predict observation *B* from observation *A*. Such a hypothesis can sometimes be refuted by Popperian falsification if one is observing *A* without observing *B*. Equivalently, if one observes the sun coming up while the birds are being silent. This is why Quine's theory of evidence is generally *behavioral/observational* but not necessarily reductive. The empirical consequences will differ from hypothesis to hypothesis, and are also a matter of degree (see, e.g., Quine 1957, 15; cf., Roth 2003).

I believe that Quine's evolutionary/historical look at scientific rationality is the key to his often misunderstood, or overemphasized, conviction that hypothetico-deductive inferences (i.e., predictive reasoning) represent *the* paradigm of science, which must be in that regard ultimately *empirical,* or one might just say experimental or observational. To put it boldly, Quine thinks, together with all empiricists, that this mode of reasoning is the only guarantee that our theories *about* the world, nested in our *intersubjective language(s)*, are somehow also *connected to* the external world in which we have evolved; it is the only guarantee that our theories have empirical contents (see, e.g., Quine 1961b, 77ff; cf., Kemp 2017).

In addition, from Quine's historical vantage point, the traditionally conceived *natural* sciences made the most of hypothetico-deductive inferences *so far* (after Darwin, also by offering plausible theories of humans qua *natural* beings and theorizers). Hence Quine in his texts sometimes uses the term "natural science" instead of just "science" to make this fact explicit.[7] However, predictive reasoning does not exhaust the modus operandi of individual scientific disciplines.

Quine substantively complicates and further crafts this inductivist picture, as is well known, especially by his *semantic holism*[8] that gradually leads to a conclusion that "our statements about the external world face the tribunal of sense experience not individually but only as a corporate body," (Quine 1961c, 41).[9] This presumption is currently labeled as the Quine-Duhem thesis.

How does the holism change the picture outlined above? "In order to deduce an observation categorical from a given hypothesis, we may have to enlist the aid of other theoretical sentences and of many common-sense platitudes that go without saying, and perhaps the aid even of arithmetic and other parts of mathematics," (Quine 1992, 13). Quine points out here that very rarely (if at all) has a single hypothesis (sentence) of a mature science the potential to imply observation categoricals independently of any other theories (sentences). Every single hypothesis/sentence is supported by many *auxiliary* hypotheses/sentences. Furthermore, Quine's fully-fledged account of science must go beyond naïve inductive steps, due to very theoretical reasons, meaning with respect to our theory-building. This is to say that mature and complex scientific theories must also encompass and work with the statements about *unobservable* and *abstract* entities and events. Quine's developed picture of science is then quite different:

> The totality of our so-called knowledge or beliefs, from the most casual matters of geography and history to the profoundest laws of atomic physics or even of pure mathematics and logic, *is a man-made fabric which impinges on experience only along the edges. Or, to change the figure, total science is like a field of force whose boundary conditions are experience*. A conflict with experience at the periphery occasions readjustments in the interior of the field. Truth values have to be redistributed over some of our statements. Re-evaluation

[7] One must also bear in mind the mere fact that, in English, the word "science" and the term "natural science" are sometimes used interchangeably.

[8] For informed analyses, see Peregrin (2016, 106ff); Roth (1983).

[9] Even for the elementary bird example above, it is essential to include a proviso, as Quine does, that our ancestors rely on the assumption that their future experiences will be the same as their past experiences. This assumption is also part of this collective, theoretical body (philosophers of science commonly refer to this fact by underlining the necessity of *auxiliary hypotheses*). We rely on these inductive generalizations all the time in our everyday life as well as in science, e.g., when we assume/hypothesize from the present that the laws of nature were the same in the past, and will be the same in the future. Stephen Jay Gould (2002, 97ff) nicely shows how this "methodological assumption of temporal invariance for laws of nature" plays a vital role in evolutionary theorizing qua historical theorizing (ibid., 102).

> of some statements entails re-evaluation of others, because of their logical interconnections – the logical laws being in turn simply certain further statements of the system, certain further elements of the field. Having re-evaluated one statement we must re-evaluate some others, whether they be statements logically connected with the first or whether they be the statements of logical connections themselves. *But the total field is so undetermined by its boundary conditions, experience, that there is much latitude of choice as to what statements to re-evaluate in the light of any single contrary experience. No particular experiences are linked with any particular statements in the interior of the field, except indirectly through considerations of equilibrium affecting the field as a whole.* (Quine 1961c, 42, emphasis mine)

Another metaphor that Quine famously uses to describe the totality of our knowledge is the "web of belief," insomuch as he sees the word "belief" in some respects better than the word "knowledge."[10] Quine's notion of *science* is thus very idiosyncratic. It does *not* represent the totality of established scientific disciplines or selected methods. Rather, it represents the totality of our knowledge/beliefs, i.e., our total theory of the world, our "total world-picture" (Quine 1961a, 17; also see again Quine 1961c, 42ff).

Quine in effect holds that our common sense, science, and rationality itself comprise one cognitive/epistemic continuum. Such a notion was at that time very original (and is often not appreciated enough even today). It makes explicit the tension between the underlying epistemic continuum of all human inquiry on the one hand, and the ordinary disciplinary/methodological pluralism parasitizing on this continuum on the other hand.[11]

This picture does not need any additional definition of natural science, nor the distinction between natural/social, hard/soft, or observational/descriptive science. A discipline fits into our total science if it adds to or accords with the rest of our system of beliefs (sentences). It seems fair to say that, according to Quine, the pivotal distinction for science cuts across disciplines and methodologies, and concerns only given bundles of *theories with direct/indirect empirical consequences*, where the theories with direct empirical consequences then enable the testing/falsification of

[10] Knowledge, in Quine's mind, strongly connotes epistemic *certainty*. See the headword "knowledge" in *Quiddities* (Quine 1987, 108ff). So, "beliefs" are for Quine simply "sentences held true."

[11] With respect to the various interpretations of Quine, it is indicative that he is often understood as arguing for the unity-of-science dictum (see, e.g., Davidson in Borradori 1994; cf. Quine 1995b). However, as I try to further argue in this book, the situation is much more complicated. Quine indeed argues for a kind of unity of "science;" but it is definitely not a unity of method nor any kind of disciplinary unity. In my reading, that is exactly the respect in which Quine is original and often misunderstood. Section 2.3 of this book discussing "historiography," qua a scientific discipline, can be then read as a specific case study of the productive tension between the unity and disunity of science, qua our total science. Ultimately, Quine might not be the first to make a case for, e.g., a strictly a posteriori approach to scientific theorizing. But he is one of the first thinkers who accepted that when science is understood as not having clear boundaries, our conception of (empirical) inquiry and its normative reflection must be re-configured altogether (so inclusive of philosophy). That is the whole point of his epistemology naturalized; it is the synergistic effect of all Quine's re-configurations that ultimately makes the real meta-theoretical difference.

these theories *taken as a whole,* since the empirical significance is "diffused through the system:"

> The scientist does indeed test a single sentence of his theory by observation conditionals, but only through having chosen to treat that sentence as vulnerable and the rest, for the time being, as firm. This is the situation when he is testing a new hypothesis with a view to adding it, if he may, to his growing system of beliefs. (Quine 1981, 71–72)

Dismantling the empirical from the theoretical (or, e.g., the analytic from the synthetic; the factual from the linguistic) is thus, according to Quine, always done against the background of a theory-driven choice made for some specified purposes (e.g., explicatory and explanatory purposes). It is a historically contingent and culturally conventional decision. In short, such a dismantling has a practical force only ad hoc.

I consider this as one of the most critical effects of Quine's re-configuration of science; one should shift the attention from *theorizing the scientific method* to *actual practices of theory-building* within and across the diverse contemporary disciplines. As Quine sarcastically remarks: "Boundaries between disciplines are useful for deans and librarians, but let us not overestimate them – the boundaries," (Quine 1976, 76). Or again in a more solemn way: "Science is neither discontinuous nor monolithic. It is variously jointed, and loose in the joints in varying degrees," (Quine 1975b, 314). If one, for any reason, thinks science benefits from some form of hierarchization, their target should be chosen bundles of theories, not disciplines and methodologies.

The substantial consequence of this modified picture is that not all scientific theories can be directly justified through some kind of experiment; not all science can be tested. The prediction serves as the ultimate checkpoint of science, not as a guiding principle of its every step. Some theories, typically the ones with very indirect empirical consequences, are built only with respect to the overall economy of our total science. "Positivistic insistence on empirical content could, if heeded, impede the progress of science," (Quine 1995a, 49). For these reasons, Quine cannot think of prediction as being the standard of scientific practice. Prediction rather creates one of the boundaries of rational space in which science operates. That is the furthest Quine can be pushed towards the demarcation criterion. In *Pursuit of Truth*, Quine explicitly employs Wittgenstein's game metaphor and states:

> When I cite predictions as the checkpoints of science, I do not see that as normative. *I see it as defining a particular language game, in Wittgenstein's phrase: the game of science, in contrast to other good language games such as fiction and poetry.* A sentence's claim to scientific status rests on what it contributes to a theory whose checkpoints are in prediction […] Prediction is not the main purpose of the science game. It is what decides the game, like runs and outs in baseball. *It is occasionally the purpose, and in primitive times it gave primitive science its survival value. But nowadays the overwhelming purposes of the science game are technology and understanding.* (Quine 1992, 20, emphasis mine)

Unfortunately, Quine never seriously elaborates on this Wittgensteinian picture of science and leaves it underdeveloped. However, I consider the game metaphor for science very illuminating, at least for three reasons. The first reason is conceptual; the concept of game is the paradigmatic instance of Wittgenstein's family resemblance, i.e., the concept of game creates a conceptual unity despite many factual and methodological differences among different games/sciences. The second reason is that such a conceptualization easily accommodates and acknowledges the historical nature of science, ipso facto, human rationality in Quinean reading.

Imagine, for instance, your favorite *sporting* game: there has always been a space to modify the rules; introduce and/or improve the equipment; the players and teams are free to create a whole range of specific strategies and their reasons for playing the game vary as well; playing to stay fit differs from playing to become the world champion. "The nature of the play into which prediction integrates is left unspecified," writes Roth aptly (1999, 106). Therefore, the standards of various scientific practices of organizing knowledge/beliefs are always tied to related, specific purposes. Quine himself explicitly mentions practical purposes (technology) and intellectual purposes (understanding). In this particular respect, Quine's notion of science/rational inquiry is undeniably one of a pragmatist (e.g., see Quine 1961c, 46; cf., Godfrey-Smith 2014).

Lastly, the third reason is that the game metaphor helps us to understand that Quine's naturalization of epistemology is not a dissolution of the normative dimension of epistemology. If one is able to articulate the purposes, the reasons why one plays a game, one can adjust one's strategies and practices accordingly. No more is needed (see Quine and Ullian 1978, 134–135). Quine merely insists that the standards of scientific inquiry are to be found, as every unit of knowledge, within our intertwined, ever-changing web of belief. That is what Quine means when he declares that naturalized epistemology is just "science self-applied" (Quine 1975a, 293; cf., Roth 1999, 104).

I contend that, against this background, one can fully appreciate the *diachronic* and *dynamic* view of science Quine, qua philosopher, sticks to when he writes things like the following often-quoted paragraph:

> I see philosophy not as an a priori propaedeutic or groundwork for science, but as continuous with science. I see philosophy and science as in the same boat – a boat which, to revert to Neurath's figure as I so often do, we can rebuild only at sea while staying afloat in it. There is no external vantage point, no first philosophy. *All scientific findings, all scientific conjectures that are at present plausible, are therefore in my view as welcome for use in philosophy as elsewhere.* (Quine 1969c, 126–127, emphasis mine)

To summarize: *science*, in Quine's reading, represents our total and contingent theory of the world. From a diachronic as well as synchronic perspective, this theory of the world is the result of our rational/pragmatic attempts to organize our experiences into a coherent whole; into the web of belief. Therefore, some beliefs in this web have empirical consequences only indirectly, and their scientific status (empirical content) is *derived* from their connection to other beliefs that have empirical consequences (that can be in principle tested/falsified). The rationality (the possibility of fallibility) of this process is grounded partially in the natural origin of our cognition, as well as in intersubjective languages (which are both understood as co-evolving with our environment); and partially in self-correcting efforts driven by our explicit practical as well as intellectual demands. *Quine simply takes for granted that science, and so all our epistemic theories, are historically dynamic.*

At this point, I want to further accentuate that, due to semantic holism and the historically dynamic conception of science to which Quine is committed, his notion of evidence/experience is *intratheoretical*: "The unit of empirical significance is the whole of science," (Quine 1961c, 42). By this move, Quine turns all the prominent (logical) empiricists on their heads, since his version of empiricism is radically anti-foundationalist. Roth interprets this stance instructively as follows:

> For a variety of scientific theories (broadly construed, so as to include the social sciences) serves to determine just which experiences count and under what conditions they count as relevant for assessment purposes. Science ultimately delimits, e.g., how many senses there are, how they function, and so what even the senses properly so-called could provide qua evidence. Both questions give rise to worries about how diffuse the notion of the empirical becomes once it cannot be restricted to terms or simple statements. (Roth 2020, 124)

Quine, in his own way, tried to dissolve these epistemic worries by pushing back against the assumption that our observations are always theory-laden through and through, and ends up talking about "degrees of observationality" (see Quine 1993, 108; Quine 1981, 71; cf., Zammito 2004, 15ff, for a critical analysis). Regardless, recollect that, for Quine, our positing of "things" in the world, meaning our projections of chosen categories/concepts on(to) the world, is indispensable for any (observational) knowledge. Our "experiences" would otherwise amount only to something akin a blooming, buzzing confusion. Put another way, there is no genuine observation without conceptualization, i.e., without language(s).

All postulated entities are our theoretical constructs, they are contingent, cultural *posits* of our theories developed to accommodate and make sense of our casual interactions with the world (Quine 1961c, 44). In this regard, Quine is an adamant anti-realist; there is no "integral reality" behind this theoretical veil: "So far as evidence goes, then, our ontology is neutral. Nor let us imagine beyond it some inaccessible reality. The very terms 'thing' and 'exist' and 'real,' after all, make no sense apart from human conceptualization. Asking after the thing in itself, apart from human conceptualization, is like asking how long the Nile really is, apart from our parochial miles or kilometers," (Quine 1993, 113). On that premise, the very conception of *nature* also emerges from within our theorizing. As Quine himself puts it:

> Our theory of nature grades off from the most concrete fact to speculations about the curvature of space-time, or the continuous creation of hydrogen atoms in an expanding universe; and our evidence grades off correspondingly, from specific to broadly systematic considerations. Existential quantifications of the philosophical sort belong to the same inclusive theory and are situated way out at the end, farthest from observable fact. (Quine 1969c, 98)

This implies that causal contexts/empirical contents do not simply force upon us conceptual contents, even in the case of our most basic concepts. The conditions of applicability of these concepts vary with respect to the cultural/linguistic communities using them. Hence, what will be considered as an observation will vary too: "One man's observation is another man's closed book or flight of fancy," (Quine 1969b, 88). However, Quine perceives this issue as a mundane fact resulting from diverse forms and degrees of cultural conditioning, which also explains his long-lasting interest in and insistence on the role of language learning and habit formation for science (for scientific language):

> For by our definition the observation sentences are the sentences on which all members of the community will agree under uniform stimulation. *And what is the criterion of membership in the same community? Simply general fluency of dialogue.* This criterion admits of degrees, and indeed we may usefully take the community more narrowly for some studies than for others. What count as observation sentences for a community of specialists would not always so count for a larger community. (ibid., 87, emphasis mine)

Also, although Quine is generally considered as the intellectual father of all current forms of naturalism, naturalism is also defended as a metaphysical/ontological doctrine nowadays (see, e.g., Bryant 2020). Quine's naturalism is, ironically enough, ontologically neutral by definition. Existential statements and specific ontological positions must, in order to be naturalistic in the Quinean sense, be formulated from within our science; from within our "provincial conceptual scheme and scientific epoch," not vice versa.

For this reason, Quine himself can embrace, e.g., physicalism as "a scientific position" (Quine 1995b, 257), hence a position cohering with and contingent on our developing total world-picture. But specific statements of physicalism are, as every other statement, prone to be re-evaluated in light of the given evidence, so physicalism could, in principle, be falsified. Quine's naturalism is conceptually as well as ontologically flexible; "If I saw indirect explanatory benefit in positing sensibilia, possibilia, spirits, a Creator, I would joyfully accord them scientific status too, on a par with such avowedly scientific posits as quarks and blackholes," (ibid., 252; the same holds even for Quine's empiricism, ipso facto, Quinean naturalism itself, see Quine 1992, 19–21). More serious attention paid to Quine's idiosyncratic, post-positivist notion of science could easily dispel the anxieties associated with his naturalizing project in epistemology. As Burton Dreben nicely underscores, Quine does not espouse any reductionist visions. Quine's physicalism is deliberately minimalistic:

> By "physical" Quine means, roughly, "materialistic as opposed to mentalistic." *He does not mean fully expressible in the vocabulary of (contemporary) theoretical physics.* His "physicalism" is what Hookway [1988, 77] has called the "disappointing" kind. Quine never intended to say anything new or interesting about "physicalism" and never attempted to fully characterize what it is for a term, notion, or concept to be "physical." (Dreben 1992, 307, emphasis mine)

In a letter responding to Hookway, Quine declares: "My basic position early and late is empiricism, and hence prediction as touchstone. Physics enters my picture only because, in my naturalism, I take the current world picture as the last word to date. If evidence mounts for telepathy or ghosts, welcome. Physicists would go back to their drawing boards. Whether to call their resulting theory physics still, on determinationist grounds, is a verbal question," (Quine in Dreben 1992, 308). Against this background, I claim that it is symptomatic of our intellectual climate that Quine is more often caricatured as a philosopher with scientistic leanings, as opposed to, e.g., relativistic ones.

In the following section, I indicate what Quine has to say about writing history from within science as he imagines it. By doing this, I also wish to illuminate Quine's stance on the pragmatic and explanatory value of all non-experimental theorizing.

2.3 Historiography and the Art of Hypothesis-Making

Historical sciences (e.g., historiography as well as evolutionary sciences) are typically considered as methodologically distinct from experimental sciences, since they paradigmatically do not rely on direct observations, predictions, law-like generalizations, let alone subsumptions under general laws. That raises serious doubts regarding the epistemic status of historical theories/explanations (for contemporary examination concerning historiography, see Kuukkanen 2015; Currie 2018; Rosenberg 2018; Roth 2020; for a historical examination, conceptual overview, and further references concerning evolutionary science, see Hubálek 2021).

As I have already pointed out in the first section, Quine does not want to exclude historical theories from our total theory of the world, even though he must be aware of the peculiar nature of historical theorizing. The fact is indicative that Quine feels obliged to mention the case of historiography, while still dealing with the headword "prediction" in *Quiddities*, from which I have already quoted. Quine's take on historiography is then also as feasible to interpret as his subtle take on non-experimental modes of inquiry.[12]

> *Hypotheses about history, even of the remote past, are part and parcel of science in the inclusive sense, and the arbiter is still prediction.* Occasionally this is evident, as when a hypothesis about ancient history implies the findings of some future excavation. More often the historical proposition will link up only very indirectly with prediction, through participating in a large, cohesive section of scholarly and scientific lore and common sense which has its ultimate links with observation categoricals and prediction somewhere outside the historical quarter. (Quine 1987, 162, emphasis mine)

In his last book, *From Stimulus to Science*, Quine performs a very similar interpretive procedure. While depicting the nature of the so-called soft sciences, he once again utilizes the case of historiography to stress his thoughts on the conception of science without direct prediction at hand.

> *In softer sciences, from psychology and economics through sociology to history (I use "science" broadly),* checkpoints are sparser and sparser, to the point where their absence becomes rather the rule than the exception. Having reasonable grounds is one thing, and implying an observation categorical is another. Observation categoricals are implicit still in the predicting of archaeological finds and the deciphering of inscriptions, *but the glories of*

[12] Of course, one might argue that there are even non-experimental modes of inquiry, such as logic, mathematics, philosophy, and arguably even some parts of physics. I agree. I simply think that Quinean scholars typically discuss them, and that some conclusions I draw for historiography will hold for some of their goals, too.

> *history would be lost if we stopped and stayed at the checkpoints.* (Quine 1995a, 49, emphasis mine)

I now take Quine's delineation of science as a starting point and extract two conclusions for historical theorizing. In addition, I would like to state here that Quine uses a vocabulary that might sound obsolete for most of today's philosophically-minded historiographers and/or philosophers of science, because he still exploits concepts like "prediction," "hypothesis," and "falsification." I use them here primarily to authentically reconstruct Quine's position. And Quine uses them in his own specific, re-configured manner where, e.g., "prediction" refers to an organism oriented in nature, as well as to exclusively human theory-building (science). Below, I say a bit more on how Quine exploits the word "hypothesis". Recall also that his notion of evidence/experience is intratheoretical. Thus, for example, when one reads the word "falsifies," one can also read the phrase, "makes less plausible."

Firstly, I present the simpler case of historical theorizing that can, in principle, include the success of predictive reasoning. Secondly, I focus on the more challenging case where historical theories connect with prediction very indirectly or seemingly not at all.

(1) Some historical theories will simply yield observation categoricals that can, in principle, be tested/falsified (by our observation). I present them in the form of their potential falsifiers.

- The theory that Vincent van Gogh was killed/murdered can be falsified by the discovery of his suicide note.
- The theory that feathers qua evolutionary traits are adaptations for flight can be falsified by the fossils indicating that feathers are older than birds.[13]

These inferences are unproblematically compatible with Quine's "game of science," as they connect the historical theories with the edges of our web of belief that can be interconnected with, and in principle falsified by, our evidence. In these cases, the hypothetico-deductive inference is principally used to answer a *what-question*[14] while inquiring into the particular matters of fact ("what was the case?;" "what happened?"). But it is not necessarily used to explain these facts, i.e., to answer a *why-question*.[15]

(2) More often, historical theories cannot be, in principle, tested/falsified by our evidence; one cannot formulate genuine prediction (predict *direct* observation). This could also mean that these theories refer to *events* (entities) that typically have their causes/reasons, as well as immediate effects, buried in the past, with no clearly discernible causes/reasons and effects in the present. Put another

[13] This is an actual evolutionary conclusion, see Gould and Vrba (1982). The fossil record indicates that feathers, as we know them today (i.e., biologically *homologous*), evolved already in terrestrial dinosaurs.

[14] Or alternatively a when-, where-, who-question.

[15] Or alternatively a how-question.

way, these historical theories are *essentially* underdetermined by available evidence. I present such events as potential historical research questions.

- Why did Vincent van Gogh cut off his ear?
- Why have birds evolved the disposition to fly?

In these cases, one is principally attempting to answer a why-question. And the explanatory challenge here is then twofold. The uniqueness and contingency of historical events detach them from the predictive and experimental inferences and practices, plus history as a temporal factor leads to fragmented or even wholly absent evidence (relevant matters of fact). These epistemic challenges related to historical research are currently widely recognized, and *scientific* explanation of the past is typically understood as a fight against the essential underdetermination of historical theories (see, e.g., Currie and Turner 2016 for an overview; cf., Gould 2002, 102).

Quine insists that science is able to accommodate these underdetermined why-questions, and maybe even must, otherwise "the glories of history would be lost". Still, the question here is "how?" A very short answer could be that theories plausibly answering historical why-questions simply give us a kind of *understanding*,[16] insomuch as Quine also lists understanding as one of the goals of our contemporary science. Quine himself does not sufficiently underscore the fact that he quite often advances answers to historical why-questions, e.g., by offering evolutionary explanations (his often-discussed essay "Natural Kinds," which I have already referred to here, is an illustrative example of that).

In the previous section, I showed that Quine naturalizes our present cognitive standards of similarity (the success of inductive inferences in general) by invoking their past survival value. He makes sense of our cognitive capacities by an evolutionary/historical explanation (cf., also Quine and Ullian 1978, 136–138, for a similar explanation related to moral capacities). In addition, there is also the obvious historical continuity of our cultural practices and conceptual frameworks. The re-evaluation of some statements/theories within our web of belief, and their subsequent re-adjustment, happens synchronically as well as diachronically. It cannot be otherwise. One must always imagine the web of belief, not only with its *horizontal* axis but also with its *vertical* axis.

Again, Quine is not often clear enough on this matter, because he predominantly discusses synchronic scientific inquiries (however, see the headword "Kinship of Words" in *Quiddities* for an example of Quine's historical sensitivity). Moreover, I also suspect that because Quine habitually thinks that to gesture towards our pragmatic choices and considerations, or to construe scientific theorizing as a product of

---

[16] However, the concept of *understanding* is not a transparent one. At a minimum, one would have to specify its relation to the concept of explanation. See, e.g., an instructive book *Understanding Scientific Understanding* written by de Regt (2017). Quine seems to use the term "understanding" rather rhetorically, and most of the time synonymously with the term "explanation," which he uses for identifying causes (see, e.g., Quine and Ullian 1978, 111–112), and at other times he seems to use the term "understanding" as an explanation with some pragmatic value.

self-conscious common sense, will do by itself. *Interpreted as such, for Quine, historical theorizing becomes indispensable for understanding/explaining how we, natural and cultural beings, wind up where we are.*

However, to stay true to Quine's own view on science, this general appeal to the concept of understanding/explanation would certainly not even satisfy our common-sense perception of scientific inquiry; one would not satisfactorily differentiate the game of science from the game of fiction in the case of historical theorizing. Put another way, there is generally no doubt that the continuity of the past to the present is formative as a temporal relation; the doubt looms over the possibility of cognitively grasping this continuity in a non-fabulative way. What is at stake is our epistemic relation to the past, especially in the essentially underdetermined cases. Without an intelligible epistemic relation to this past, significant parts of it would become a historical screen on which anything can be projected (or at least on which anything could be intentionally distorted, for instance, politically). Thus, one must go a little further with Quine.

What does Quine mean when he says that historical theories are connected to prediction indirectly "through participating in a large, cohesive section of scholarly and scientific lore and common sense which has its ultimate links with observation categoricals and prediction somewhere outside the historical quarter?" Or when he contrasts "having reasonable grounds" with "implying observation categorical?" These claims are, I suggest, an illustration of the normative part of naturalized epistemology/science which, in turn, becomes: "the art of guessing, or framing hypotheses [...]. [But as Quine adds], creating good hypotheses is an imaginative art, not a science. It is the art of *science*. Normative epistemology is the art or technology not only of science, in the austere sense of the word, *but of rational belief generally*," (Quine 1995a, 49–50, emphasis mine). By referring to the "artistic" nature of hypothesizing, Quine makes it clear that there is no algorithm for creating good hypotheses.

Hypothesis-making is an integral, imaginative part of the constant and historically continuous process of developing our web of belief. True, to today's ears, the term "hypothesis" connotes with hypothesis-testing, and Quine, of course, *primarily* works with the term "hypothesis" in this modern and technical manner too, as I exposed in a previous section. That being said, Quine never seems to deviate seriously from his idiosyncratic view on science, which is driven, inter alia, "by our efforts to organize beliefs into a coherent whole"—which one might paraphrase at this point as "by the scientific art of hypothesizing." So, *secondarily*, Quine also uses "hypothesis" with its original connotation as "supposition"/"uncertain belief." In *The Web of Belief*, he together with Joseph Ullian, state the following: "In a word, hypothesis is guesswork; but it can be enlightened guesswork," (Quine and Ullian 1978, 65).

Quine and Ullian then catalog broad epistemic virtues that can abet our hypothesizing *in relation* to our total world-picture (see ibid., 66–80). For instance, they identify the virtue of *modesty*; hypothesis *A* is more modest than *A* and *B*, taken as a united hypothesis (i.e., where *A* does not entail *B*). Also, hypothesis *A* is more modest than *B* if hypothesis *A* assumes happening(s) that are more probable or usual with respect to our web of belief (cf., ibid., 68). Hence, the first possible step to enlighten our guesswork is to narrow down the list of plausible hypotheses by evaluating them by their own structural efficacy, and the second possible step is to evaluate the hypotheses through the lens of our total science, i.e., through our standards of "rational belief generally."

Besides that, as Quine and Ullian ingeniously argue, ordinary beliefs become hypotheses by the very reasons for which one entertains them. One engages in hypothesis-making as one presupposes that, if a given hypothesis is proved testable, or just complementary with our total world-picture, it would also explain some of the circumstantial matters of fact one already believes in (ibid., 66).

> Hypothesis, where successful, is a two-way street, extending back to explain the past and forward to predict the future. What we try to do in framing hypotheses is to explain some otherwise unexplained happenings *by inventing a plausible story, a plausible description or history of relevant portions of the world.* (ibid., emphasis mine)

To step down on a more concrete level, the very reasons for formulating hypotheses draw on or are deduced from other previously held hypotheses/beliefs; one simply cannot think (or build theories) in a cultural and conceptual vacuum. And, of course, some of these "hypothetical beliefs" can be connected with either other beliefs with indirect empirical consequences or directly with our observations. This is why the diachronic and dynamic aspect of human science is vital. Our partially inherited and partially invented—when viewed from a time-anchored perspective—total world theory always serves as the "reasonable ground" from which one must start reasoning, and simultaneously, as the very set of standards for our reasoning. This once again underscores Quine's all-inclusive conception of science that is in mutual relation to our everyday reasoning (common sense). Consider in this light the following passage:

> We talk of framing hypotheses. Actually we inherit the main ones, growing up as we do in a going culture. The continuity of belief is due to the retention, at each particular time, of most beliefs. In this retentiveness science even at its most progressive is notably conservative. […] But the going culture goes on, and each of us participates in adding and dropping hypotheses. Continuity makes the changes manageable. Disruptions that are at all sizable are the work of scientists, but we all modify the fabric in our small way, as when we conclude on indirect evidence that the schools will be closed and the planes grounded or that an umbrella thought to have been forgotten by one person was really forgotten by another. (ibid., 81–82)

This all clearly holds for the case of historical research as well. This is something that is standardly done when historical research is being conducted. Historians attempt to explain Van Gogh's actions not only in the consistency of already existing information about him and about his milieu, but also in the light of our developing total science, e.g., in the light of chosen psychological/psychiatric/medical theories, or other relevant theories produced by disciplines conducting synchronic research (for useful overview and further references, see Turnbull 2022; see also Arenberg et al. 2020; Voskuil 2020; Dasgupta et al. 2022; the amount of literature about Van Gogh is massive, so I deliberately choose some recent works here).

The same applies in the case of the evolution of birds. Evolutionary scientists base their inferences about evolutionary history on facts/theories from genetics, paleontology, comparative morphology, and similar disciplines, as well as on other more specific evolutionary theories. From where we stand now, that is from the perspective of our contemporary total science, the dominant meta-theory (paradigm) in evolutionary theorizing is still Neo-Darwinism, with its emphasis on natural selection, and the concepts of adaptation, exaptation, and spandrel.[17]

So, when done right, the diachronic and synchronic development of our web of belief is in a dialectical relationship. When answering any what- and why-question, one is guided by diachronically upheld beliefs and standards, as well as by direct or indirect synchronic observations and/or experiments. But, in principle, every belief is prone to be re-evaluated, e.g., in the light of given falsifying evidence; and which experiences will count as evidence can be re-considered by re-evaluating the very assessment standards. Our scientific hypothesizing (including our historical guesses) is then in this particular sense always empirically and theoretically guided.

Nonetheless, on the one hand, the fact that our (historical) hypotheses with indirect empirical consequences can be connected/made compatible with other beliefs from different disciplines, or areas within our web of belief, and even potentially interconnected with beliefs having direct empirical consequences, is *not* by itself an argument for their truthfulness, in any philosophically interesting sense of the term. The essential underdetermination of the vast majority of historical theories will remain unbeaten by these procedures, too.[18] Thus, for naturalists, it is no surprise that alternative competing historical explanations of the same past happenings exist.

On a more positive note, these synthetizing procedures create the conditions of possibility to determine which hypotheses about the past are more plausible and which are less plausible (in the same cases also, e.g., which are more and which less probable). They allow historical theories to participate in the Quinean rational game of science; they prove that it is possible to write history *also* from within science.

---

[17] More complicated, but simultaneously interesting and important, is the case of human evolution and/or history, since the possibility to conceptualize cultural history using evolutionary theories is still a debated one. For a classic work on this topic, see Fracchia and Lewontin (1999); for an unorthodox yet biologically motivated piece, see Portmann (1990); and to track the current debate, see Richerson and Christiansen (2013); and Heyes (2018). Also, Stovall (2015, 16ff) offers a focused analysis of Darwin's own "hypothesis" about the evolutionary past with respect to his theory of natural selection.

[18] This does not seem that problematic, since a certain degree of underdetermination is characteristic even for non-historical inquiries, which is taken for granted in the current philosophy of science and Quine, of course, acknowledges it, too. Quine considers our total theory of the world underdetermined as a whole. See, e.g., Quine's reply to Chomsky in Quine (1975a, 302ff).

2.4 Quine's Scientism in Reverse

Does Quine have any scientistic leanings? I argued that his notion of science does not privilege *natural* sciences per se, meaning experimental methods and observations, as the only acceptable and appropriate way of producing knowledge (of scientific theory-building). Quite the opposite, as I have demonstrated, Quine actively denies this assumption. In the collection of essays published in 2000 and put together on the occasion of his 90th birthday, Quine responds to a text commenting on his epistemology and laments: "It is awkward that 'science', unlike *scientia* and *Wissenschaft*, so strongly connotes natural science nowadays," (Quine 2000, 411).

The main aim of Sect. 2.2 and Sect. 2.3 was to reconstruct Quine's conception of science, and his subsequent views on writing history. My main motivation for this interpretative exercise was that Quine is often considered an intellectual father of contemporary forms of naturalism, which is sometimes understood and defended as a scientistic and/or physicalist doctrine. However, my analysis shows that Quine has a very inclusive and idiosyncratic conception of science that, inter alia, incorporates all the so-called soft and social sciences, and is ontologically neutral.

I have also argued that Quine's conception of science as our total world-picture can accommodate historical theories, inclusive of the essentially underdetermined historical theories. Thus, these two sections can also be read as a case study analysis of Quine's take on all non-experimental modes of scientific inquiry. To encapsulate, without understanding Quine's specific conception of science, one cannot properly understand his natural*ism* or pragmat*ism*, or so I argue.

However, in keeping with Quine, is there a way to acquire factual knowledge apart from science qua our total theory of the world? No, there is not. But it is rather an acknowledgment of the limitations of our science, not the opposite. It is primarily the acknowledgment that science is our best tool for rationalizing beliefs. Science simply exemplifies rationality (especially inductive but *also* deductive and abductive inferences). Put another way, if our science has co-evolved with our rationality (culture and conceptual frameworks), then science is a man-made, cultural tool to manage and manipulate our environment. It cannot be a tool to review the man-independent nature of our past and present world. Or, to be less presumptuous, claiming so would require different, additional arguments. I am of the view that to simply assume that just because "science works" (i.e., is technologically very effective), it therefore must be in touch with something like "the reality in itself" is against the idea of Occam's Razor.

At any rate, this is both a limitation and a liberation of our epistemic condition (cf., Quine 1995b, 256–257). Liberalizing is the fact that naturalized epistemology does not need to fix any method or logic of science. *Naturalism is a methodologically and metaphysically opportunistic epistemological project.* That is, at the same time, its limitation insomuch as Quinean naturalism intentionally steps back from identifying a historically invariant nature of science and (scientific) knowledge. According to Quine, there are no epistemic certainties nor unconceptualized givens. However, despite that, our common sense as well as science rely on the historically accumulated belief system guiding and grounding our epistemic cultures. Absent these "certainties," empirical science would not be possible. "If experience is the ground of our certainty, then naturally it is past experience," (Wittgenstein 1969, 275). Still, history also proves that such a ground is, in fact, shifting ground. This all re-configures and also creates wholly new problems and challenges for our scientific theory-building.

I suspect that, for many thinkers, the distinction between fact and fabulation (or even literary fiction) one is left with after Quine's naturalization of epistemology will not be sharp enough. For instance, one might be able, in principle, to compose an alternative fictitious historical theory/explanation of a given event that fits the available evidence and/or is just compatible with our current total world-picture. Quinean naturalists can, however, embrace this problem as an empirical, and in fact pragmatic, challenge which can be dealt with only from within the historiographical practices working with anything our ongoing total science has to offer. True, building essentially underdetermined theories about historical what- and why-questions is a difficult task, but no more difficult than building any other theories with indirect empirical consequences.

I thus contend that historians, like any other specialists, must first answer a more difficult question: what is the ultimate purpose of their inquiry? Or as Hayden White puts it in his seminal paper, *The Burden of History* (where he, among many other things, also makes a plea for "methodological and stylistic cosmopolitanism" in historiography):

> Is there any reason why we ought to study things under the aspect of their past-ness rather than under the aspect of their present-ness, which is the aspect under which everything offers itself for contemplation immediately? (White 1966, 132, cf., 130)

Although, with Quine's naturalism, we all lose the *philosophical* antidote to bad forms of theorizing, apart from the standard practice of citing what one takes the facts to be and what is taken to be most plausible, no one seems to be theoretically or pragmatically paralyzed in this epistemic condition; naturalistic historians can benefit from including and experimenting with diverse methods and styles of inquiry, and naturalistic scientists can freely utilize various historical theories (for example, the histories of their respective disciplines).

That is where the normative part of the very process of naturalizing epistemology itself helps but also stops. It is the therapy one no longer needs when appreciating that there is no extra-scientific perspective, and that further specific inquiries, inclusive of their normative implications, must be conducted and confronted from an intra-scientific perspective.[19] Put another way, further inquiries, e.g., into the specifics of historical theorizing, such as specific theory choices and specific analyses of historical explanations/evidence, are a going concern of respective scientific disciplines and their theoretical as well as practical aims.

[19] I frame here the issue analogically to Roth (1999, 100, 105). Intra-scientific perspective here also simply means *assumption-laden* perspective. Equivalently, one could take extra-scientific perspective to be some kind of assumption-free perspective.

However, the word "science" must be read with Quine in its re-configured, all-inclusive sense, viz., as our total science/total world-picture. And not simply as an interchangeable term for "natural science." What we need is scientism in reverse; we need to be methodologically opportunistic and disciplinarily intersectional in practice.[20] In this sense, Quine's naturalism does both humanize and historicize the whole domain of rational inquiry.

References

Ankersmit, Frank. 2013. Reply to Professor Roth: On how antidogmatism bred dogmatism. *Rethinking History* 17 (4): 570–585.

Arenberg, Israel K., Vincent J. Di Maio, and Michael M. Baden. 2020. A reevaluation of the death of Vincent van Gogh: Suicide or murder? The need for a definitive autopsy. *The American Journal of Forensic Medicine and Pathology* 41 (4): 291–298.

Borradori, Giovanna. 1994. *The american philosopher: Conversations with Quine, Davidson, Putnam, Nozick, Danto, Rorty, Cavell, MacIntyre, Kuhn*. Trans. Rosanna Crocitto. Chicago: University of Chicago Press.

Bryant, Amanda. 2020. Naturalisms. *Think* 19 (56): 35–50.

Ciupe, Alexandru. 2007. Quine and postmodernism. *Lingua B. Cultură şi Civilizaţie* (VI–VII):25–32.

Currie, Adrian. 2018. *Rock, bone, and ruin: An pptimist's guide to the Historical Sciences*. Cambridge, MA: MIT Press.

Currie, Adrian, and Derek Turner. 2016. Introduction: Scientific knowledge of the deep past. *Studies in History and Philosophy of Science A* 55:43–46.

Dasgupta, Shouvik, Rik Vanspauwen, Eylem A. Guneri, and Matteo Mandala. 2022. Vincent van Gogh and the elusive diagnosis of vestibular migraine. *Medical Hypotheses* 159:110747.

Dreben, Burton. 1992. Putnam, Quine—and the facts. *Philosophical Topics*, 20(1), 293–315.

Fracchia, Joseph, and Richard C. Lewontin. 1999. Does culture evolve? *History and Theory* 38 (4): 52–78.

Gibson, Roger F. 1992. The key to interpreting Quine. *The Southern Journal of Philosophy* 30 (4): 17–30.

Godfrey-Smith, Peter. 2014. Quine and pragmatism. In *A companion to W. V. O. Quine*, ed. Gilbert Harman and Ernest Lepore, 54–68. Hoboken: Wiley-Blackwell.

Gould, Stephen Jay. 2002. *The structure of evolutionary theory*. Cambridge, MA: Harvard University Press.

[20] Moreover, together with Ullian, Quine writes that there are obviously other language games, such as fiction and poetry, that also "teach us" and "bring enlightenment," (see Quine and Ullian 1978, 4). So, although, according to Quine, *science* has a monopoly on the fact-stating discourse, he does *not* deny, contra a traditionally conceived scientism, that one can learn something about the world from other forms of discourse. To examine the specific reasons for such a claim is beyond the scope of this chapter. However, the general reasons are present here; Quine dismisses any *formal* demarcation criterion of science. That naturally leads to complex and context dependent overlaps between, e.g., the factual and non-factual discourse. Science is thus not the only rational language game in town. As I have also emphasized here, Quine thinks that philosophy itself is (methodologically) continuous with science, but that does not mean that philosophy does not pursue different aims. Quine does not say much on this issue but when he does, he seems to opt for this possibility. That is to say, Quine does not deny that philosophy, qua an intellectual discipline, still can have a certain kind of autonomy.

Gould, Stephen Jay, and Elisabeth S. Vrba. 1982. Exaptation – A missing term in the science of form. *Paleobiology* 8 (1): 4–15.

Haack, Susan. 1993. The two faces of Quine's naturalism. *Synthese* 94:335–356.

Hacking, Ian. 1983. *Representing and intervening*. Cambridge, UK: Cambridge University Press.

Heyes, Cecilia. 2018. *Cognitive gadgets: The cultural evolution of thinking*. Cambridge, MA: Harvard University Press.

Hookway, Christopher. 1988. *Language, experience and reality*. Oxford: Blackwell Press.

Hubálek, Michal. 2021. A brief (hi)story of just-so stories in evolutionary science. *Philosophy of the Social Sciences* 51 (5): 447–468.

Hylton, Peter. 2007. *Quine*. Abingdon: Routledge.

Kemp, Gary. 2017. Quine, publicity, and pre-established harmony. *ProtoSociology* 34:59–72.

Kuukkanen, Jouni-Matti. 2015. *Postnarrativist philosophy of historiography*. New York: Palgrave Macmillan.

Patterson, Dennis M. 1996. *Law and truth*. Oxford: Oxford University Press.

Peregrin, Jaroslav. 2016. *Meaning and structure: Structuralism of (post)analytic philosophers*. New York: Routledge.

Portmann, Adolf. 1990. *A zoologist looks at humankind*. Trans. James Schaefer. New York: Columbia University Press.

Putnam, Hilary. 1992 (1988). *The greatest logical positivist*. In Realism with a human face, ed. James Conant, 268–277. Cambridge, MA: Harvard University Press.

Quine, Willard Van Orman. 1957. The scope and language of science. *The British Journal for the Philosophy of Science* 8 (29): 1–17.

Quine, Willard Van Orman. 1961a. On what there is. In *From a logical point of view: Logico-philosophical essays*, 1–19. New York: Harper and Row, Publishers. (Originally published 1948).

Quine, Willard Van Orman. 1961b. Identity, ostension, and hypostasis. In *From a logical point of view: Logico-philosophical essays*, 65–79. New York: Harper and Row, Publishers. (Originally published 1950).

Quine, Willard Van Orman. 1961c. Two dogmas of empiricism. In *From a logical point of view: Logico-philosophical essays*, 20–46. New York: Harper and Row, Publishers. (Originally published 1951).

Quine, Willard Van Orman. 1969a. Speaking of objects. In *Ontological relativity*, 1–25. New York: Columbia University Press.

Quine, Willard Van Orman. 1969b. Epistemology naturalized. In *Ontological relativity*, 69–90. New York: Columbia University Press.

Quine, Willard Van Orman. 1969c. Natural kinds. In *Ontological relativity*, 114–138. New York: Columbia University Press.

Quine, Willard Van Orman. 1975a. Replies. In *Words and objections*, ed. Donald Davidson and Jaakko Hintikka, 292–352. Dordrecht: D. Reidel. (Originally published 1969).

Quine, Willard Van Orman. 1975b. On empirically equivalent systems of the world. *Erkenntnis*, 9(3), 313–328.

Quine, Willard Van Orman. 1982. Fives milestones of empiricism. In *Theories and things*, 67–72. Cambridge, MA: Harvard University Press. (Originally published 1981).

Quine, Willard Van Orman. 1987. *Quiddities: An intermittently philosophical dictionary*. Cambridge, MA: Harvard University Press.

Quine, Willard Van Orman. 1992. *Pursuit of truth*. 2nd ed. Cambridge, MA: Harvard University Press. (Originally published 1990).

Quine, Willard Van Orman. 1993. In praise of observation sentences. *The Journal of Philosophy* 90 (3): 107–116.

Quine, Willard Van Orman. 1995a. *From stimulus to science*. Cambridge, MA: Harvard University Press.

Quine, Willard Van Orman. 1995b. Naturalism; or, living within one's means. *Dialectica* 49 (2–4): 251–263.

Quine, Willard Van Orman. 2000. Quine's responses. In *Knowledge, language and logic: Questions for Quine*, ed. Alex Orenstein and Petr Kotatko, 407–430. Berlin: Springer Science & Business Media.

Quine, Willard Van Orman. 2013. *Word and object*. Cambridge, MA: MIT Press. (Originally published 1960).

Quine, Willard Van Orman. 1976. *Necessary truth*. In The ways of paradox and other essays, 68–76. Revised and enlarged edition. Cambridge, MA: Harvard University Press.

Quine, Willard Van Orman, and Joseph S. Ullian. 1978. *The web of belief*. 2nd ed. New York: Random. (Originally published 1970).

de Regt, Henk W. 2017. *Understanding scientific understanding*. Oxford: Oxford University Press.

Richerson, Peter J., and Morten H. Christiansen, eds. 2013. *Cultural evolution: Society, technology, language, and religion*. Cambridge, MA: MIT Press.

Rorty, Richard. 2021. *Pragmatism as anti-authoritarianism*. Cambridge, MA: Harvard University Press.

Rosenberg, Alexander. 2018. *How history gets things wrong: The neuroscience of our addiction to stories*. Cambridge, MA: MIT Press.

Roth, Paul A. 1983. Reconstructing Quine: The troubles with a tradition. *Metaphilosophy* 14 (3/4): 249–266.

Roth, Paul A. 1999. The epistemology of "Epistemology Naturalized". *Dialectica* 53 (2): 87–110.

Roth, Paul A. 2003. Why there is nothing rather than something: Quine on behaviorism, meaning, and indeterminacy. In *Philosophy, psychology, and psychologism: Critical and historical readings on the psychological turn in philosophy*, ed. Dale Jacquette, 263–287. Dordrecht: Springer.

Roth, Paul A. 2008. The epistemology of science after Quine. In *The routledge companion to philosophy of science*, ed. Stathis Psillos and Martin Curd, 3–14. Abingdon: Routledge.

Roth, Paul A. 2020. *The philosophical structure of historical explanation*. Evanston: Northwestern University Press.

Stovall, Preston. 2015. Inference by analogy and the progress of knowledge: From reflection to determination in judgements of natural purpose. *British Journal for the History of Philosophy* 23 (4): 681–709.

Turnbull, Alan. 2022. *Who shot Van Gogh?: Facts and counterfacts about the world's most famous artist*. London: Thames & Hudson.

Verhaegh, Sander. 2018. *Working from within: The nature and development of Quine's naturalism*. Oxford: Oxford University Press.

Voskuil, Pieter. 2020. Vincent van Gogh and his illness: A reflection on a posthumous diagnostic exercise. *Epilepsy & Behavior* 111:107258.

White, Hayden V. 1966. The burden of history. *History and Theory* 5 (2): 111–134.

Wittgenstein, Ludwig. 1969. On certainty. Ed. G. E. M. Anscombe and G. H. von Wright. Oxford: Blackwell Publishers Ltd.

Zammito, John H. 2004. *A nice derangement of epistemes: Post-positivism in the study of science from Quine to Latour*. Chicago: University of Chicago Press.

Chapter 3
Quinean Explication of Reality and Naturalistic Philosophy

Abstract This chapter deals with two often underappreciated Quinean naturalistic explications. The first explication concerns the metaphysical status of reality under naturalism, by examining what can be viewed as Quine's attempt to dissolve realism and idealism into each other; the second explication concerns the metaphilosophical dimension of naturalized epistemology. I aim to demonstrate that Quine's way of approaching these two concepts turns out to be not only complementary, but also highly indicative of his overall naturalistic outlooks as well as professional preferences. Namely, they are approached by Quine's characteristic *from within* approach, which highlights the intra-theoretic and intra-linguistic nature of all human empirical inquiry.

Keywords Quine · Explication · Reality · Naturalistic philosophy · Language games

3.1 Introduction: Defending Quine's Naturalistic Legacy

As I have also tried to indicate in Chap. 1, naturalism is nowadays without a doubt understood in many different ways (see again Bryant 2020; Rosenberg 1996; cf., Danto 1967). Part of the reason for this is arguably the fact that the subscription to at least a very broad naturalism is a standard for present-day philosophers. As David Papineau pointedly states: "For better or worse, 'naturalism' is widely viewed as a positive term in philosophical circles – only a minority of philosophers nowadays are happy to announce themselves as 'non-naturalists,'" (Papineau 2021). So, in recent years, we have witnessed several instructive attempts to (re)define naturalism anew (see again Price 2011 and De Caro and Macarthur 2004).

Many of these attempts go, whether directly or indirectly, in the footsteps of the two well-known naturalists writing in the second half of the twentieth century, W. V. O. Quine and Wilfrid Sellars (see, e.g., Peregrin 2023; Roth 2020). The aim of this chapter, then—apart from supplementing the material from the previous two

M. Hubálek, *Dissolving Naturalism and Historicism Into Each Other*, Studies in History and Philosophy of Science 67,
https://doi.org/10.1007/978-3-032-07535-2_3

chapters—is to contribute to past and contemporary debates about naturalism by examining two often underappreciated Quinean naturalistic explications.

In the first part of this chapter, I focus on the first explication which concerns the metaphysical status of "reality" under naturalism, by examining what can be viewed as Quine's own attempt to dissolve realism and idealism into each other. In the second part, I move on to parse the second explication which concerns the metaphilosophical dimension of naturalized epistemology.

Moreover, as will gradually become evident, I present and defend Quine's naturalistic legacy in a manner that makes Quine much more complementary to thinkers such as John Dewey, Ludwig Wittgenstein, Wilfrid Sellars, Stephen Jay Gould, or Richard Rorty than is usually acknowledged in print. Put another way, although there are certainly many differences between Quine and the thinkers just mentioned, the similarities highlighted in this chapter, however local they might seem, challenge the widespread view that these thinkers radically broke away from Quine's very own naturalistic framework.

3.2 Dissolving Realism and Idealism into Each Other

> Natural selection the great solvent. Dissolved final cause into efficient cause. Dissolves realism and idealism into each other. Dissolves disparateness of neural input into a like output. Kinds are projected for terms, and terms are vague, applying in degrees. Relative similarity is what is objective; not natural kinds. Biological species are atypically clean-cut. Similarity is projected on the world from our associations.
>
> —W. V. O. Quine[1]

3.2.1 *A Realist Quine*

A lot has been written on Quine and real*ism* and no wonder (see, e.g., Jaksland 2024; Pils 2020; Keskinen 2012; Dreben 1992; Rorty 1976).[2] Let me start right in the middle and introduce the basics of Quine's position on this matter. Quine's naturalistic, intra-theoretic position holds that the "objects" laymen as well as scientists

---

[1] (Quine 2008, 179–180).

[2] I purposely refrain from providing the reader with a concrete delimitation of realism and anti-realism here to effectively highlight the fact that Quine's own position is quite idiosyncratic. A very minimalistic delimitation should do; *The Oxford Companion to Philosophy* instructively states: "[Realism and anti-realism are] primarily directions, not positions. To assert that something is somehow mind-independent is to move in the realist direction; to deny it is to move in the opposite direction. No sane position is reached at either extreme. Not everything is in every way independent of minds; if there were no minds, there would be no pain. Not everything depends in every way on minds; if I forget that Halley's comet exists, it does not cease to exist. Many philosophical questions have the general form: Is such-and-such mind-independent in so-and-so way? Given specifications of such-and-such and so-and-so, one may call someone who answers 'Yes' a realist.

talk about (e.g., atoms, genes, stones, bodies, and numbers) are to be regarded as linguistic posits. Qua posits, objects of our discourse are then seen as conceptual tools we develop and utilize to accommodate and make sense of our experiences (of our causal interactions with the world). Then, for both laymen as well as scientists, any idea of "reality" must be worked out in a theory-internal way.

We inherit and invent concepts and theories about the world which commit us to a certain ontology; they commit us to positing certain kinds of objects. For us humans, the existence of specific objects with their specific characteristics is thus always theory-dependent. To ask what objects there are independently of any concepts and theories is hopeless, according to Quine. Without theories there are no things-in-themselves, but only blooming, buzzing confusion (Quine 1995a, 225). Also, due to Quine's well-known commitment to holism, the identity conditions for particular objects are dependent on a bundle of theories taken as a whole; they depend on *science as a whole*. Quine often also speaks of our "web of belief" or "total theory of the world" (Quine 1961a, 1961b; cf., Quine and Ullian 1978 and Quine 1987, 108ff). Note that this all applies to "physical objects" as well. In this sense, physical objects are also science-dependent:

> Physical objects are conceptually imported into the situation as convenient intermediaries – *not by definition in terms of experience*, but simply as irreducible posits comparable, epistemologically, to the gods of Homer. Let me interject that for my part I do, qua lay physicist, believe in physical objects and not in Homer's gods; and I consider it a scientific error to believe otherwise. *But in point of epistemological footing the physical objects and the gods differ only in degree and not in kind. Both sorts of entities enter our conception only as cultural posits.* The myth of physical objects is epistemologically superior to most in that it has proved more efficacious than other myths as a device for working a manageable structure into the flux of experience. [...]
>
> Total science, mathematical and natural and human, is [...] extremely underdetermined by experience. *The edge of the system must be kept squared with experience; the rest, with all its elaborate myths or fictions, has as its objective the simplicity of laws.* (Quine 1961a, 44, 45, emphasis mine)

For many thinkers, these postulations have anti-realist or even idealist undertones. Nevertheless, Quine still wants to be an empiricist; he holds that for an experience to be *evidence of/for* something there must be some intersubjective, observable/behavioral criteria at work. So, the objects, to whose existence we are committed, are in fact dependent not only on our theories but also on "the world" (given causal contexts) (see Quine 1969a, 1992, 5ff; cf., 1987, 159–161). Quine of course never denies this, ipso facto, he never denies that *the world exists independently of us*. The only difficulty here is that, according to him, our human knowledge of this world is co-determined by our conceptual choices and the facts of the matter at the same time, with no unambiguous way of separating these two elements. This is a result of dismantling the fundamental distinction between analytic and synthetic beliefs/truths (cf., Roth 2020, 115, 169, footnote 2; and Danto 1967).

Since different philosophers take different specifications for granted, the word 'realism' is used in a bewildering variety of senses [...]," (Honderich 1995, 746–747).

One of the early commentators dealing with Quine's holistic view of language and scientific knowledge, Albert Hofstadter, who moreover pointedly labels Quine's position as "holistic pragmatism," contemplates that, in Quine's theoretical system, there is no room for a philosophically significant distinction between scientific invention and discovery. And although Hofstadter in the end to a large degree diverges from Quine on this point, he instructively summarizes Quine's outlook as follows: "The consequence for the invention-discovery distinction is that while science depends upon both human and non-human factors, one can not distinguish the two factors within science. [...] Both operate *in* a kind of functional relation to each other and *on* the whole of the language of knowledge. There are no empirical tests or conventional decisions regarding statements individually. Every test and every decision reverberates through the whole, affects and is affected by the whole, so that in every empirical test the whole of our knowledge is tested and in every decision the whole fabric is decided on. *Or perhaps better, the distinction between test and decision also breaks down*," (Hofstadter 1954, 401; cf., 412ff). One could say for now that, for Quine, our world, the world we live and theorize in, is both invented and discovered; made and found.

However, especially due to Quine's insistence on the observable/behavioral notion of evidence, is not Quine then committed to a position taking, e.g., observable "items" (however holistically conceived) as in some sense primary, or even as the ones that are *really real*? The answer to this question is negative, as Quine explicitly subscribes to a view that all objects are theoretical objects. If there is a difference between positing the existence of moles and molecules, it is a difference in degree and not in kind (e.g., Quine 1961c, 18; Quine 1969b, 87). Again, without our sentences (or generally conceptual frameworks), there would not be any objects or facts for us to talk about (see, e.g., Quine 1981a, 20). But, interestingly enough, this does not discourage Quine from declaring, in the next breath, that he endorses "robust realism," i.e., "unswerving belief in external things – people, nerve endings, sticks, stones." And then continues by stating: "I believe also, if less firmly, in atoms and electrons and in classes," (Quine 1981b, 21). But the belief in physical objects is firmer, among other things, only insofar as the words for them are acquired ontogenetically *earlier* than for abstract objects when one acquires a first language, so physical objects are "more deeply rooted in our formative past." In this particular regard, the vocabulary of physical objects is considered primitive, as it is directly tied to our interactions with the causal realm (Quine 2013, 215).

3.2.2 Naturalistic Reconfiguration of "Reality"

With respect to the traditional disputes about the philosophical forms of realism and anti-realism, Quine finds himself in an outlandish situation insomuch as he prima facie saves the most fundamental beliefs of both of these isms; Quine does not deny that the world itself (the causal realm) exists independently of us, and simultaneously, grants us significant creative freedom in imposing our conceptual frameworks on this world.

Nonetheless, insomuch as Quine is resolute in denying the theory-*independent* existence of objects for us human cognizers and agents, I contend that he is much closer to anti-realism than to realism, at least to any naïve forms of realism; for Quine, the identity conditions and/or truth values for categorizing and describing objects are theory-*dependent*. Also, I think that the vast majority of realists would want to hear more, for example, that the world possesses its own unifying and/or knowable structure as well as rules/principles. In other words: its own integrity (cf., Wright 1993).

I thus want to proceed by defending Quine, even if somewhat indirectly, against the accusation that his holism forces him to embrace a form of idealism. Hofstadter, for example, adverts Quine's "holistic pragmatism" as "an empiricist equivalent of the Idealistic Absolute" (Hofstadter 1954, 416). Doing this will, I hope, help to establish what kind of anti-realist Quine actually is. Because indeed, according to Quine, the *reality* we recognize and live in is co-created by us, because "all ascriptions of *reality* must come [...] from within one's theory of the world; it is incoherent otherwise," (Quine 1981a, 21). Quine a bit later sums up his views in this way:

> *The objectivity of our knowledge of the external world remains rooted in our contact with the external world, hence in our neural intake and the observation sentences that respond to it.* We begin with the monolithic sentence, not the term. A lesson of proxy functions is that our ontology, like grammar, is part of our own conceptual contribution to our theory of the world. *Man proposes; the world disposes*, but only by holophrastic yes-or-no verdicts on the observation sentences that embody man's predictions. (Quine 1992, 36, emphasis mine)

It is obvious, and fully comprehensible from a naturalistic vantage point, that Quine's "solution" (I use quotation marks, as Quine himself never addresses this issue directly) to the problem of "reality" is motivated by understanding our human position in the world; *our human epistemic condition*. We are always *in* the world and all our beliefs about this world are always determined by our cognitive faculties and vocabularies interacting with the world. After all, that is what our current science (numerous branches of biology and anthropology) tells us. We are animals/beings sharing the world with other animals/beings. We "antecedently acknowledge" the external world in everything we do, hence also when inquiring into and describing the world (Quine 1992, 19).

One must always take into account that Quine thought of himself as a Darwinian (see, e.g., Quine 1996). However, in this particular regard, Quine is also very close to the pragmatism/naturalism of John Dewey who, inter alia, forcefully advanced an anti-representational view of our knowledge of the world (cf., Quine 1969d, 27ff). Dewey was convinced that many thinkers, and philosophers especially (see Dewey 1911), make a pragmatically impotent and theoretically unnecessary question from the relationship between our cognition and the world. Louis Menand fruitfully summarizes Dewey on this point as follows:

> The pragmatist response to this question is to point out that nobody has ever made a problem about the relation between, for example, the hand and the world. The function of the

> hand is to help the organism cope with the environment; in situations in which a hand doesn't work, we try something else, such as a foot, or a fishhook, or an editorial. [...] *Dewey thought that ideas and beliefs are the same as hands: instruments for coping. An idea has no greater metaphysical stature than, say, a fork.* [...] Dewey's point was that "mind" and "reality," like "stimulus" and "response," name nonexistent entities: they are abstractions from a single, indivisible process. It therefore makes as little sense to talk about a "split" that needs to be overcome between the mind and the world as it does to talk about a "split" between the hand and the environment, or the fork and the soup. (Menand 2002, 360–361, emphasis mine)

This is a broadly naturalistic approach, since it openly draws on the Darwinian insight that we are ultimately organisms *coping* with our environments, and not necessarily *copying* its metaphysical structures (cf., Rorty 1998, 2021). With this starting point, it is very hard to conceive an argument claiming that we might not be in touch with the (external/real) world. The only possibility seems to be claiming that we are not in touch with the world *as it is in itself*. But I do not think that there actually is a conclusive counter-argument to this Kantian move, i.e., an argument that makes a pragmatic difference to our everyday life and/or various forms of theorizing. It will always be possible to insist on the possibility that there is a "really real" world we cannot reach for some reason or other.

In view of this, I argue that one cannot satisfactorily answer the question "what is the metaphysical status of reality in Quine's theoretical system?," unless one takes seriously his naturalistic viewpoints. That is to say, our relationship to the world is, in principle, empirically explainable, meaning causally and/or historically (genealogically). In contemporary idiom, one could say that the inquiries into our phylogeny, ontogeny, and cultural history offer the most fundamental understanding of the human condition in the world.[3]

Just like Darwin, Quine does not strive to identify an essence of human beings/cognizers; his strategy is ethological and ecological; he is interested in behavioral patterns and principles typical for us as a biological and social species, which are the result of our modus vivendi (of our living conditions). This directly implies that

[3] However, this is not to say that these inquiries produce some "objectively" fundamental starting points. They are obviously also scientific theories. In other words, e.g., while developing specific evolutionary explanations, scientists make many conceptual choices and commitments, at minimum with respect to "science" and "evolution." Also, as the survey study of Tuomisto et al. (2018) tentatively shows, the inquiry into the origin of exclusively human traits across the scientific fields such as paleoanthropology, paleontology, ecology, evolution, and human biology is very diverse, meaning that it is possible to identify popular hypotheses, but there is no universal consensus. Joseph Heath and Catherin Rioux in the same vein reflect on the current state of affairs in evolutionary ethics and write: *A common mistake made by philosophers working under the banner of "evolutionary ethics" has been to suppose that the evolutionary science on this question is settled*—that we have an answer to the question, not just how human morality is possible, but even how it evolved. *If this were correct, then the only task remaining for the philosopher would be to draw out the normative implications of this body of science. Unfortunately, there is no such scientific consensus. Indeed, the evolution of human ultrasociality—whether it be altruistic or cooperative—is one of the most important unanswered questions in the life sciences*," (Heath and Rioux 2018, 1–2, emphasis mine). Of course, it is also necessary to ask whether the consensus on these matters is possible/desirable, but I leave aside that here.

we, as humans, always approach the world *from within our theories about it*. There is neither a profound nor fundamental level of the world. We cannot go beyond a posited world to understand *who we are* and *what world* we actually inhabit. This is a conceptual leap most of the naïve realists will simply not accept.[4] This leap, moreover, undermines any philosophical, and I would also say any pragmatic, substance of the distinction between *the world in itself* and the *world for us*.

For Quine himself, the most valuable naturalistic insight into "the nature" of the world comes from the study of our acquisition of first language and of our language use in general, since any genuine observation is a linguistic matter (it presupposes inductive agreement on our basic categories). Inter alia for this reason, our common sense, science, and rationality itself comprise one cognitive and epistemic continuum. As is well known, Quine is not interested in any ahistorical demarcation criterion of science (see especially Quine 1957, 1995a).

That is the proper context for understanding Quine's much-quoted passages on ontology from his notorious essay "On What There Is." Consider now these samples:

> *To be assumed as an entity is, purely and simply, to be reckoned as the value of a variable.* In terms of the categories of traditional grammar, this amounts roughly to saying that *to be is to be in the range of reference of a pronoun*. Pronouns are the basic media of reference; nouns might better have been named propronouns. (Quine 1961c, 13, emphasis mine)

> We can very easily involve ourselves in ontological commitments by saying, for example, that *there is something* (bound variable) which red houses and sunsets have in common; or that *there is something* which is a prime number larger than a million. But, this is, essentially, the *only* way we can involve ourselves in ontological commitments: by our use of bound variables. (ibid., 12)

> Viewed from within the phenomenalistic conceptual scheme, the ontologies of physical objects and mathematical objects are myths. *The quality of myth, however, is relative; relative, in this case, to the epistemological point of view. This point of view is one among various, corresponding to one among our various interests and purposes.* (ibid., 19)

However, it is easy to underestimate, or just forget, that Quine's theoretical interests and interpretations are constantly driven by pragmatic hence primarily epistemic concerns. Ontology follows epistemology, not vice versa. Or, as one of the two epigraphs for *Word and Object* (2013) has it; "ontology recapitulates philology."[5] Quine memorably opens his essay "Ontological Relativity" by this announcement: "With Dewey I hold that knowledge, mind, and meaning are part of the same world that they have to do with, and that they are to be studied in the same empirical spirit that animates natural science," (Quine 1969d, 26). This announcement, however, can be read in several ways, each of which is relative to a particular understanding of the key concepts ("the world," "empirical," and "natural science") as well as to

[4] See, for example, the recent attempt to endorse naïve realism by Mitrović (2022). Typically, as treated again in *The Oxford Companion to Philosophy*, naïve realism amounts to "a theory [...] that holds that our ordinary perception of physical objects is direct, unmediated by awareness of subjective entities, and that, in normal perceptual conditions, these objects have the properties they appear to have [...]," (Honderich 1995, 602).

[5] The author of the epigraph is an American biologist James Grier Miller.

our interests and purposes. Put another way, although it is not obvious on first reading, or on contextless reading, it is a conceptually hence ontologically plastic announcement.

The ontological relativity, as Quine construes it, arises from the appreciation that "there be no ultimate sense in which [the] universe can have been specified. [...] *Ontology is indeed doubly relative*. Specifying the universe of a theory makes sense only relative to some background theory, and only relative to some choice of a manual of translation of the one theory into the other. [...] We cannot know what something is without knowing how it is marked off from other things. *Identity is thus of a piece with ontology*," sums up Quine (1969c, 50, 54–55, emphasis mine). Also, nothing prevents us in these settings from utilizing the Wittgensteinian conception of "language games," as Quine himself has done occasionally since the 1990s (see, e.g., Quine 1992, 20).

Now, to rebut the reading of Quine as being an idealist of a solipsistic kind, one must ask how the causal realm (the world) constrains our linguistic practices of posit*ing* objects. One also must agree with Davidson that: "It is reasonable to call something a posit if it can be contrasted with something that is not. [For Quine] the something that is not is sensory experience – at least that is the idea," (Davidson 1973–1974, 16). But since Quine denies that there is a possibility to unambiguously detect the empirical content of individual fact-stating terms and sentences, one must reformulate the question.

Let me try. For starters, it would be more useful to ask, "what makes some posited objects *better* than others and/or what makes them *genuinely scientific*?" (meaning simply what distinguishes them from our everyday/common-sense positing).

The straightforward (but only partial) answer would be, of course, that the answer to this question will always remain relative to our interests and purposes, but Quine in fact offers more than that. And, in this particular sense, Quine's dealing with the problem of positing is a special instance of (scientific) hypothesizing (cf., Quine 1995b, 49–50). Quine appreciated very soon that abductive inferences, or what is sometimes called "the inference to the best explanation," is a very useful and in fact indispensable tool for our hypothesis making. "Hypothesis, where successful, is a two-way street, extending back to explain the past and forward to predict the future. What we try to do in framing hypotheses is to explain some otherwise unexplained happenings by inventing a plausible story, a plausible description or history of relevant portions of the world," write cogently Quine and Ullian (1978, 66).

An instructive example at several levels is Quine's dealing with the concepts of "meanings" and "mental states" (believing, wishing, intending, etc.). Nowadays, it is hardly controversial that Quine refuses mind-body dualisms of Descartes' type and subscribes to "anomalous monism" (see Davidson 2002): "The point of anomalous monism is just that our mentalistic predicate imposes on bodily states and events a grouping that cannot be defined in the special vocabulary of physiology. Each of those individual states and events is physiologically describable, we presume, given all pertinent information," explains Quine (1995b, 88; cf., Quine 1992,

71ff).[6] However, *as a naturalist*, Quine cannot simply dogmatically start with this kind of physicalist ontology. Such an ontology must be a result of empirical/scientific considerations in order to be a naturalistic ontology.

So, the question here for us ultimately is: "why does Quine not think that meanings and mental states should be considered as (scientific) posits?" Roth offers a lucid summary of Quine's take on this matter in the form of an argument:

Pr. 1: For posits to have claim to reality, they must be justified as part of an explanatory theory within a naturalized epistemology.

Pr. 2: In order to be a justified part of an explanatory theory, the posits must either be necessary for (constitutive of) stimulations being evidence (as in quality spaces), or must provide an observable mark marking them as the things they are, i.e., there must be an objective basis for ascertaining that some behavior has been rightly or wrongly categorized a behavior of a certain type,

Pr. 3: But mental states and concepts have yet to be shown to be necessary for or even robustly explanatory of observed stabilities in behaviors; in addition, nothing marks a behavior as objectively categorized correctly or incorrectly,

C: Meanings and mental states have no claim to reality (because no claim to any genuine explanatory function). Whatever has no place in our best scientific scheme of explanations is not real. Whatever is not real does not belong to the realm of facts, i.e., constitute a fact of the matter. (Roth 2003, 275)

Premise number two has principal significance for understanding Quine's concept of posit. "Posits can be constitutive of the possibility of stimulations being evidence, or they can be kinds made objective by accepted theoretical explanations," (ibid.). The only objective, better said intersubjective, criteria for ascribing mental states to people are observable, i.e., physicalist/behavioral, criteria.[7] Note that this is merely an alternative way of acknowledging the "reality" of sensory experience, as Davidson alludes to in the quote above.

---

[6] Quine is very consistent on this point, consider this telling passage of his: "Descartes' dualism between mind and body is called metaphysics, but it could as well be reckoned as science, however false. He even had a causal theory of the interaction of mind and body through the pineal gland. *If I saw indirect explanatory benefit in positing sensibilia, possibilia, spirits, a Creator, I would joyfully accord them scientific status too, on a par with such avowedly scientific posits as quarks and black holes*," (Quine 1995a, 252, emphasis mine).

[7] "Quine charts what falls by the way in an advancing scientific picture of what there is. His view of epistemology as science self-applied, and his corresponding conclusion about the explanatory utility of meanings, has interesting parallels here to Kantian themes. *For at least one important link that connects a tradition that runs from Kant to Quine studies how human minds come to constitute a shared and mutually intelligible world. Kant, in this regard, assigns the inquiring mind a strongly constitutive role. But how concepts and percepts come together to form understanding remains an unsolved problem, a problem which finds its modern incarnation in questions of finding rules that guide behavior. Quine despairs of the task of using philosophical analysis to find such rules. He proposes, instead, to turn that job over to empirical psychology*," (ibid., 278, emphasis mine).

What we then actually do when speaking about mental states is *imposing mentalistic predicates on bodies*. Quine's argument is thus not reductionist, nor *primarily* ontological. It is a pragmatic and explanatory argument. So, according to Quine, to say that "ontology recapitulates philology" also means that *ontology recapitulates epistemology*.

To put it crudely, Quine argues that when explaining or interpreting human behavior, one does not need to assume "meanings" and "mental states" as ontologically distinct entities existing somewhere in the causal world (or in our heads, for example). However, he does not directly reject the idea that mentalistic or intentional vocabulary can sometimes serve as a satisfactory means for explaining and interpreting particular cases.

That is the key pragmatic difference here; to work with mental states (ascribing them to people) might be useful for interpreting human action, but positing a sui generis ontology of the mental does not, in itself, add to their explanatory force (unless, of course, positing an autonomous realm of the mental proves to be explanatorily beneficial, too). Recall that, Quine ultimately holds that all vocabularies function "merely" as conceptual tools for navigating our experiences of the world. This stance certainly differentiates Quine from other influential naturalists, e.g., from Wilfrid Sellars, when it comes to its ontological implications. However, how much difference this stance makes for our conception of empirical inquiry as such remains an open and often-discussed question (to learn more, one can start with Rosenberg 2007; Borradori 2008; and Roth 2023).

For the purposes of this chapter, the key point is that Quine's epistemology-first stance intricately aligns ontology with empirical and pragmatic concerns.

3.2.3 *Fusing Idealism with Instrumentalism*

At this point, another way of resuscitating the realist element of Quine's ontology reappears. Quine could be seen as a *scientific* realist. It could be argued that, because science is ultimately very successful technologically, it must provide us with at least some truths about the real nature of the world around us.

However, in Quine's intra-theoretic/intra-linguistic system, such a realism loses its metaphysical bite and is still only a form of instrumentalism, meaning a *methodological* position emphasizing the usefulness of posits/concepts qua tools. Be that as it may, Quine is very far from admitting the meaningfulness of the Platonic assumption of seeing our theories as "carving nature at its joints." Strictly speaking, for Quine, one can only "know" (intelligibly state) that there is nature (independent causal realm); whether nature has *its own* joints cannot be known unambiguously. It is simply hard to imagine what it would mean to know that *in pragmatic terms*. And to proclaim our "best" scientific posits as really real is at least problematic, due to all kinds of semantic indeterminacies and evidential underdeterminations Quine famously introduces (cf., e.g., Quine 1975a, 302ff). If the history of science teaches us anything, it is the fact that from both synchronic and diachronic perspectives, there is more than one way of conceptually carving the world.

To summarize: according to Quine, our view of the world with its particular objects is always interwoven with our theoretical and other pragmatic choices.

> *To call a posit a posit is not to patronize it.* […] *Everything to which we concede existence is a posit from the standpoint of a description of the theory-building process and simultaneously real from the standpoint of the theory that is being built. Nor let us look down on the standpoint of the theory as make-believe; for we can never do better than occupy the standpoint of some theory or other, the best we can muster at the time.* What reality is like is the business of scientists, in the broadest sense, painstakingly to surmise; and what there is, what is real, is part of that question. The question how we know what there is is simply part of the question […]. *The last arbiter is so-called scientific method, however amorphous.* (Quine 2013, 20–21, emphasis mine)

On my reading then, Quine performs a flight *from realism,* but it is not a flight *from reality* (hence nor *to any kind of idealism*). The metaphysical nature of our world and its objects are indeed "only" hypothesized; but the hypothesizing is our practical doing, it is hypothesizing via and against our interactions with the world/the causal contexts.

So, despite Davidson's intermittent suspicions, Quine would quite happily condone that: "In giving up dependence on the concept of an uninterpreted reality, something outside all schemes and science, we do not relinquish the notion of objective truth – quite the contrary. […] *In giving up the dualism of scheme and world, we do not give up the world, but re-establish unmediated touch with the familiar objects whose antics make our sentences and opinions true or false,*" (Davidson 1973–1974, 20, emphasis mine; cf., Quine 1984).

True, Quine's holistic notion of evidence implies, inter alia, that no single experience or scientific experiment will ever resolve an ontological dispute (cf., Quine 2013, 254). But why would one even expect that? It is certainly not licensed, for example, by our everyday experiences. Quite the opposite suggests itself. Consider the following quote from Roth, who briefly but pointedly explicates, in the Quinean/naturalistic attitude, what depending on theoretical and pragmatic choices means for our lives and our (scientific) understanding of the world:

> Background beliefs regarding social status or religious affiliation might influence which individual beliefs count or how they count. In addition, which beliefs might be open to revision will be determined by perceptions regarding how those beliefs connect to religious or political views deemed important. *Consideration such as these makes the "unit of empirical significance" culture-size.* (Roth 2020, 125)

Of course, Quine's, in a sense instrumental, belief in/accepting of physicalism is also of this parochial kind. Physicalism, in Quine's view, keeps our science going; it provides science with many sometimes verifiable and sometimes falsifiable *hypotheses* on which we then gradually construct the rest of our web of belief/our total theory of the world.

Thus, although "we are now seeing ontology as a more utterly human option than we used to," the options are not infinite, precisely because we are regulated by the world and limited by our natural capabilities (Quine 1995a, 260). Naturalism in fact embraces these limitations. "Naturalism need not cast aspersions on irresponsible metaphysics, however deserved, much less on soft sciences or on the speculative reaches of the hard ones, except insofar as a firmer basis is claimed for the experimental method itself," (ibid., 252, "experimental method" here simply stands for the intersubjective/third-person perspective).

Therefore, contra Hofstadter's interpretation, nothing in Quine justifies claiming that "whatever view I want is a view of the Universe." A naturalistic corrective of this idealistic vision of the world could be: "There is the view of the universe which sees it as it is."[8] However, an obligatory note for those still under the spell of Kantian visions: here, "as it is" refers to "as it is for us, the beings of the universe," rather than "as it is in itself."

3.3 Naturalizing Philosophical Inquiry

> I see philosophy not as an a priori propaedeutic or groundwork for science, but as continuous with science. I see philosophy and science as in the same boat – a boat which, to revert to Neurath's figure as I so often do, we can rebuild only at sea while staying afloat in it. There is no external vantage point, no first philosophy. All scientific findings, all scientific conjectures that are at present plausible, are therefore in my view as welcome for use in philosophy as elsewhere.
>
> —W. V. O. Quine[9]

3.3.1 The Continuity of Philosophy and Science

Despite many misunderstandings, Quine's naturalistic approach to various forms of inquiry is not by any means hostile towards philosophy itself. Even philosophy is part and parcel of his naturalized epistemology. Philosophy only loses its special, grounding status towards science as Quine does not think that there is any purely philosophical method, meaning *epistemologically privileged philosophical perspective*, such as first-person perspective, introspective insight, etc. Methodologically-wise, *philosophy is continuous with science* (e.g., Quine 1969a, 1995a). As I also revealed in the previous section, for Quine, all forms of explication and explanation are science-dependent, or one could say result-driven; they must be formulated from within a third-person perspective, otherwise the intersubjective, direct or indirect empirical checks, would simply not be available.

Philosophy thus also relies on the synchronic and diachronic developments of our web of belief; of our total theory of the world. So, Quine generally seems to oscillate between saying that philosophy and science are continuous (when he speaks in a disciplinary and methodological sense) and saying that philosophy is a

---

[8] I borrow these two characterizations from Peregrin (1999).

[9] (Quine 1969a, 126–127).

part of science (when he means by science "our total theory of the world"). Consider in this light these two paragraphs:

> The naturalistic philosopher *begins his reasoning within the inherited world theory as a going concern*. He tentatively believes all of it, but believes also that some unidentified portions are wrong. *He tries to improve, clarify, and understand the system from within*. He is the busy sailor adrift on Neurath's boat. (Quine 1981a, 72, emphasis mine)

> *Naturalistic philosophy is continuous with natural science. It undertakes to clarify, organize, and simplify the broadest and most basic concepts, and to analyze scientific method and evidence within the framework of science itself.* The boundary between naturalistic philosophy and the rest of science is just a vague matter of degree. (Quine 1995a, 256–257, emphasis mine).

However, the pragmatic and intellectual purposes of philosophy as a distinct discipline are not specified in greater detail by Quine himself. As is clear from the quotes above and from one of the epigraphs I have chosen for this section, Quine often only casually lists the general purposes of explication (clarifying and organizing concepts), methodological considerations, and aiming at understanding our web of belief/our total theory of the world as a whole. All the same, Quine can be read simply as emphasizing that philosophy can continue doing what it has always been doing as long as it respects the practices and results of other relevant disciplines and their theories, i.e., Quine can be read as not restricting the specific aims of philosophy in naturalism in advance of formulating given pragmatic or intellectual needs (cf., Churchland in Quine 2013, xiv).[10]

On the other hand, I think that one must be very careful and always distinguish between the very consequences of Quine's naturalism for philosophy, and Quine's own philosophical preferences and personal tastes. From a certain perspective, Quine is a conservative thinker immersed in analyzing a very specific subset of scientific and philosophical problems, where the latter are separated from the former only by their higher degree of generality and abstractness. Quine is simply not a philosopher who would engage in substantial debates concerning, for instance, the nature of morality or the problem of free will in our deterministic world. That being said, one has good reasons to suppose that Quine considered these problems as important empirical problems, but just not problems he is particularly invested in as a professional philosopher.

Another clue to interpreting Quine on this matter is provided by the BBC television interview with Bryan Magee in 1978 (see Philosophy Overdose 2021).[11] Magee

---

[10] Quine at one point also likens his idea of *philosophical explication* to Wittgenstein's vision of philosophy. He writes: "According to an influential doctrine of Wittgenstein's, the task of philosophy is not to solve problems but to dissolve them by showing that there were really none there. *This doctrine has its limitations, but it aptly fits explication.* For when explication banishes a problem it does so by showing it to be in an important sense unreal; viz., in the sense of proceeding only from needless usages," (Quine 2013, 240, emphasis mine). I see this as yet another argument for interpreting Quine's view of philosophy as a very general view, adaptable to many different aims and purposes.

[11] Quine in this interview even briefly addresses the problem of free will, thanks to Magee's interest in the broader consequences of his philosophical system.

asks Quine if he includes or excludes from philosophy "the age-old questions" like "how the world got here in the first place" or "how life began". Quine quickly and resolutely answers "I exclude these from philosophy." For Quine, the question of the beginning of the world is a question for physicists and astronomists and their "conjectures" and the question how life began is a question for biologists. No surprise here. However, then Quine adds that these two questions are in fact "pseudo-questions" (after that, the camera moves to Magee and his mischievous smile). Quine perceives these questions as meaningless as he "can't imagine what an answer would look like."

According to Quine, meaningful questions and their respective answers are articulated from within a theoretical framework, so what "world" or "life" is depends on the chosen frameworks. Put another way, we cannot speak of life without already having some beliefs, i.e., assumptions, about what life is, or generally what it is to exist. In "The Limits of Knowledge," Quine writes: "We have to work within some conceptual scheme or other; we can switch schemes, but we cannot stand apart from all of them. It is meaningless, while working within a theory, to question the reality of its objects or the truth of its laws, unless in so doing we are thinking of abandoning the theory and adopting another," (Quine 1976a, 65).[12]

Therefore, one could also say that the more internal a question to a conceptual framework is being asked, the more specific an answer will follow. Consider this example to illustrate the sense of gradability proposed here; a question "what causes genetic mutations?" will be answered from within the conceptual frameworks of genetics/evolutionary biology. For all that, the question "what is the nature of causality?" requires a more abstract answer that will not be associated with only one particular framework. One can clearly understand how the abstract answers impact the more specific ones. So, one clearly sees the place of philosophy on the continuum with science; indeed, to ask and answer abstract and general questions is a substantial contribution to (re-)wiring our web of belief, too. It could be argued that the answers to the most important general questions comprise the very wires of this web. Therefore, I claim that to deny a unique epistemic position of philosophy does not mean to deny the cognitive division of labor between traditionally recognized intellectual disciplines. And we, the post-Quine philosophers, can obviously exploit this division of labor in ways that were not available or desirable to Quine.

I thus also think that Quine's understanding of naturalistic philosophy is perfectly compatible with the synoptic and synthetizing aims of philosophy as (in) famously introduced by Wilfrid Sellars: It is […] the "eye on the whole" which

[12] "Meaningless" must thus be read very loosely in this context, *and with the knowledge that it is Quine's pragmatism that displaces the residues of positivistic thinking from his philosophy.* Meaningfulness is not, for Quine, necessarily linked directly to empirically verifiable conditions (see Quine 1995b, 49). So, Quine is here simply being at one with Wittgenstein and emphasizes that some question may sound "meaningful," or even "philosophical," and yet be strictly speaking pseudo-questions in the sense that they push our language to its very limits, where it is not possible to make clear sense of our words. Let alone to specify how some of these questions are related to our technological and intellectual goals (I thank Paul Roth for forcing me to comment on this issue, and for reminding me of the parallel with Wittgenstein here).

distinguishes the philosophical enterprise," (Sellars 2007, 371). Or more forcefully and distinctively put:

> *The aim of philosophy, abstractly formulated, is to understand how things in the broadest possible sense of the term hang together in the broadest possible sense of the term.* Under "things in the broadest possible sense" I include such radically different items as not only "cabbages and kings", but numbers and duties, possibilities and finger snaps, aesthetic experience and death. To achieve success in philosophy would be, to use a contemporary turn of phrase, to "know one's way around" with respect to all these things, not in that unreflective way in which the centipede of the story knew its way around before it faced the question, "how do I walk?," but in that reflective way which means that no intellectual holds are barred. (Sellars 2007, 369, emphasis mine)

For Quine, the synthetizing aspect of philosophy is inevitable. Absent a special method or the possibility to step out of all the conceptual frameworks, philosophers depend on the beliefs/knowledge generated, inter alia, by the special sciences. That is the principal assumption of Quine's intratheoretical—*from within*—approach. And, although Quine is a much more prosaic thinker than Sellars, he endorses the synoptic vision of philosophy. See again the quotes concerning philosophy above. If one reads the word "science" or the phrase "the conceptual scheme" in these passages as "the total theory of the world," one gets very close to Sellars' eye-on-the-whole designation of philosophy.

I would argue that this similarity is obscured, among many other things, by the fact that Quine often unfortunately hides in his writing style and rhetoric the assumption that he conceives science very broadly and unorthodoxly. One possible example is that, most of the time, Quine uses the terms "natural science," "hypothetico-deductive method," "empirical checks," or "prediction" actually interchangeably. Or he also casually uses the adjectives "soft" and "hard" for sciences. But he himself does not offer any special demarcation of "natural" and "hard" science (cf., Quine 1976b, 76; Quine 1975b, 314).

In the BBC interview I referred to earlier, Magee effectively translates from Quine's austere language and takes from Quine the following synoptic message:

> So, in other words, you regard the central tasks of philosophy as the analysis and elucidation of concepts *that are central to various fields of human activity*, and also, in particular, notions like what it is for something to be a cause of something else, what it is for something to exist, what it is for something to be, shall we say, a scientific law. *The most general notions that are, as it were, the connecting tissue of thought and that we have to use and have to employ in the specific activities that people like scientists – or it could even be politicians, lawyers and so on – are engaged in.* Is that a correct way of putting your view?
>
> [Quine laconically replies:] Yes, yes. I would agree with that. (Philosophy Overdose 2021, transcribed and emphasized by me)

In light of what has been said so far, I claim that even Quine is the child of the twentieth century (philosophy of) science when it comes to the vocabulary.

3.3.2 Science Is Not the Only Game in Town

If one also seriously takes into account Quine's inclination to view science as a Wittgensteinian language game—one among several such games—it follows that other human linguistic practices, such as poetry and fiction writing, exemplify other and legitimate language games. Additionally, for Quine, our modern science game has already moved past its *primary* evolutionary purpose of predicting our experiences, and its two dominant purposes are currently "technology" and "understanding," meaning specific pragmatic purposes and specific intellectual purposes (Quine 1992, 20). I will put aside the technological purposes, as I think no one in their right mind would deny the technological advances of our present-day sciences.

However, if one takes on board the plurality of language games, one is entitled to ask how to account, in the Quinean spirit, for the difference between "understanding" from within science and, for instance, from within fiction writing (and, for that matter, from within any kinds of writings that were deemed non-scientific at the time). If fiction qua a form of art also "teaches" and "enlightens" us, as Quine unproblematically holds (Quine and Ullian 1978, 4), when should we adhere to science and when, for instance, to a novel? Again, these sorts of questions are off Quine's radar most of the time, so I am aware that my answers cannot be supported by Quine's explicit reflections. However, I am not particularly interested in interpreting Quine as in interpreting his naturalistic legacy. Put another way, from now on, I engage in the Quinean naturalistic explication, without necessarily arguing that it is an explication Quine himself would have defended.

The main purpose of this explication is to clarify the relationship between science, the arts, and various forms of second-order inquiry—often referred to as "philosophy" but also by other names, such as "ethics." Fiction writing thus represents the limit case, as Quine does not go so far as to claim that even fiction is *cognitively* worthless. So, secondarily, the explication should clarify why and in what sense Quine does *not* have scientistic inclinations.

I will start in a roundabout way by deliberating on the difference between factual and non-factual discourse. Then, I will briefly discuss fictional discourse in the artistic sense. Stephen Jay Gould argued some time ago that, if perceived properly, science and religion are not in any conceptual or explanatory conflict.

> No such conflict should exist because each subject has a legitimate magisterium, or domain of teaching authority – and these magisteria do not overlap (the principle that I would like to designate as NOMA, or "nonoverlapping magisteria"). *The net of science covers the empirical universe: what is it made of (fact) and why does it work this way (theory). The net of religion extends over questions of moral meaning and value. These two magisteria do not overlap, nor do they encompass all inquiry (consider, for starters, the magisterium of art and the meaning of beauty*. (Gould 2011a [1997], 274, emphasis mine)

> The attainment of wisdom in a full life requires extensive attention to both domains – *for a great book tells us that the truth can make us free and that we will live in optimal harmony with our fellows when we learn to do justly, love mercy, and walk humbly*. (ibid., 271, emphasis mine)

As an evolutionary scientist and left-wing liberal, Gould was sympathetic to many kinds of intellectual and ideological movements; Gould understood very well that to agree with NOMA is easier than to actually practice it. The vast majority of questions central to our everyday lives cannot be moved too far away from the boundaries of the given magisteria. In this respect, for Gould, the friction surfaces between science and religion represent the paradigm of NOMA in practice. Therefore, although Gould primarily speaks of science and religion, he undoubtedly intends to extrapolate from these two to say something general about our human epistemic condition. In this sense, I regard NOMA as a naturalistic thesis.[13]

> *Many of our deepest questions call upon aspects of both for different parts of a full answer – and the sorting of legitimate domains can become quite complex and difficult.* To cite just two broad questions involving both evolutionary facts and moral arguments: Since evolution made us the only earthly creatures with advanced consciousness, what responsibilities are so entailed for our relations with other species? *What do our genealogical ties with other organisms imply about the meaning of human life?* (ibid., 274, emphasis mine)

Now, at face value, Quine and all like-minded naturalists would certainly agree that science is a theory-driven, fact-stating discourse (a language game). Similarly, naturalists have no problem recognizing what Gould hints at here; scientific discourse does not, in itself, imply any normative guidelines. By explaining our origins in evolutionary terms, scientists are not prescribing anything. Their explanations do not directly translate into value-stating discourse; in other words, these explanations are not direct value claims. The "ought" cannot be derived from the "is" directly without further background assumptions and aims, such as chosen ethical frameworks or societal values that guide and ground our interpretations.

For the type of holistic naturalism I derive from Quine, this follows, inter alia, since the very constituting and communicating "facts" are already inseparable from our historical cultures/scientific communities, i.e., inseparable from many other (background) beliefs, interests, and purposes. Let alone any further attempts to explain or theorize the facts. "Only theologians and unthinking ideologues imagine that normative conclusions can be read off some recitation of the facts," (Roth in Domańska et al. 2019, 544; cf., Gould 2011a [1997], 282).

At any rate, Quine and Gould are in agreement that "science" has a monopoly in determining what counts as facts, i.e., monopoly in generating meaningful fact-stating statements. This monopoly, however, is not tied to any methodological ownership. Nor to any unity-of-science trusts. In Quine's view, science has no ahistorical demarcation criterion from other commonly recognized *non-fictional* practices and discourses. The monopoly is rather grounded in the appreciation that the general intersubjective criteria (of individual sciences) has been the most successful tool for settling, i.e., rationalizing, our beliefs about the world, and the most effective tool for manipulating the world.

[13] I use "naturalistic" here in its most basic sense, referring simply to the aim of reconfiguring the relationship between scientific discourse and other forms of discourse. However, I consider this sense to be a properly understood interpretation of Quine's perspective as well (see again, e.g., Quine 1995a).

So, the claim so far is that as long as religion/theology, or any other non-fictional discourse for that matter, does not claim to be more fundamental or accurate than science with respect to factual matters, no conflict arises. For the naturalist of the Quinean kind, this does not necessarily mean that theology must abruptly accept, e.g., all the conclusions and conjectures of up-to-date physics. In this regard, respect for the facts is enough, regardless of how loosely one demarcates them. Take as an example the concept of "soul." This implies that understanding the concept of the soul dualistically (as a sui generis non-material entity) would need to be based on some form of intersubjective criterion and/or epistemologically-pragmatic purposes. The dualism of body and soul could, in principle, be empirically vindicated if it would give us some explanatory or technological benefits (for example, in sociology, medicine, etc.). The same would hold for concepts such as "prayer," "miracle," "revelation," or even "God."

Gould, however, seems to primarily target an additional issue; the nature of non-factual discourse, i.e., typically ethical/value discourse. As he further clarifies: "NOMA also cuts both ways. If religion can no longer dictate the nature of factual conclusions properly under the magisterium of science, *then scientists cannot claim higher insight into moral truth from any superior knowledge of the world's empirical constitution. This mutual humility has important practical consequences in a world of such diverse passions*," (Gould 2011a [1997], 281–282, emphasis mine). It is obvious that theology exemplifies only one of the possible ethical and value discourses in Gould's understanding of NOMA (cf., Gould 2011b [1999]). So, if anything, Gould can be charged with the intention of narrowing down the role of religion and theology exclusively to their ethical dimensions, and interpreting the existential (ipso facto factual) statements of theology in a *non-literal* way. But that is a problem for another day. Now, against this background, instead of science and religion, one could just speak of fact-stating discourse and ethical/value discourse, or generally of factual and non-factual discourse.

NOMA then certainly does not imply, as for example Richard Dawkins (1998) incorrectly infers, that the non-factual (ethical) discourse of theology can "unproblematically" work with *transcendental* concepts or notions (e.g., heaven, life after death, etc.). That would indeed amount to making *super*-natural existential claims, however implicitly. On the other hand, NOMA does not prevent us from working with transcendental concepts in metaphorical ways, e.g., operationalizing transcendental concepts qua Kantian regulative ideas.

This fact also seems to be one of the main motivations for Gould when introducing NOMA: "While I cannot personally accept the Catholic view of souls, *I surely honor the metaphorical value of such a concept both for grounding moral discussion and for expressing what we most value about human potentiality: our decency, care, and all the ethical and intellectual struggles that the evolution of consciousness imposed upon us*," (Gould 2011a [1997], 282, emphasis mine, cf., 271). Interestingly, Quine, when parsing the notion of the "absolute truth," also states that "metaphor is perhaps a handy category in which to accommodate transcendental concepts, from a naturalistic point of view," (Quine 1995a, 261). In general, it could be said that *non*-factual use of language simply serves different purposes and tackles different problems in which one is interested. It addresses different kinds of questions, e.g., "what is worth living for?," or "why should I stay healthy?" Therefore, even though Rorty actually thinks otherwise, I argue that Quine allows for considering our diverse uses of factual as well as non-factual language as tools, or in Rorty's own words, "techniques of problem solving" (Rorty 2021, 184). On my reading then, Quine's *functional* view of our human discoursing also leads to the pragmatist stance Rorty himself sharply depicts this way:

> Democritus, Newton, and Dalton solved some problems with particles and laws. Darwin, Gibbon, and Hegel solved others with narratives. Carpenters solve others with hammers and nails, and soldiers still others with guns. *Philosophers' problems are about how to prevent the words used by some of these problem-solvers from getting in the way of other words used by other problem-solvers.* (ibid., emphasis mine; cf., Thomasson 2024 for a more focused take on this matter)

I am also convinced that the functional view of discourse gives us a suitable conceptual bridge for comprehending the cognitive component of the arts. I briefly focus on fiction writing/novels. It is common sense that fiction writers can always use the language to confront us with "what ifs" or "as ifs" propositions related to our "real" lives. They can always imaginatively (re-)construct dialogues between two "real" persons or simply compose entirely fictional worlds and events to educate and/or entertain us. Also, this nicely illustrates that the first-person perspectives, although strictly speaking not scientifically useful in the naturalistic sense, can often be valid and valuable with respect to our human needs, interests, and purposes.

So far so good, but one might still wonder how to satisfactorily differentiate between artistic, and in some sense scientific usage of language. Take, for example, someone like Kendall L. Walton and his methodically naturalistic explicatory attitude towards all the art forms. Walton conceives of the artistic modes of expression as "games of make-believe:" "I take seriously the association with children's games – with playing house and school, cops and robbers, cowboys and Indians, with fantasies built around dolls, teddy bears, and toy trucks. We can learn a lot about novels, paintings, theater, and film by pursuing analogies with make-believe activities like these," (Walton 1990, 4).

Construed as such, the games of make-believe cannot be limited only to arts (or solely to discursive activities), but are rather "*a pervasive element of human experience.* […] There is nothing distinctively 'aesthetic' about make-believe itself at all. […] I suspect that make-believe may be crucially involved as well in certain religious practices, in the role of sports in our culture, in the institution of morality, *in the postulation of 'theoretical entities' in science, and in other areas in which issues of metaphysical 'realism' are prominent* […]," (ibid., 7, emphasis mine). This approach is moreover congenial to Quine's metaphysically *anti-representational* view of reality. As I have already established, Quine often stresses that the objects science postulates are, from the epistemological vantage point, useful myths or fictions (see again the previous section). And it also seems to be a quite expectable conclusion when the demarcation criterion of science is being abandoned.

Therefore, even though one cannot hope to identify a universal and unambiguous demarcation between scientific works and, e.g., novels, this does not mean that there are not pragmatically useful criteria for adjudicating individual instances. As naturalism dictates; the demarcation of scientific propositions will always have to be done empirically and intratheoretically, i.e., always ad hoc. Walton himself offers an illuminating exemplification of this type of constraint. Consider in this light the following passages:

> An important symptom of the difference between *The Origin of Species* and works like *Gulliver's Travels* [...] is that what is said in *The Origin of Species* does not of itself warrant assertions like "Species evolved by means of natural selection." *It justifies such assertions only insofar as it provides good reason to think they are true.* But the sentences in *Gulliver's Travels* warrant the assertive utterance "A war was fought over how to break eggs," *quite apart from whether they give us reason to think such a war actually was fought.* (Walton 1990, 71, emphasis mine)

> If we are to believe the theory of evolution, it is because that theory is true, or because there is good evidence for it, not because it is expressed in *The Origin of Species* – although of course *The Origin of Species* might convince us of the theory's truth or inform us of evidence for it. *Darwin's book itself does not prescribe believings. So we cannot conclude that it prescribes imaginings, even if believing involves imagining.* (ibid., 70–71, emphasis mine)

Consequently, the very purposes of scientific works and novels can be, in principle, very similar or even identical. They can aim at the same goals via different means. Again, no principled reasons exist for identifying the universal purposes of either scientific or artistic works.[14] Whether or not a statement has a scientific status must be decided against the background of other beliefs; it must be done from within our web of belief, to use the Quinean expression once again. Let me return to the example of the novel; it is obvious that every writer works, explicitly or implicitly, with the purposes of their work in manifold, and in principle, open-ended ways. And these ways can intersect (or be wholly interwoven) with other purposes of other intellectual endeavors and disciplines. It seems to be an understandable consequence of our human condition and of our linguistic modalities. To give one specific example, Milan Kundera in his *The Art of the Novel* explicates the "spirit of the novel" at one point as follows:

> *As a model of [the] Western world, grounded in the relativity and ambiguity of things human, the novel is incompatible with the totalitarian universe.* This incompatibility is deeper than the one that separates a dissident from an apparatchik, or a human-rights cam-

[14] Especially not for evaluating single, isolated statements. It is also a dialectical process; dispositions as well as expectations of the audience matter. Walton addresses this point via speech-act theory: "*The fundamental disanalogy between illocutionary actions and acts of fiction making comes out in differences in the roles of agents' intentions.* A crucial question for a person on the receiving end of an illocutionary action is almost always, Did he mean it? Did he intend to assert this, to promise that, to issue such and such an order or apology? But one may well read a story or contemplate a (fictional) picture without wondering which fictional truths the author or artist meant to generate. Photographers, especially, can easily be unaware of fictional truths generated by their works. Authors and other artists may be surprised at where extrapolation from the fictional truths they intentionally generated leads. *This need not make any particular difference to the appreciator—unless he is concerned with what the artist might be asserting in producing the fiction, what illocutionary actions she might be performing in the process of, and in addition to, producing it,*" (ibid., 88, emphasis mine).

> paigner from a torturer, because it is not only political or moral but *ontological*. By which I mean: *The world of one single Truth and the relative, ambiguous world of the novel are molded of entirely different substances. Totalitarian Truth excludes relativity, doubt, questioning*; it can never accommodate what I would call the *spirit of the novel*. (Kundera 2020, 10, long emphasis mine)

Kundera also thinks that the novel is inherently historical. "Novels take place in a time that has a date and is thoroughly historical," (ibid., 20). But one must carefully distinguish between "*the historical dimension of human existence*," and "*the illustration of a historical situation*;" meaning "popularizations that translate non-novelistic knowledge into the language of the novel." Kundera then concludes: "Well, I'll never tire of repeating: The novel's sole *raison d'etre* is to say what only the novel can say," (ibid.). So, for him, there are several principles distinguishing novels from scientific (e.g., historiographic) works. One of them being a different point of focus:

> *Historiography writes the history of society, not of man. That is why the historical events my novels talk about are often forgotten by historiography.* Example: In the years that followed the 1968 Russian invasion of Czechoslovakia, the reign of terror against the public was preceded by officially organized massacres of dogs. *An episode totally forgotten and without importance for a historian, for a political scientist, but of the utmost anthropological significance!* By this one episode alone I suggested the historical climate of *The Farewell Party*. Another example: At the crucial point of *Life Is Elsewhere,* History intervenes in the form of an inelegant and shabby pair of undershorts; there were no others to be had at the time; faced with the loveliest erotic occasion of his life, Jaromil, for fear of looking ridiculous in his shorts, dares not undress and takes flight instead. *Inelegance! Another historical circumstance forgotten, and yet how important for the person obliged to live under a Communist regime.* (ibid., 21, long emphasis mine)

This is one of the many possible approaches, compatible with the Quinean approach I develop and defend here, to distinguishing between the purposes and cognitive components of scientific works (in our case above historiographical works), and novels (in our case above of novels dealing with the history of humanity).

When taken together, it represents an inquiry into the human "soul" in a manner that would be appreciated by not only Gould but Quine as well. True, Quine writes at the very end of his autobiography: "*I have little bent for soul-searching.* [...] My way of coping with the spells of nostalgia, loneliness, anxiety, or boredom over the years has been to escape into my projects," (Quine 1985, 475, emphasis mine).[15] However, his naturalizing project, which emphasizes the role of the third-person perspective over the first-person one, nevertheless creates a space that preserves all

[15] Quine then continues: "I am deeply moved by occasional passages of poetry, and so, characteristically, I read a little of it. I respond similarly to passages of grand opera, and this is due to the libretto as much as to the music. Otherwise I have a poor memory for fiction, for it resists integration with my system of the world. I appreciate style more than plot. I enjoy the eighteenth-century brand of humor that we get in Fielding's prefaces and belatedly in Dickens; also the humor of W. S. Gilbert and Wodehouse and, in moderation, zany S. J. Perelman. Encouraged by the indulgent interest of readers who have read all the way to here, I am evidently accomplishing some soul-searching after all," (ibid., 476).

the typically acknowledged and appreciated purposes and cognitive functions of the arts, or so I have argued with the assistance of Walton and Kundera above.

To conclude, in this section, I have proposed that Quine's functional view of our language use (hence also scientific, philosophical, ethical, and artistic discourse) empowers naturalists to see human linguistic practices as tools for specific pragmatic and intellectual purposes. Such a view also sheds light on the relationship between and different functions of third-person perspective (paradigmatic for but not exclusive to science); and first-person perspective (utilized often but not uniquely within the arts).

In addition, my own attempt to interpret Quine on this matter exemplifies a case of philosophical exemplification under natural*ism*.

References

Borradori, Giovanna. 2008. *The american philosopher: Conversations with Quine, Davidson, Putnam, Nozick, Danto, Rorty, Cavell, MacIntyre, Kuhn*. Trans. Robert Crocitto. Chicago: University of Chicago Press.

Bryant, Amanda. 2020. Naturalisms. *Think* 19 (56): 35–50.

Danto, Arthur Coleman. 1967. Naturalism. In *The encyclopedia of philosophy*, ed. Paul Edwards, vol. 5, 448–450. New York: The Macmillan Company and The Free Press.

Davidson, Donald. 1973–1974. On the very idea of a conceptual scheme. *Proceedings and Addresses of the American Philosophical Association* 47:5–20.

Davidson, Donald. 2002. Mental events. In *Contemporary materialism: A reader*, ed. Patricia Kitcher Moser and John D. Trout, 122–137. London: Routledge. (Originally published 1970).

Dawkins, Richard. 1998. When religion steps on science's turf: The alleged separation between the two is not so tidy. *Free Inquiry* 18 (2): 18–19.

De Caro, Mario, and David Macarthur, eds. 2004. *Naturalism in question*. Cambridge, MA: Harvard University Press.

Dewey, John. 1911. Brief studies in realism. II. *The Journal of Philosophy, Psychology and Scientific Methods* 8 (20): 546–554.

Domańska, Ewa, María Inés La Greca, Paul A. Roth, Xiaoxing Chen, Valentina Tozzi Thompson, and Kalle Pihlainen. 2019. Globalizing Hayden White. *Rethinking History* 23 (4): 533–581.

Dreben, Burton. 1992. Putnam, Quine – and the facts. *Philosophical Topics* 20 (1): 293–315.

Gould, Stephen Jay. 2011a. Non-overlapping magisteria. In *Leonardo's mountain of clams and the diet of worms: Essays on natural history*, 269–284. Cambridge, MA: Harvard University Press. (Originally published 1997).

Gould, Stephen Jay. 2011b. *Rocks of ages: Science and religion in the fullness of life*. New York: Ballantine Books. (Originally published 1999).

Heath, Joseph, and Catherine Rioux. 2018. Recent trends in evolutionary ethics: Greenbeards! *Biology & Philosophy* 33 (1): 1–20.

Hofstadter, Albert. 1954. The myth of the whole: A consideration of Quine's view of knowledge. *The Journal of Philosophy* 51 (14): 397–417.

Honderich, Ted, ed. 1995. *The oxford companion to philosophy*. Oxford: Oxford University Press.

Jaksland, Rasmus. 2024. Naturalized metaphysics in the image of Roy Wood Sellars and not Willard Van Orman Quine. *Metaphilosophy* 55 (2): 214–230.

Keskinen, Antti. 2012. Quine on objects: Realism or anti-realism? *Theoria* 78 (2): 128–145.

Kundera, Milan. 2020. *The art of the novel*. London: Faber & Faber. (Ebook).

Menand, Louis. 2002. *The metaphysical club: A story of ideas in america*. New York: Macmillan.

Mitrović, Branko. 2022. A naïve realist rumination on the Roth-and-Dewulf versus Currie-and-Swaim exchange. *Journal of the Philosophy of History* 16 (3): 295–307.

Papineau, David. 2021. Naturalism. In *The Stanford encyclopedia of philosophy*, ed. Edward N. Zalta.

Peregrin, Jaroslav. 1999. *Zen & the art of (Davidsonian) analytic philosophy*. Jarda.Peregrin.

Peregrin, Jaroslav. 2023. *Normative species: How naturalized inferentialism explains us*. New York: Routledge.

Philosophy Overdose. 2021, December 16. *The ideas of Quine – Bryan Magee & Quine (1978)* [Video]. YouTube.

Pils, Raimund. 2020. Quine's scientific realism revisited. *Theoria* 86 (5): 612–642.

Price, Huw. 2011. *Naturalism without mirrors*. New York: Oxford University Press.

Quine, Willard Van Orman. 1957. The scope and language of science. *The British Journal for the Philosophy of Science* 8 (29): 1–17.

Quine, Willard Van Orman. 1961a. Two dogmas of empiricism. In *From a logical point of view: Logico-philosophical essays*, 20–46. New York: Harper and Row, Publishers. (Originally published 1951).

Quine, Willard Van Orman. 1961b. On what there is. In *From a logical point of view: Logico-philosophical essays*, 1–19. New York: Harper and Row, Publishers. (Originally published 1951).

Quine, Willard Van Orman. 1961c. Identity, ostension, and hypostasis. In *From a logical point of view: Logico-philosophical essays*, 65–79. New York: Harper and Row, Publishers. (Originally published 1950).

Quine, Willard Van Orman. 1969a. Natural kinds. In *Ontological relativity*, 114–138. New York: Columbia University Press.

Quine, Willard Van Orman. 1969b. Epistemology naturalized. In *Ontological relativity*, 69–90. New York: Columbia University Press.

Quine, Willard Van Orman. 1969c. Ontological relativity. In *Ontological relativity*, 26–68. New York: Columbia University Press.

Quine, Willard Van Orman. 1969d. Speaking of objects. In *Ontological relativity*, 1–25. New York: Columbia University Press.

Quine, Willard Van Orman. 1975a. Replies. In *Words and objections*, ed. Donald Davidson and Jaakko Hintikka, 292–352. Dordrecht: D. Reidel.

Quine, Willard Van Orman. 1975b. On empirically equivalent systems of the world. *Erkenntnis* 9 (3): 313–328.

Quine, Willard Van Orman. 1976a. The limits of knowledge. In *The ways of paradox and other essays*, 59–67. Revised and enlarged edition. Cambridge, MA: Harvard University Press.

Quine, Willard Van Orman. 1976b. Necessary truth. In *The ways of paradox and other essays*, 68–76. Revised and enlarged edition. Cambridge, MA: Harvard University Press.

Quine, Willard Van Orman. 1981a. Things and their place in theories. In *Theories and things*, 1–23. Cambridge, MA: Harvard University Press.

Quine, Willard Van Orman. 1981b. Five milestones of empiricism. In *Theories and things*, 67–72. Cambridge, MA: Harvard University Press.

Quine, Willard Van Orman. 1984. Relativism and absolutism. *The Monist* 67 (3): 293–296.

Quine, Willard Van Orman. 1985. *The time of my life: An autobiography*. Cambridge, MA: MIT Press.

Quine, Willard Van Orman. 1987. *Quiddities: An intermittently philosophical dictionary*. Cambridge, MA: Harvard University Press.

Quine, Willard Van Orman. 1992. *Pursuit of truth*. 2nd ed. Cambridge, MA: Harvard University Press. (Originally published 1990).

Quine, Willard Van Orman. 1995a. Naturalism; or, living within one's means. *Dialectica* 49 (2–4): 251–263.

Quine, Willard Van Orman. 1995b. *From stimulus to science*. Cambridge, MA: Harvard University Press.

Quine, Willard Van Orman. 1996. Progress on two fronts. *The Journal of Philosophy* 93 (4): 159–163.
Quine, Willard Van Orman. 2008. The innate foundational endowments. In *Confessions of a confirmed extensionalist and other essays*, ed. Dagfinn Føllesdal and Daniel B. Quine, 176–181. Cambridge, MA: Harvard University Press.
Quine, Willard Van Orman. 2013. *Word and object*. Cambridge, MA: MIT Press. (Originally published 1960).
Quine, Willard Van Orman, and Joseph S. Ullian. 1978. *The web of belief*. 2nd ed. New York: Random House. (Originally published 1970).
Rorty, Richard. 1976. Realism and reference. *The Monist* 59 (3): 321–340.
Rorty, Richard. 1998. *Truth and progress: Philosophical papers*. Vol. 3. New York: Cambridge University Press.
Rorty, Richard. 2021. *Pragmatism as anti-authoritarianism*. Cambridge, MA: Harvard University Press.
Rosenberg, Alexander. 1996. A field guide to recent species of naturalism. *The British Journal for the Philosophy of Science* 47 (1): 1–29.
Rosenberg, Jonathan F. 2007. *Wilfrid Sellars: Fusing the images*. Oxford: Oxford University Press.
Roth, Paul A. 2003. Why there is nothing rather than something: Quine on behaviorism, meaning, and indeterminacy. In *Philosophy, psychology, and psychologism: Critical and historical readings on the psychological turn in philosophy*, ed. Daniel Jacquette, 263–287. Dordrecht: Springer.
Roth, Paul A. 2020. *The philosophical structure of historical explanation*. Evanston: Northwestern University Press.
Roth, Paul A. 2023. What does translation translate? Quine, Carnap, and the emergence of indeterminacy. In *The philosophical project of Carnap and Quine*, ed. Stewart Morris, 154–176. Cambridge: Cambridge University Press.
Sellars, Wilfrid. 2007. Philosophy and the scientific image of man. In *In the space of reasons: Selected essays of Wilfrid Sellars*, ed. Robert Brandom and Kevin Scharp, 369–408. Cambridge, MA: Harvard University Press. (Originally published 1962).
Thomasson, Amie L. 2024. How should we think about linguistic function? *Inquiry* 67 (3): 840–871.
Tuomisto, Hanna, Matleena Tuomisto, and Jouni T. Tuomisto. 2018. How scientists perceive the evolutionary origin of human traits: Results of a survey study. *Ecology and Evolution* 8 (6): 3518–3533.
Walton, Kendall L. 1990. *Mimesis as make-believe: On the foundations of the representational arts*. Cambridge, MA: Harvard University Press.
Wright, Crispin. 1993. *Realism, meaning, and truth*. 2nd ed. Cambridge, MA: Blackwell.

Chapter 4
Roth and Historiography Naturalized

Abstract In *The Philosophical Structure of Historical Explanation* (2020), Roth advances a naturalistic case for integrating historical/narrative explanations into the canon of scientifically recognized modes of explanation. He simultaneously challenges the traditional dichotomy between explanation and understanding and proposes an "irrealist" middle ground that navigates between scientific realism and anti-realism. Moreover, Roth offers a novel reinterpretation of Kuhn's *Structure*, arguing that Kuhn establishes historical explanation as indispensable not only for delineating what counts as science but also for rational understanding in general. This interpretation consequently renders history essential to any comprehensive account of human rationality. And since Roth himself subscribes to Quinean naturalism, this chapter seeks to introduce his work in detail and critically engage with it, with particular attention to its implications for naturalism as developed and defended herein.

Keywords Roth · Narrative · Historical explanation · Irrealism · Naturalism

4.1 Introduction: Historical Explanation as a Philosophical Problem

> It sometimes is said that the task of philosophy is not to think or talk about the world, but rather to analyze the ways in which the world is thought and talked of. But since we plainly have no access to the world apart from our ways of thinking and talking about it, we scarcely, even in restricting ourselves to thought and talk, can avoid saying things about the world. The philosophical analysis of our ways of thinking and talking about the world becomes, in the end, a general description of the world as we are obliged to conceive of it, given that we think and talk as we do. Analysis, in short, yields a descriptive metaphysic when systematically executed.
>
> —Arthur C. Danto[1]

[1] (Danto 1985, xv).

M. Hubálek, *Dissolving Naturalism and Historicism Into Each Other*, Studies in History and Philosophy of Science 67,
https://doi.org/10.1007/978-3-032-07535-2_4

> Since the second half of the nineteenth century, history has become increasingly the refuge of all of those "sane" men who excel at finding the simple in the complex and the familiar in the strange. This was all very well for an earlier age, but if the present generation needs anything at all it is a willingness to confront heroically the dynamic and disruptive forces in contemporary life. The historian serves no one well by constructing a specious continuity between the present world and that which preceded it. On the contrary, we require a history that will educate us to discontinuity more than ever before; for discontinuity, disruption, and chaos is our lot.
>
> —Hayden White[2]

This chapter introduces and critically examines Roth's book *The Philosophical Structure of Historical Explanation* (2020, hereafter PSHE). This book is a culmination of Roth's long-lasting interests in and investigations of the concept of historical explanation (or, as he often prefers to say, "narrative explanation"), the philosophy of (social) science, and naturalism. Roth is, moreover, a Quinean naturalist exactly in the sense I developed in Chap. 2 (see his own delimitation and defense of naturalism with additional references in Roth 2020, 113ff). Roth thus sees naturalism as an a posteriori and intratheoretical epistemological/scientific position: "Naturalism in epistemology can be characterized negatively by its eschewal of any notions of analytic or a priori truths. Positively, naturalism asserts a normative and methodological continuity between epistemological and scientific inquiry. The techniques endemic to the former are only a subset of the historically received and contingently held norms and methods of the latter," (Roth 2020, 115). Consequently, for Roth just like for Quine, there is ultimately no substantial epistemological difference between explaining/theorizing "the past" and "the present:"

> At the most general epistemological and metaphysical levels, *no principled distinction emerges between empirical knowledge generally and knowledge of the past.* The forms of inference required to have empirical knowledge at all – inductive, abductive, and deductive – arise for all such cases of knowledge. *Once the presumption of givenness with regard to evidence or of shared conceptual schemes is renounced, the "shape of the past" and the "shape of the present" receive their form under fundamentally similar holist constraints.* Temporal distance may accentuate problems of making sense of others and what they did, but the problems posed turn out not to be at all unique. Only in a theory do things – for example, facts, events, kinds, actions – exist and have explanations. (Roth 2020, 61, first emphasis mine).

From the general perspective of philosophy of science, Roth's approach to the issues of historical explanation and to the scientific status of historiography[3] can be labeled as post-Quinean and, as will become clearer and more crucial, also post-Kuhnian (see Sect. 4.2.3). Thus, in a word, post-positivist. Roth's approach to the issues directly related to historiography is then also instructive for tackling all the usual problems related to post-positivist conceptions of science and scientific explanation; especially again the problem of determining how "experience" relates to "theory;" how theory-ladenness of observations relates to our conception of "evidence."

---

[2] (White 1966, 134).

[3] I use the word "historiography" here in this chapter when I want to refer to historical inquiry qua scientific inquiry into "the past."

Roth's work is, from within this perspective, motivated epistemologically, meaning by normative and evaluative concerns common to the (post-)analytic tradition of philosophy (Roth 2020, xiii–xiv).

From the general perspective of philosophy of history, Roth's approach can be labeled as post-Hempelian, as he intentionally contributes to, in fact attempts to revive, the field of "analytical philosophy of history," which he perceives historically as an offshoot of the early Anglo-American philosophy of science. Here is how Roth perceives the genesis of the analytical philosophy of history:

> Birth [of analytical philosophy of history] could plausibly be dated to coincide with the publication of Hempel's (1942) classic article, "The Function of General Laws in History" [...]. Hempel's explication effectively mandates the de facto exile of academic history from the realm of the legitimate sciences. Analytical philosophy of history, for the twenty-five years following Hempel's article, by and large consists in critiques or defenses of the applicability of this model to historical explanation. Indeed, Hempel's essay became a near exclusive focus of discussion both in and out of philosophy by (in)famously insisting that historical explanations in their usual guise constitute at best "explanation sketches." Such sketches then have to be completed by citing some law or law-like connection between the explanandum statement and those putatively serving as explanantia. (Roth 2020, 4, cf., also xi–ix)
>
> In the subsequent three decades of debate triggered by Hempel's essay, one of the most notable responses can be found in Danto's (1965) important work, Analytical Philosophy of History. It is in Danto's work that I first find the phrase "narrative explanation" (Danto 1965, 237). What makes this point noteworthy is that it has come to name that form of explanation specific to history and connotes for our purposes those differences already in play prior to Hempel's article. Danto's specific ways of motivating consideration of this term and Mink's (1987) subsequent crucial modifications and elaborations of Danto's insights remain central to any serious consideration of this topic. (Roth 2020, 4–5).

However, as Roth quickly and rightly notes, the phrase "narrative explanation" (or even the very word "narrative") is today used very loosely; "the term 'narrative explanation' as now used has been evacuated of the substantive content it has in Danto and in Mink," (Roth 2020, 5, cf., 7). So, an important part of Roth's agenda is to demonstrate the philosophically and pragmatically substantive content of the concept of narrative explanation (qua historical explanation), which Danto introduces in theoretical opposition to Hempelian/Popperian attempts to grasp history via a deductive-nomological model of explanation (cf., Danto 1985, 201; and Uebel 2019).

Also, Roth openly states that his work "also endorses the primacy of narrative form in the case of historical explanation, and thus the constructivist position long advocated by Hayden White and the narrativist theorizing that follows from his work," (Roth 2020, xiii). Relatedly, as I show in Sect. 4.3, Roth argues that the dichotomy between historical realism and anti-realism is a false dichotomy by making a case for irrealism—a position he finds already in the works of Nelson Goodman and Ian Hacking, but further crafts with respect to historical reasoning. By this maneuver, Roth, among other things, also commits himself to elaborating on the metaphysical status of historical events and entities, i.e., historical reality.

The strength of Roth's approach lies precisely in his innovative synthesis of these two perspectives (i.e., the perspectives of philosophy of science and philosophy of history). He primarily strives to show how the epistemic standards developed within the currently commonly recognized sciences hang together with the "narrativist" and "constructivist" structure of historical explanation. Put another way, Roth strives to show that narratives indeed explain. In his own words: "Philosophy of history in the sense that primarily interests me connects to issues that concern the nature of historical explanation. These arose in discussions originating in the philosophical literature in the nineteenth century and in terms that still dominate. Varieties of positivism, whether of Comtean or Vienna Circle style, advocated for forms of explanation modeled on their idealization of the natural sciences," (Roth 2020, 3). Roth eventually argues that from within naturalism (at any rate from within the kind of naturalism he himself defends and which I take to be the one Quine developed), historiography is a form of empirical inquiry, i.e., historiography is a science (Roth 2020, 145).

Thus, with respect to the aims of this book, Roth supplements the general Quinean argument for the scientific status of historiography provided in Chap. 2 with his detailed and distinct analysis of narrative explanation, and its subsequent inclusion into the pool of scientific methods. Secondarily, and perhaps prima facie more controversially, Roth interprets Kuhn's widely celebrated historicizing of science (historicizing of the scientific "progress") as providing, in fact, a narrative account of science, and so Roth then in PSHE makes a case for a deep interdependence between natural science and history, arguing that a narrative history of natural science ironically proves critical to providing a basis for characterizing what now passes as a science. [...] Inasmuch as narrative explanation turns out to be required for explaining what theories come to have status as scientific, this book maintains that the issue of how to tie narrative historical explanation back into the general catalog of accounts of what happens cannot be avoided. (Roth 2020, xiv–xv)

This interpretation aspires to entirely transform our understanding of science, because it "effectively reverses the received order of epistemic authority. Prior to Kuhn, a work of history, in order to count as providing a legitimate explanation, needed to conform to a certain standard determined by an ahistorical account of science. Post-Kuhn, philosophers fashion histories to account for which explanatory forms come to prevail and why," (Roth 2020, 86). I unpack these claims in Sect. 4.2.3. Here I just want to underscore that I take these conclusions again as supplementary to the already debated Quinean conclusions implying that the very standards of reasoning (hence scientific reasoning) are always established, and so determined, historically.

The Quinean web of belief/our total science is erected diachronically, and only then can serve as a default ground for our reasoning (see Sects. 2.2.1 and 2.3). Construed as follows, however, one still does not know what "fashioning histories" in the narrativist and constructivist mode amounts to epistemologically. And that is what Roth's work brings to the table. As John Zammito in his review of PSHE also recognizes, for Roth, "explicating historical explanation becomes part of a general account of empirical knowledge. That is its important philosophical relevance," (Zammito 2020). On this account, Roth's parsing of historical/narrative explanation is a vital contribution to our understanding of naturalism (in philosophy).

Since Roth's work on historical reasoning and explanation is, at least in its compact form, of recent vintage, I analyze it purposefully in great detail, relying on many of Roth's own expressions and preferred interpretative structures. I thus attempt to lay out the basics of his position in the most authentic yet enriching manner. The enrichment consists, I hope, of presenting and highlighting Roth's rich *naturalistic* edifice. However, while I consider selected reviews or critiques of Roth's work which are now available in print (see especially Sect. 4.4), my own evaluative work of Roth's framework is done, in the naturalistic spirit, mostly indirectly via the two individual case studies submitted in Chaps. 5 and 6 of this book.

4.2 Explicating Narrative Explanation

Aware of the many confusions in the usage and understanding of the concept of narrative, Roth opts for its very minimalistic definition, which he considers already commonly accepted and accommodated within various disciplines (outside of historiography, also literary theory and psychoanalysis). Firstly, he appreciates Daniel Little's nearly all-purpose understanding of a narrative as "an account of the unfolding of a series of events, along with an effort to explain how and why these processes and events came to be. A narrative is intended to provide an account of how a complex historical event unfolded and why [...]," (Little 2010, 28).[4] And then he comes up with his own sharper versions: "Narratives are stories, a telling that something happened," (Roth 2020, 22); "a narrative involves an 'unfolding', sequencing of a series of events that accounts for a development," (Roth 2020, 66). However, as is clear to Roth (2020, 65), these minimalistic definitions say very little about its explanatory value, let alone about its scientific status. After all, defined as such, narratives are part and parcel of fiction writing as well. To parse the narrativist structure determinative for historical explanation, Roth turns to the work of Danto and Mink.

So now, together with Roth, we must pay some attention to Danto's ingenious and intellectually provoking thought experiment of an Ideal Chronicler and Ideal

[4] Consider another characterization of a narrative that Roth quotes from Humphrey Morris (a psychoanalytic theorist): "A 'narrative' [...] is a particular language form that is organised according to a fundamentally temporal principle, that is, according to some variation on a 'beginning-middle-end' structure. Narrative, in this structural sense, is self-evidently 'historical,'" (Morris in Roth 2020, 66).

Chronicle, respectively, and consequently to his notion of narrative sentences (Danto 1985, 149ff).[5] Danto invites us to imagine the following ideal scenario:

> [An Ideal Chronicler] knows whatever happens the moment it happens, even in other minds. He is also to have the gift of instantaneous transcription: everything that happens across the whole forward rim of the Past is set down by him, as it happens, the way it happens. The resultant running account I shall term Ideal Chronicle (hereafter referred as I.C.). Once *E* is safely in the Past, its full description is in the I.C. We may now think of the various parts of the I.C. as accounts to which practicing historians endeavor to approximate their own accounts. (Danto 1985, 149)

A selected part of the I.C. would then represent what one might consider as "a description which really is a full description, which tells everything and is perfectly isomorphic with an event. Such a description then will be *definitive*: it shows the event *wie es eigentlich gewesen ist* [how it actually was]," (Danto 1985, 148). And as Roth emphasizes, "the purpose of [an Ideal Chronicler and an I.C.] is to suggest a case in which the factual record of the past is as complete as can be imagined *at the moment at which events occur*," (Roth 2020, 27). For this reason, an Ideal Chronicler could be also conceived as an "Ideal Witness" (see Danto 1985, 151).

However, Danto thinks that I.C. will not meet the expectations one standardly associates with historical accounts. This is to say that I.C. cannot replace historians in their work, because I.C. is blind to what historians, and arguably also laymen, typically consider as historical significance, viz., how past relates to present and future. Danto identifies a kind of statement (sentence, truth) that, in principle, cannot be recorded in I.C. because of its specific relation to a later time/later events. The clearest example he offers goes as follows: an Ideal Witness could not have seen and an Ideal Chronicler could not have recorded in 1618 that "The Thirty Years War" began (see Danto 1985, 151–152). Roth makes the logic of these sentences fully explicit:

> Danto [...] demonstrates that there are statements true of some time *t* in the past which cannot have been known to be true at that time. [...] Examples are easy to generate. Simply formulate descriptions known to be true of persons at a time later than *t* and use them to refer to those persons at *t*. The result – what Danto calls "narrative sentences" – will be sentences true at *t* but which could not have been known at *t*, and so escape even the Ideal Chronicler. (Roth 2020, 27, see also 28).

Another, slightly more complicated, narrative sentence could be: the movie The Terminator (released 1984) stars a former Governor of California, Arnold Schwarzenegger (in office from November 17, 2003 until January 3, 2011). This statement is truthful only from January 4, 2011. The only difference here is an insertion of another future ("a *former* Governor"), for us maybe better said present, event/perspective (I write this in 2024, and you are reading this in 2025, at the earliest).

[5] I choose to refer generally to Danto's later book *Narration and Knowledge,* which contains his influential and, for this book, relevant work *Analytical Philosophy of History* (Danto 1965). It is also worth mentioning that *Analytical Philosophy of History* contains Danto's "Narrative sentences" (1962).

The slight complication in this example does not as such add anything new to the argument; it should only highlight the role future events/perspectives play in our statements about the past. After all, an event is a future event only in relation to a past event, and vice versa (cf., Danto 1985, 199).

Danto considers this role of "future contingencies" as critical for our understanding of the past, insomuch as if there are future contingencies, there are also "past contingencies." Past contingencies amount to "incompatible descriptions, so to speak, hovering over a given past event, unable to establish definite semantical relations with it until something happens in the future," (Danto 1985, 195). What is important to note about this kind of past contingency is that, in Roth's words, "knowledge of the past remains contingent, but a contingency that not arises because of any lack of evidence about the past," (Roth 2020, 55). Again, for Danto and Roth, past contingency is not an issue an Ideal Witness/Ideal Chronicler could help us to solve.

The core of the argument is that the truth-value of some statements will depend on a given time that one asserts them. Consider the following sentence (I modify Danto's 1985, 193ff; and Roth's 2020, 55ff, example and considerations):

(a) The Terminator starred Arnold Schwarzenegger, and Arnold Schwarzenegger was the Governor of California.

This narrative sentence is now true (both its conjuncts are true now). But that was not always the case. Say, in 1985, the first conjunct is true but the second is false. That creates a logical paradox since, formally viewed, the truth-value of this sentence should be derivable from both of its conjuncts (from their respective truth-values). Danto suggests that, in cases like these, one may consider making these kinds of sentences "time-true" (Danto 1985, 195). Meaning to restate (a) so it respects the time difference and is true (or at least time-true). The sentence I composed would afterwards look like this:

(b) The Terminator starred Arnold Schwarzenegger, and Arnold Schwarzenegger will be the Governor of California.

Although possibly time-true in 1985, the second conjunct "Arnold Schwarzenegger will be the Governor of California" seems to be neither true nor false, as it concerns future and not past or present matters. Put another way, one simply does not know in 1985,[6] and formal logic will not help here either. According to Danto, one is free to choose whether (b) is in fact false or without a truth-value. It seems like there is

[6] One may think that, with regard to the future, one might be able to predict at least some events, or at least to filter probable and/or possible future events. That strikes me as a correct conclusion. But it does not change the argument much, as the argument concerns the past not the future. It is indeed improbable that Arnold Schwarzenegger will win any Mr. Olympia title again, or impossible that he will become the President of the USA (as of now, one is required to be a natural-born citizen of the USA to run for President). However, it does not change the fact that Arnold Schwarzenegger as a person who starred in *The Terminator* and was the Governor of California has already become and will yet become describable in ways not known, thus not truthful, in 1984, 2011, or even 2025.

no satisfactory way of squaring time-true parts of narrative sentences with "true" in some absolute sense/atemporal language. However, to state the obvious once again, those entertaining the perspective from and later than January 4, 2011 can assert (a) truthfully. And that is the difference that makes the epistemological difference for historical inquiry.[7] To account for truths embedded in or, in fact, emerging via narrative sentences, one needs either a suitable temporal standpoint or a temporal language.

Thus, Danto not only discerns a peculiar kind of temporal sentence he tags narrative, but also demonstrates that, due to the openness of the future, the past is also, in principle, open to indefinite re-descriptions and, in effect, that our epistemic relation to history is inevitably dynamic (Roth 2020, 56, 11). As Roth also fittingly puts it, "a narrative sentence *adds* a truth to an earlier time because related to some subsequent occurrence," (Roth 2020, 68). This is a small but decisive step towards appreciating that narratives have a certain constructivist effect on our accounts of the past. Danto in his own way indirectly suggests this in the concluding passage of the chapter titled "History and Chronicle" which is worth quoting at length:

> I shall say now, that you cannot give a complete description of any event which does not use narratives. Completely to describe an event is to locate it in all the right stories, and this we cannot do. We cannot because we are temporally provincial with regard to the future. [...] The complete description then presupposes a narrative organization, and narrative organization is something that we do. Not merely that, but the imposition of a narrative organization logically involves us with an inexpungable subjective factor. *There is an element of sheer arbitrariness in it. We organize events relative to some events which we find significant in a sense not touched upon here. It is a sense of significance common, however, to all narratives, and is determined by the topical interests of this human being or that. The relativists are accordingly right.* (Danto 1985, 142, emphasis mine)

Roth, while commenting on this passage, summarizes Danto's influence on his own conception of historical explanation as follows: "In short, historians look to justify an explanation of an event under a particular description, a description tied to a retrospective and so narrative perspective," (Roth 2020, 9, emphasis mine). Now, I can finally add that narrative explanation qua historical explanation,[8] for Roth, "concerns itself with a developmental process that emerges only in retrospect," (Roth 2020, 20). Put another way, "narrative explanations explain narrative sentences," (Roth 2020, 65). However, in Roth's view, only the later reception and rethinking of theoretical consequences of Dantonian narrative sentences by Mink (1987) fully bring the constructivist nature of historical knowledge to the forefront. So, in the next section, I focus on Roth's innovative fusion of Danto's and Mink's insights.

---

[7] However, this is not to say that all the subject matters practicing historians deal with are constituted and/or structured narratively. For example, sometimes "merely" inquiring into the matters of fact (answering a what-question and not necessarily a why-question) can be of historical significance too. *The claim here is that narrative structures are tied to historical discourse as opposed to, for instance, physicalist discourse* (cf., Roth 2020, 62ff).

[8] From now on, especially in the context of Roth's thinking, I use the phrases "historical explanation" and "narrative explanation" interchangeably.

4.2.1 *Three Nons: Non-Standardization, Non-Detachability, Non-Aggregativity*

Led among other things by Mink's reading of Danto, Roth identifies three constitutive features of narrative explanations: non-standardization, non-detachability, and non-aggregativity (hereafter *non-S*, *non-D*, *non-A*; or together *non-SDA*). These features at the same time constitute the difference between historical and other parts of scientific discourse (especially chemistry and arguably some parts of physics, see Roth 2020, 11, 62). Let me briefly summarize the contents of these features as Roth introduces them, together with his mentions/and chosen quotes of Mink's thoughts.

(non-S) Roth states that historical events (event kinds, e.g., wars, revolutions, famines, genocides, as well as particular historical events, e.g., the American Civil War, the Second World War) lack theoretically standardized descriptions, and so historians cannot constitute a "normal science" in Kuhn's sense (Roth 2020, 9–14, 63–67, 74).

According to Roth: "Events simpliciter cannot be shown to exist; they are not known to be of nature's making rather than of ours. Events exist only by proxy. [...] Knowledge of events is restricted to happenings isolated under descriptions provided by interested parties," (Roth 2020, 30); "Events do not come prepacked, like elements on some periodic table, and so determinately conjoinable," (Roth 2020, 130).

Compare with Mink: "A scientific account of an event determines a standard description of the event, by counting, say, statements of the mass and velocity of a moving body as relevant descriptions, and statements about its color [...] as irrelevant. History, on the other hand, reports how descriptions change over time, that is, how 'The author of Rameau's Nephew was born in 1713' came to be true. [...] There can be a history of science, that is, of changes in the kinds of descriptions accepted as standard at different times, but no science of history, that is, a complete description of events which includes or subsumes all possible descriptions," (Mink 1987, 139).

This boosts Danto's insights in at least one important regard. As Mink insists, when historians construct theories about and (narrative) explanations of the past, they often rely on concepts in a temporarily asymmetrical way: "For just as we cannot tell a story whose descriptions refer to future events which we cannot predict, so we cannot tell a story whose descriptions depend on concepts which we do not yet possess," (Mink 1987, 142, emphasis mine). Roth selects one example of this asymmetry from Mink: "the unpropertied citizens of Rome constituted the first urban proletariat," (Mink in Roth 2020, 10). And he himself suggests as other examples the vocabularies of critical race theory and feminism. Therefore, since the narrative sentences calling for explanation are of our own making, it follows that historians in fact construct their own explananda (Roth 2020, 10–12).

(non-D) If one accepts *non-S*, one accepts that historical events come to existence via our narrativizing and conceptualizing efforts. Historical events are thus also nondetachable from historical theories and (narrative) explanations positing them (Roth 2020, 14ff).

Roth argues: "As a consequence of this feature narratives explain only by virtue of the narrative order itself. Not, of course, because the world it depicts does not exist; rather, the narrative selectively orders materials. That results, I have maintained, from the fact that the events to be explained, and the events used to explain it, turn out to be part and parcel of the narrative to which they belong. A narrative constructs both the explanans and the explanandum," (Roth 2020, 14).

Compare with Mink: "But despite the fact that an historian may 'summarize' conclusions in his final chapter, it seems clear that these are seldom or never detachable conclusions. [...] The significant conclusions [...] are ingredient[s] in the argument itself [...] in the sense that they are represented by the narrative order itself. As ingredient conclusions they are exhibited rather than demonstrated," (Mink 1987, 79).

Taken together, *non-S* and *non-D* imply, according to Roth, that one cannot make any good sense from the idea that the past has some underlying and unifying structure on its own, which can be discovered and then described by us. In other words, *non-S* and *non-D* go "against any idea that there exists just one human past," (Roth 2020, 14, cf., 14ff). This, in the end, undermines even the very plausibility of Danto's I.C.

(non-A) If there exist more than one past, one must simply give up the idea that our histories can aggregate into a one all-inclusive and consistent explanation/meta-narrative of the past.

Roth writes: "Aggregation presupposes that all the events could belong to some one narrative, an implied unifying perspective. But there can no more be a single story than there can be an Ideal Chronicle, for new and different events and new and different stories constantly come into being," (Roth 2020, 14, emphasis mine; Roth here in the text refers to Mink 1987, 197); "Nothing now licenses an assumption of The Past conceived as an untold or partially told story, but always nonetheless the same story, a human past narratable *sub specie aeternitatis*," (Roth 2020, 15).

Compare with Mink: "The determinateness of the past is part of common-sense ontology; it is not a theory but a presupposition of unreflective common experience. [...] "We could not conceive or imagine an Ideal Chronicle at all unless we already had the concept of a totality of 'what really happened.' We reject the possibility of a historiographical representation of this totality, but the very rejection presupposes the concept of the totality itself. It is in that presupposition that the idea of Universal History lives on," (Mink 1987, 195); "If we accept that the description of events is a function of particular narrative structure, we cannot at the same time suppose that the actuality of the past is an untold story. There can in fact be no untold stories [...]. There can only be past facts not yet described in

> a context of narrative form," (Mink 1987, 201). "There are hopes, plans, battles, and ideas, but only in retrospective stories are hopes unfulfilled, plans miscarried, battles decisive, and ideas seminal," (Mink 1987, 60; cf., Roth 2020, 94)

In Sect. 4.3, I shall focus again in more detail on *non-S*, I.C., and Mink's negative concept of Universal History (the idea of the past as an untold story) and their implications for the metaphysical status of historical entities and events hence our idea of "past reality."

Now, as *non-SDA* further clarifies, historical explanations that Roth targets are narrative *essentially* insomuch as the narrative/retrospective perspective comprises the very explanandum. Thus, only historical explanation qua essentially narrative explanation can explain truths emerging in retrospect (contained in narrative sentences). Roth claims:

> That is, in those cases where the available antecedents provide no basis for rationally accounting for the outcome [e.g., assuming that in 1984 it could not be predicted that Arnold Schwarzenegger, who starred in The Terminator, would be elected the Governor of California in 2003], then a sequencing of events that has the later event emerge as a consequence of the earlier (i.e., a narrative) provides the only sort of explanatory account one could have for such cases. (Roth 2020, 68, rewritten to respect my example)[9]

It is not possible to account for these kinds of truths in a non-narrative form, because they are entrenched in developments that cannot be outlined by laws. Roth's *non-SDA* also highlights two general features of historical explanation that I shall expand upon in this chapter. Firstly, historical explanations cannot be translated to the other forms of explanations, i.e., mathematical equation, statistical probability or simply any law-like generalization. Logically speaking, one cannot produce an explicative formula: *X* happened in virtue of general rule *Y*. Secondly, historical explanations will always be shaped by subjective considerations (speculations, if you will) since they map the landscape of immense complexity, contingency and uniqueness.

These are the reasons why historical explanations viewed as "hypotheses" cannot be "tested" in the traditional sense of the term (verified/falsified). The cause-effect components of the given narrative explanations are then fundamentally fallible and probabilistic. These features are the epistemic limitations of narratives qua the category of explanation.

[9] As a counter-example, consider this passage: "What proves critical to clarifying narrative as a form of explanation involves whether or not an explanation in this form can also be nonnarratively structured. That is, does it allow for paraphrase into some other, nonnarrative explanatory form? For if so, then whatever explanatory import such a narrative seemingly possesses—revealing how things at the beginning of a series came to be what they later were—turns out to be inessential for purposes of explanation. Hempel's well-known example of a radiator bursting provides a case in point. A story explaining why it burst—e.g., one's failure to put in antifreeze—can be recast and given instead classic deductive-nomological form," (Roth 2020, 66).

4.2.2 Two Examples: Constructing and Evaluation Holocaust and Evolution

Roth obviously reveals the epistemic impacts of *non-SDA* by discussing actual historical explanations. Here, for reasons of space, I am going to touch upon only two of them. The first concerns the evergreen of ethical and theoretical debates and disagreements related to historical theorizing—the Holocaust. Roth also chooses this example purposefully, among other things, with the ambition of pointing out the nuances of his position that certainly has a constructivist and relativist flavor (see Roth 2020, 16ff). However, back to the example, Roth sketches the working definition of the Holocaust as follows: "the pursuit by the Nazis of a deliberate policy of extermination against the Jews and selected others," (Roth 2020, 16). And then discloses:

> Indeed, my view does understand the Holocaust as a social construction, at least in the following sense. Were it not, I would claim, for the publication of Raul Hilberg's (1985) *The Destruction of the European Jews*, it would be highly unlikely that discussion of Nazi policy with respect to the Jews would have received the scholarly attention it does. Hilberg does not in that work use that term, but his work arguably proves central to the creation of a historiographical focus that does. (Roth 2020, 16)

Roth's argument is not about denying any aspect of the "reality" of the Holocaust, i.e., any facts (however exactly demarcated), such as the number of people killed, the conditions in concentration camps/extermination centers, the existence and purposes of the Nuremberg Laws, etc. He only proposes that Hilberg's work isolates certain facts in a manner that eventually brings into existence the whole field of Holocaust studies with its own discourse and which inevitably seeps into our everyday understanding, not only of our past but of our present as well: "In short, discussion of the Holocaust as an event in its own right, as one that grimly foreshadows what happens repetitively going forward in the twentieth century, as a basis for rethinking the limits of nation-states for the purpose of judging war crimes, etc., depends in many key respects on how Hilberg bequeaths to us a way of thinking about that time," (Roth 2020, 16). Roth then asks us to reflect on the following passages from Hilberg:

> The Germans killed five million Jews. The onslaught did not come from the void; it was brought into being because it had meaning to its perpetrators. [...] In retrospect, it may be possible to view the entire design as a mosaic of small pieces, each commonplace and lusterless by itself. Yet this progression of everyday activities, these file notes, memoranda, and telegrams, embedded in habit routine, and tradition, were fashioned into a massive destruction process. Ordinary men were to perform extraordinary tasks. A phalanx of functionaries in public offices and private enterprises was reaching for the ultimate. (Hilberg 1985, 993–994)

> The destruction of the Jews was thus no accident. When in the early days of 1933 the first [German] civil servant wrote the first definition of "non-Aryan" into a civil service ordinance, the fate of European Jewry was sealed. (Hilberg 1985, 1044)

The first passage, putting aside the fact that Hilberg himself acknowledges the power of retrospective insight, primarily manifests the still paradigmatic way in which Hilberg explained the Holocaust (Roth 2020, 17, 140). As Roth puts the matter: "He crafted this mosaic; to him goes the credit of providing the world a new way of seeing familiar pieces. His 'mosaic' depicts that destruction as a deliberate, careful policy," (Roth 2020, 17). Put another way, in Hilberg's explanation, the main agents of the Holocaust are "bureaucracies" and "institutions" operated by "ordinary man" (Roth 2020, 73, 140).

The second passage, in fact, discloses the explanandum Hilberg invents and intends to explain. "Somewhat more prosaically, Hilberg's statement may be reworded as a narrative sentence: The Holocaust began in 1933," (Roth 2020, 72). This is what Roth means by social/historical constructivism that is characteristic for our human relation to historical events. He remarks: "Narratives 'create' what they simultaneously set out to explain, not because they 'make things up,' but precisely for the reason that narrativists such as Hayden White have for so long insisted: only by this means does a historian provide meaning and structure to a morass of details that otherwise has neither," (Roth 2020, 75).

To put it baldly, there could be numerous historical accounts of the Second World War or of the Third Reich leaving "the Holocaust" out of the picture, or at least out of its center.[10] Also, no fact of the matter directly forces Hilberg, just to give one example, to shift his attention from the victims and various atrocities of "the Holocaust" to the administrative, and in certain sense very rational, processes of mass destruction. So, alternatively put, to the attempt to explain "the Holocaust" as a gradual transition from "ordinary" to "extraordinary" patterns of human behavior (Roth 2020, 16–17, 72–74). Roth, while keeping in mind his constitutive features of narrative explanation, concludes:

> The Destruction of the European Jews fits the mold of a narrative explanation developed above. Its conclusion can be stated as a narrative sentence, one that the text explains by providing a beginning-middle-end structure that presents a story line detailing the causes of that event, but where "causes" can be identified only by offering specific steps in an extended developmental sequence. No laws underwrite this sequencing. And while other genocides happen both before and after, "genocide" does not name a scientifically standardized event type [*non-S*]. The result will be explanatory, an answer to an important "Why?" question that depends essentially (in my sense of the term) on a temporal sequencing of certain statements of fact. [...] The event explained – what "the mosaic of small pieces" depicts – moreover cannot be detached from the narrative that presents it [*non-D*]. [...] And as reactions to and subsequent scholarship reveals, the narratives concerning what happens over this time span do not aggregate [*non-A*]. (Roth 2020, 74–75; see Roth 2004 for addressing the issue of *non-A* with regard to the Holocaust in greater detail).

[10] Consider in this context a remark from a distinguished Holocaust historian, Christopher Browning, that Roth quotes in his book: "In Hilberg's portrayal, this event was a vast bureaucratic and administrative process employing a cross-section of German society, not the aberrational accomplishment of a few demented individuals. The Holocaust comes to be marked as an *independent event*, one whose workings have implications that extend beyond situating it as a historical aberration, a freak, pathological event on the margin of German and European history," (Browning in Roth 2020, 73, emphasis added by Roth).

The second and last example I want to examine here in this section concerns narrative explanations and their relevance for evolutionary science.

This example thus also exemplifies the relationship of narrative explanation to a historical science which at the same time works with a non-trivial degree of theoretical standardization. As I have already indicated in Sect. 2.3, the Neo-Darwinist paradigm in evolutionary biology is still determining for evolutionary theorizing as such. Thus, in cases of ultimate evolutionary explanations, the theoretical arsenal in question would typically consist of the principle of natural selection and correlated concepts of adaptation, exaptation, and spandrel. Roth builds his example on a paper written by John Beatty and Isabel Carrera (2011), in which they together argue for the need of narrative explanations in evolutionary biology in cases where history matters.

Beatty and Carrera discuss and further develop Steven Jay Gould's well-known conjecture that, due to genetic variation, which is the sine qua none of biological evolution, one must accept that historical contingency plays a central role in the very creation of organisms. As Roth elegantly paraphrases it: "Following Gould, they call this 'replaying life's tape.' (Beatty and Carrera 2011, 472–473). In this thought experiment, Gould maintains that if one could erase 'life's tape' going back to some particular point, and then 'replay' it from that point (start over with the state description of that time *t*, so to speak), Gould doubted that from a biological/evolutionary standpoint that the tape would 'play out' exactly as before," (Roth 2020, 76). Thereafter, Beatty and Carrera refer to an experiment that, in essential respects, substantiates Gould's suggestive conjecture:

> The basic (ongoing) experimental setup involves the investigation of twelve, initially identical (cloned) populations of the bacterium E. coli, as they evolve in identical (and identically altered) environments. The investigators have detected a number of differences in evolutionary outcomes among the twelve lines, differences that cannot be attributed to differential selection pressures (since the groups have faced identical selection pressures in their identical environments), but that seem instead to be causally dependent on chance differences in the variations (and order of variations) that have arisen in the different lineages. (Beatty and Carrera 2011, 488)

Roth continues in retelling the course of the experiment: "After 31,500 generations, one lineage exhibited an extremely rare but highly adaptive mutation. The question addressed by the researchers in line with the 'replay' scenario concerned whether or not this mutation would occur in the other populations as well," (Roth 2020, 77); or as Beatty and Carrera pose the question: "Whether the population in question had by that time, through a series of contingencies, evolved to become uniquely capable of taking the final evolutionary steps in the direction of citrate metabolism," (Beatty and Carrera 2011, 488).

Insomuch as the researchers conducting this particular experiment confirmed by subsequent tests that indeed only the one lineage in question did evolve the ability to metabolize citrate, one can conclude that, although initial conditions were identical for all the E. coli, only 1 out of the 12 lineages developed such an adaptive trait. "This suggests the conclusion that, starting from a genetically identical initial state and holding the environment constant, in some worlds (at least one, anyway) a specific mutation emerges, and in some (indeed, most) it does not. There will be, moreover, no predicting that this mutation might ultimately emerge because 'in the beginning' all these 'possible worlds' share a point of origin," sums up Roth (2020, 77–78). *Put simply, history matters even in biological evolution as, for instance, the emergence of adaptive designs is unpredictable in individual cases. Each individual evolutionary case represents an individual historical scenario.*

Beatty and Carrera then emphasize that narratives are indispensable for explaining these scenarios, i.e., "situations where history matters: where a particular past had to happen in order to realize a particular future, and when the past that had to happen (in order to realize that future) was not bound to happen, but did," (Beatty and Carrera 2011, 491).

However, Roth asserts that, when put this way, "history matters" rather because after the fact one knows that something unpredictable happened. Only after the lineage in question evolved the ability to metabolize citrate, can one truthfully assign evolutionary significance to the earlier adaptive mutation. "And knowledge so gained through hindsight can be used to fashion a narrative that then charts a developmental course from a beginning to an end. This may also have been unpredictable [...]. But unpredictability just serves to emphasize that this insight could be obtained only retrospectively," (Roth 2020, 78–79). In a phrase, one can identify a beginning only by knowing an end *that for some reason matters to us* (cf., Roth 2020, 79). Roth thus thinks that this example also harmonizes with his constitutive elements of narrative explanation:

> *What counts as an adaptation remains contextually defined and retrospectively identified.* To the extent that context remains ineliminable and so an adaptive mutation proves only retrospectively specifiable, then the explanandum event will in turn also be nonstandardized in the sense relevant to how essentially narrative explanations have been characterized. For under that description, the event has no nonnarrative explanation. Those other, more standardized aspects will enter into the sequencing, but they can neither displace nor replace a need to narrativize. Put another way, the relation of explanans and explanandum (whatever the content of statements in the explanans) remains essentially narrativized. (Roth 2020, 80, emphasis mine)

As the two examples above also nicely illustrate, Roth's position is, as a good naturalistic position, in itself non-committal with respect to certain conceptual, ontological, and explanatory frameworks.

To give some examples, the emphasis on retrospective and narrative insights and even *non-SDA* does not require that the best strategy to explain cultural history is focusing on actions and decisions of individual historical actors, nor does it prescribe employing intentional vocabulary for constructing explananda and explanans. By the same token, Roth's position, of course, does not discount these as viable possibilities *in advance*.[11] What is important, for Roth, is the fact that only from

[11] Viewed as such, narrative explanations structurally resemble abductive inferences or an explanatory strategy nowadays commonly called *inference to the best explanation* (see Roth 2020, 51, 143ff). Roth quotes Peter Lipton's depiction of inference to the best explanation (IBE): "Beginning with the evidence available to us, we infer what would, if true, provide the best explanation of that

within chosen descriptions—which ultimately implicitly or explicitly express specific human values—does a "morass of details" (or data, or facts, if you prefer) have significance as (historical) events. Historical knowledge is thus a product of our moral, political, and epistemic cultures and their interests.

This, naturally, has some bearing on the rational possibilities of evaluating narrative explanations. And it is again this respect in which Roth's conclusions resonate with the much debated and disputed work of Hayden White, who systematically highlights the moral and political choices and responsibilities associated with our narrativizing efforts. "In short, choices of form impact content," (Roth 2020, xv, cf., 141ff).

Especially due to *non-D* and *non-A* of historical events, "evaluating explanations that have narrative form essentially [...] will primarily be a function of assessing competing explanations, and so draw on evaluative criteria more akin to theory appraisal than to hypothesis confirmation," (Roth 2020, 66). Therefore, in general, evaluation of historical explanations will be principally comparative on several levels. As Roth clearly summarizes:

> Although there will be factors both internal to a narrative explanation (assessing the sequencing) and external to it (comparison with competing narratives, if any), I suggest that evaluation in the end can only be on a case-by-case basis. The fact that narratives cannot be expected to aggregate will be a limiting factor; that the events explained have been standardized, at least to some extent, will abet comparative evaluation. In other words, nonaggregativity and nonstandardization force this result; no general test can be had. (Roth 2020, 75)

In Chap. 6, I present my own examples of evaluating internal factors of ultimate evolutionary explanations qua historical explanations; see Roth (1998 and 2004) for his informative attempts to demonstrate how to evaluate historical explanation with respect to external factors, i.e., with respect to other competing historical explanations.

I am convinced that Roth's position will not be considered controversial for theoretically reflective scientists and scholars, for example, that they will not find it constructivist in some solipsist sense. And if so, one must simply keep in mind that Roth's position is motivated naturalistically; now at least in the sense that he avoids any a priori truths and commitments, and conceive thus scientific evidence and even our ordinary experiences intratheoretically. "The general point to keep in mind

---

evidence," (Lipton in Roth 2020, 144). According to Douven (2011), Peirce characterizes abduction in this, for our matters here, very instructive, way:

"The surprising fact, *C*, is observed.

But if *A* were true, *C* would be a matter of course.

Hence, there is reason to suspect that *A* is true."

These inferences then obviously work only hand in hand with our interests and prior theoretical commitments (cf., Roth 2020, 144). Roth in this context says: "Thus quite unlike the Hempelian model, explanation in this case is not a consequence of mimicking a certain inferential form. Rather, it could be said, the explanandum determines what to count as explanans by virtue of providing an explanation for those factors. In this important respect, IBEs share the key feature of nondetachability characteristic of essentially narrative explanations," (ibid.).

involves the fact that scientific reasoning begins as parasitic on 'common sense,' that is, whatever passes as received knowledge for a time and place. Science certainly refines such views and may in time transcend and transform 'common sense.' But reasoning begins from within some set of received views – Quine's 'web of belief,'" (Roth 2020, 159). Roth in the same context also mentions an illuminative, and characteristically post-positivist, quote from Danto: "Narrative structures penetrate our consciousness of events in ways parallel to those in which [...] theories penetrate observations in science," (Danto 1985, xii). For similar reasons, Mink thinks of narratives as cognitive instruments; as a "primary and irreducible form of human comprehension, an article in the constitution of common sense," (Mink 1987, 186). I think that, in these details, Danto and Mink think naturalistically as well.

Narrative structures then manifest our peculiar human relation to past events; they manifest the configuration of the historical discourse qua an important area of our human epistemic condition. "There is nothing wrong with being wise after the event; it is just that we can't be wise after the event, before the event," helpfully quips Mink (1987, 145). For better or worse, naturalism leaves us without any strict demarcation between our everyday reasoning and scientific theorizing. In his review of Roth's PSHE, Maksymilian Del Mar seems to appreciate all of this and suitably comments:

> Notice how, on this view of history, one thing that emerges as both significant and fascinating about historical practice are the disagreements between historians as to what count as events, as well as events worth describing. *Indeed, on this view, there are many different kinds or degrees of historical disagreement*: e.g., first-level disagreements as to what to treat as a unit of description (e.g., whether to talk at all of "events" or "actions"); second-level disagreements as to whether to describe something as an event (or action); and third-level disagreements, i.e., on just how to characterize and describe the events (or actions) in question, even when there is second-level agreement on what the relevant events (or actions) are.
>
> The history of history, on this view, becomes a fascinating aesthetic, ethical, and political battle: *the work of historians now appears as a series of emotionally charged disputes about what is relevant and important to describe, and how*. (Del Mar 2020, emphasis mine)

In the next section, I briefly concentrate on Roth's re-reading of Thomas Kuhn's *The Structure of Scientific Revolutions* (2012 [1962], hereafter SSR) through the lens of his views on historical discourse and historical explanation.

This re-reading has a threefold impact on Roth's meta-historiographical position: it (1) instantiates another example of employing narrative explanation. Then, by arguing that Kuhn explained the development of science narratively, Roth (2) naturalizes narrative explanations (he shows that they are necessary for reconstructing what a science is), and also (3) provides naturalism with a general approach of establishing in a historical, yet epistemically justified, manner which rational practices are to be counted as scientific, i.e., Roth provides us with a way of establishing how and why given explanatory practices have stood the test of time without any prior conceptual commitments to what science or rationality is or should be.

4.2.3 Kuhn and His Story of Science

Roth's exegetical excursion into SSR does not aim to rebut or reconsider any of the now commonly accepted Kuhnian conclusions, with the key being the historicizing of scientific progress, and so rationality as such.[12] Or as Roth himself at one point puts it, "that what to count as science has no historically stable boundaries," (Roth 2020, 91). Quite the opposite, as I shall demonstrate in this section, Roth forcefully supports these highly appreciated conclusions by nesting Kuhn's historical account of science into his own irrealist and naturalist frameworks.

Also, remember that in Chap. 2 (Sect. 2.4) I concluded that naturalism, as Quine developed it anyway, maintains that "our common-sense as well as science rely on the historically accumulated belief system guiding and grounding our epistemic cultures." Naturalists do not have any atemporal invariant means of distinguishing scientific from non-scientific practices. Naturalists need to first establish what science is, in order to gain any normative means of evaluation. Well aware of this epistemic predicament, Roth employs Kuhn's historicizing approach to science as a possible naturalistic strategy for explaining why given explanatory practices now commonly accepted as scientific should be perceived as such (see Roth 2020, 108ff).

Thus, Roth's pivotal interest lies in SSR's methodology in particular, and Kuhn's historiographic self-reflection in general. "Kuhn certainly thought of himself as a historian […]," affirms Roth (2020, 98). Generally, as Roth stipulates, Kuhn's understanding of historical inquiry is, at least from a certain period, explicitly very close to the one he himself advocates for. In the book *The Essential Tension*, published 15 years after SSR, but containing philosophical elaborations written instantly after its publication, Kuhn asserts:

> The final product of most historical research is a narrative, a story, about particulars of the past. In part it is a description of what occurred […] Its success, however, depends not only on accuracy but also on structure. The historical narrative must render plausible and comprehensible the events it describes. In a sense to which I shall later return, history is an explanatory enterprise, yet its explanatory functions are achieved with almost no recourse to explicit generalizations. (I may point out here, for later exploitation, that when philosophers discuss the role of covering laws in history, they characteristically draw their examples from the work of economists and sociologists, not of historians. In the writings of the latter, lawlike generalizations are extraordinarily hard to find.) (Kuhn 1977, 5, the quote is reconstructed by Roth, and he also documents that this view of historiography is arguably influenced by reading Mink, see Roth 2020, 91–94)

> Theories, as the historian knows them, cannot be decomposed into constituent elements for purposes of direct comparison either with nature or with each other. […] For the historian, therefore, or at least for this one, theories are in certain essential respects holistic. (Kuhn 1977, 19–20, reconstructed by Roth 2020, 93)

[12] It might be objected that the latter is a stronger claim, definitely not accepted commonly, and that is perhaps correct. I can only say here that my construal of naturalism in this book assembles many arguments for this claim.

Kuhn thus clearly regards narratives as explanatory devices and subscribes to a holistic view of historical events and entities; Roth understands it as a subscription to his *non-D* of historical events (Roth 2020, 93). So, in effect, Roth's endeavor to read SSR through his own historiographic framework receives additional justification, regardless of whether one believes that Kuhn wrote SSR already with these assumptions in mind.

Roth's main explicatory target is then Kuhn's concept of "normal science." Alexander Bird's concise description of Kuhnian normal science changing arm in arm with given scientific paradigm will do for me here: "A mature science, according to Kuhn, experiences alternating phases of normal science and revolutions. In normal science the key theories, instruments, values and metaphysical assumptions that comprise the disciplinary matrix are kept fixed, permitting the cumulative generation of puzzle-solutions, whereas in a scientific revolution the disciplinary matrix undergoes revision, in order to permit the solution of the more serious anomalous puzzles that disturbed the preceding period of normal science," (Bird 2022). Now, although the persistent interpretative controversies associated with the (im)proper usage of the term "paradigm" certainly deserve some attention, they are not directly related to Roth's arguments. Henceforth, simply regard the above-mentioned "disciplinary matrix" as equivalent term for "paradigm" (cf., Roth 2020, 111–112).

Roth argues that Kuhn's own historical explanation in SSR of how and why a normal science develops within a certain period (after a revolution) exemplifies essentially narrative explanation (it thus also manifests the *non-S* and *non-A* elements of narrative explanation as Roth states them).

> SSR does not examine events that come in some standardized format, and it provides a narrative of the history of science that does not aggregate either with then current histories or as the history of a stable something called "science." Regarding standardization, the "revolutions" that Kuhn identifies possess a "structure" only in a figurative sense, inasmuch as much of his account goes to show that there can be neither any specifying of exactly when one paradigm will be overwhelmed by anomalies nor exactly what considerations induce scientists to shift from one theory to the next. The very contrast between "normal" and "revolutionary" science – "complementary notions" as Kuhn [2012, 8] terms them – itself proves contingent and contextual. As Kuhn also notes, what counts as "revolutionary" change may be invisible to all but a limited group engaged in a highly specialized undertaking [Kuhn 2012, 180]. The events exist only as described retrospectively, for that alone reveals them as normal or revolutionary.
>
> As a narrative about science, the history that SSR presents aggregates neither internally – thus incommensurability – nor with standard "textbook" histories of science. Indeed, Kuhn emphatically sets his account in opposition to those histories that maintain that accounts of scientific change do aggregate. (Roth 2020, 101, rewritten to respect my style of referencing)

It is not without interest that, at the very end of SSR, in the last chapter titled "Progress through Revolutions," Kuhn briefly ventures to model the non-cumulative concept of scientific progress on Darwin's non-teleological concept of biological evolution. Kuhn proposes that, in a very loose sense, one might consider given scientific practices also to be selected for. However, the selection must in this case have both dimensions; the natural (it is a selection happening in and against the world) and the social/cultural (it concerns people/scientists working on the theories of the world) (see Kuhn 2012, 169ff). There will thus not be any definitive unscrambling of these two dimensions, as both the formative concepts ("nature"/"culture") require a theoretical reconstruction.

At one point, Kuhn illuminatively writes: "*The resolution of revolutions is the selection by conflict within the scientific community of the fittest way to practice future science.* The net result of a sequence of such revolutionary selections, separated by periods of normal research, is the wonderfully adapted set of instruments we call modern scientific knowledge. Successive stages in that developmental process are marked by an increase in articulation and specialization. *And the entire process may have occurred, as we now suppose biological evolution did, without benefit of a set goal, a permanent fixed scientific truth, of which each stage in the development of scientific knowledge is a better exemplar*," (Kuhn 2012, 171–172, emphasis mine; cf., Popper 1972; Hull 1988). So, to put it simply, what is being selected is actually the fittest scientific community. It cannot be any other way if there is no "fixed scientific truth"/"a stable something called science." And to insist that what is being selected are, e.g., the theories and/or metaphysical assumptions (the disciplinary matrix), would be putting the cart before the horse. It would again be putting reason before history.[13]

As Roth (2020, 102ff) cleverly points out, Kuhn reconfigures the whole explanatory goal of the general history of science from something like what counts as a science to what counts as a scientific community/"who counts as a scientist" (these questions must, of course, be asked and answered within a particular time *t*). "This sets (or perhaps I should say resets) the problem. For narrating a history of science (or of reason) that is not also a history of Science (or of Reason) proves to be a tricky matter indeed. In particular, Kuhn's quicksilver view of science foists on him a severe rhetorical challenge. What remains for a history of science to be a history of? Kuhn's awareness of the rhetorical challenge – What does a history of science now take as its object? – pushes him to fashion a novel answer: how a certain type of community constitutes what counts as science," (Roth 2020, 104; cf., Kuhn 2012, especially 179). The result is a transformed picture of science and its practitioners:

[13] Consider also this reflection by Kuhn: "*I say that we must learn to see development and progress in science as something which is not drawn by a goal, as the way things really are at the far end, but as simple evolution from a starting point, like biological evolutions.* Therefore I am very much sympathetic to Foucault's notion of *epistemes*. [...] On the other hand, [...] what I found destructive in Foucault was the account of the way one went from one *episteme* to another. The developmental process, the evolutionary process, all has to be part of a story that emphasizes *epistemes* and *coupures. On the contrary, Foucault states that epistemological fractures can be represented but not explained, for they embody the "other" in itself.* The interpretation of this otherness constituted Foucault's political involvement, which I never had. *For Foucault it was less essential to talk about how one goes from one place to another, than to talk about where to go.* [...] *If the process is driven from behind, you then have to say what the response is, what direction it moves in as a result of a drive from behind. But if you believe, like Foucault, that history is being pulled from in front, you don't have to understand the process, changes become more trivial*," (Kuhn in Borradori 1994, 161, long emphasis mine).

> The textbook image has that community as constituted derivatively, by reference to some prior and historically independent standard. Scientists become scientists by virtue of instantiating that standard. But Kuhn detaches that which constitutes a community of science from any timeless standard. [...] All that matters for purposes of historicizing the notion of science involves the fact that identifying what science is for Kuhn cannot happen until and unless one identifies what counts as normal science at a given time. What will emerge as normal science cannot be predicted on the basis of what passed previously as such. Until the proverbial dust from a revolution settles, no clear view can be had regarding what counts as normal science. (Roth 2020, 106)

In the end, what counts as a normal science at a time *t*, is always relative to what counted as a normal science before *t* (before a revolution). Normal science is an essentially historical concept. Therefore, according to Roth (2020, 111), to account for the developments in "science," hence developments in our notion of "rationality," one needs a historical explanation. One needs to narratively describe the transition, within a given mature science, from normal science *S1* to normal science *S2*. In other words, to put forward a developmental sequence. Roth thus ultimately reads Kuhn as constructing *a historical counter-narrative* to all *ahistorical accounts* of science which, moreover, explains what is and should be regarded as a (normal) scientific practice in the narrative mode.

Viewed as such, it is a strategy having significant practical/explicatory force for naturalism, viz., it, most importantly, provides a naturalistic account of (a) science. "That strategy requires only a narrative of discontinuities, one in which community practices and membership shift in certain key ways during the transition periods of interest," (Roth 2020, 112). Roth thus naturalizes simultaneously both the concept of science and the concept of historical explanation. Without retrospective histories of scientific evolution, one cannot make sense of a (normal) science. "Understanding Kuhn's work as a narrative naturalizes narrative explanation through a form of mutual containment – since narrative helps constitute what counts as normal science, narrative cannot be separated from an understanding of what science is," (Roth 2020, 162, footnote 4, emphasis mine). It is thus the particular (hi)story of a (normal) science that matters in the end:

> That a discipline has successive paradigms related in certain ways – has a particular history – creates a lineage for physics and chemistry as sciences but not, e.g., for astrology and alchemy. Kuhn's narrative also exhibits that what passes as scientific rationality has a history. The "argument" for his history of reason consists in showing how different cases of theoretical transition also alter what makes for proper procedures within a science. Kuhn's emplotment of these successive transitions shows them resistant to any overarching analysis of scientific rationality. (Roth 2020, 19, cf., 126ff)

Roth's re-reading of Kuhn truly "reverses the received order of epistemic authority," insomuch as all the epistemic and generally rational criteria must be understood from within the historical discourse, and not vice versa (cf., Danto 1995, 72).

In the next section, I focus on how Roth's overall delineation of the historical discourse from the epistemological vantage point relates to metaphysical issues at stake. I start with his additional deconstruction of the very idea of I.C., and then move to elaborate on his alternative position to historical realism and anti-realism which he calls *irrealism*.

4.3 Historiography and the Metaphysical Multiplicity of the Past(s)

While discussing the issue of *non-A* of historical events in Sect. 4.2.1, we learnt that Roth together with Mink reject the idea of *one unalterable past*. Or, in Mink's words: the idea of Universal History; the totality of what really happened in the past; the totality of the past as *untold or partially untold but still one and the same story*. In this section, I wish to further elaborate on the consequences of this rejection for our general notion of historical *reality*.

Moreover, Roth himself believes that to spell out the metaphysical consequences of his conception of historical explanation is, at the same time, another argument for its epistemic license. "For imagined as timelessly fixed and unchangeable, i.e., a past composed of a determinate inventory of events, there appears to be no reason why histories should necessarily have any of the distinctive qualities [of narrative explanation]. That is, a fixed past should plausibly contain [...] explanatory sequences that aggregate, standardize, and detach," (Roth 2020, 35). Also, recall that for Roth and Mink, the negative model of historical reality is a version of Danto's I.C.: "If the past is fixed, if it is a story waiting to be told, then it must be logically possible to have some chronicle of it of the sort Danto imagines," (Roth 2020, 28). That is to say, any variety of (naïve) realism which takes for granted the *investigation-independent*[14] nature of historical events, ipso facto, historical explananda, would constitute an argument against the explanatory necessity and value of narrative explanations as Roth delimits them.

Roth is well aware that the Dantonian observation that retrospective insight opens the past to infinite re-descriptions is not by itself an argument against (naïve) realism in historiography. One could, in principle, accept that the temporal distance and/or conceptual asymmetries enable new (re)descriptions of chosen historical happenings/facts, and still hold the idea of Universal History (Roth 2020, 47). However, as I have also specified in Sect. 4.2.1, Roth's constitutive elements of narrative explanation (*non-SDA*) point towards anything but an investigation-independent nature of historical events, hence theorizing. For Roth, there is no room for historical events outside of a language (i.e., our conceptual frameworks). As he himself puts it at one point in italics: "*An eventful historical past exists only as a result of human theorizing. History becomes an artifact of a disciplined disciplinary imagination*," (Roth 2020, 51).

To debunk the realist metaphysical picture implicit in the idea of Universal History, Roth challenges the rational plausibility of the notion of I.C. Let me therefore first focus in more detail than in the previous section on Roth's deconstruction of I.C. The get-go of his deconstruction rests in pressing the issue of *non-S* to its very limits. Roth tries to put aside the issue that an Ideal Chronicler/Ideal Witness

[14] I borrow the phrase "investigation-independent" in the context of real*ism* from Crispin Wright, on whose thoughts Roth also builds his views (Wright in Roth 2020, 47).

does not possess narrative sentences and asks whether their job could be regarded as recording historical events in any relevant sense of the term.

> Imagine the Ideal Chronicler at work. What does the Ideal Chronicler record? Danto's suggestion is everything, and at once at that. But in agreeing that the Ideal Chronicler can write anything at all, we have, in a Wittgensteinian sense, been tricked; the very first step is the fatal one. The conjuring trick is complete once one concedes that there is anything for an Ideal Chronicler to record.
>
> What is the basic unit of the posited perfect record? It is events of every sort: visits home, heartbeats, a first kiss, the jump of an electron from one orbital position to another. But, as we know, events may be sliced thick or thin; a glance may be identified as an isolated event or as an instance in an event. What the unit-event is depends on the telling of it. Given the instructions to record "everything that happens, as it happens," the problem is not that there is too much for an Ideal Chronicler to record; the irony is that there are no things in the abstract to be recorded. An Ideal Chronicler never gets started because there are no ideal events to chronicle. (Roth 2020, 29)

Roth's intention here is not to simply suggest that events are not as real as, for example, sticks and stones. Again, Roth is a naturalist, so all entities/objects are for him ultimately theoretical posits. As long as the vocabulary of "events" proves, for instance, useful for explicating and explaining human action, Roth is happy to accept "events" as "legitimate objects of discourse" (ibid.).[15] Or I would at least argue that he should as a naturalist. Nonetheless, treating events as "individuated objects" does not solve the problem at hand which is, in fact, to somehow detect "ideal" events.[16] In other words, the plausibility of something akin to I.C. presupposes a description-independent view of events, and that is certainly not an epistemic presupposition. To start with an epistemically ungrounded ontology of "events" is not only against naturalism, as characterized in this book, but also does not help us in any way to identify the kinds of events that should end up in I.C. It is an epistemically groundless and pragmatically worthless presupposition.

Roth (2020, 30–31) subsequently also tries to make the idea of I.C. plausible by understanding it as a sum of all available descriptions. But this will not suffice either. Danto's I.C. must contain "everything that happens as it happens, the way it happens." The descriptions in I.C. must be "objective" in the sense of "impartial." Can this version of I.C. be considered as plausible? No, if one allows all possible

[15] "A reason for worrying whether events exist in some philosophically relevant sense of that term—that is, whether they count as legitimate objects of discourse—is that assuming their existence proves a convenience for the purposes of explicating the logical form of sentences about actions. Countenancing events facilitates the ability to draw permissible inferences which otherwise cannot be readily managed if events are ruled out as individuated objects," (ibid.).

[16] Some might feel that a solution could be to count only the physical events as ideal. Firstly, how do we decide that physical equals ideal in each and every case of historical explanation? Secondarily, granted that particularly defined physical events are ideal events, one still faces the problem of underdetermination of theory by evidence. So, then, the task becomes discerning the *really* ideal physical theory from among a set of physical theories deemed non-ideal (Roth 2020, 29–30).

descriptions to enter the I.C., there will inevitably be inconsistencies hardly compatible with any notion of impartiality. Roth offers this example:

> Consider the events depicted in Kurosawa's film *Rashomon*. The story of what happened in the forest is successively retold from the perspectives of the husband, the wife, and the robber. None tells the same story; indeed, their stories are inconsistent. One tells a tale of rape and humiliation and a husband's cowardice; in other tales, one or more of these descriptions is contradicted. The audience sees what happens each time through the narrator's eyes; it is just that the narrators see different things. Events off the screen certainly have this quality as well. If the conjunction of all possible descriptions is included in the chronicle, the imagined purpose of the chronicle is defeated. Since it contains inconsistencies, it is no longer the hoped-for objective record of what actually happened. But if some descriptions of events are excluded, then the chronicle is incomplete, contrary to its intended purpose. Hence, if complete, then inconsistent, and if consistent, then incomplete. There can be no Ideal Chronicle. (Roth 2020, 30–31)

Another attempt by Roth to save the plausibility of some version of I.C. is thus to imagine a god-like Ideal Chronicler who would be absolutely impartial in our human sense. "Boethius imagined that God saw everything at once; all actions at all times stood revealed simultaneously to God. Certainly this is a way of capturing all that happens," (Roth 2020, 31). Roth, however, rejects the resulting god-like picture of I.C. "A Boethian chronicler" also fails to provide us with historical events because if a temporarily bounded portion of the past is disclosed to a Boethian chronicler all at once, he can provide us actually with "just one event, the total picture at *t*," (ibid.). Consider this telling clarification from Roth:

> The past so pictured presents not a chronicle, moreover, but a Jamesian buzzing, blooming confusion. Put another way, the identification of events from the Boethian tapestry of the past requires separating the simultaneous presentation of happenings which Boethius imagined into particular strands, the ones that interest us. God may see everything at once; an Ideal Chronicler, within a temporal limit, may do the same [...]. But this chronicle gives us less than we have even now. It is not just that there might be a need to factor in cultural conditioning and personal quirks when discussing what we see; seeing is just not perceiving, not in any simple sense.
>
> The basic problem is more elementary than that. When we view a snapshot or read a page of a book, if the object is not at the proper distance from our eyes, in appropriate light, and so on, we cannot see what we want to see. If someone pushes the book or picture up so it touches our nose, we see something – but not, for example, the picture of the picnic or the story of the latest political gaffe. Given the Boethian view, the Ideal Chronicler is in just this position, or leaves us in this position when consulting the resulting tapestry of happenings. The Boethian Chronicler has no natural point of focus. But without a focus, either nothing appears – the blooming, buzzing confusion – or God-knows-what looms before us, like the photo pressed too close for one to view. Total information gives us less than we need to know. (Roth 2020, 31–32, emphasis mine)

And this is not the final exploration of the possibilities of articulating the notion of I.C. for Roth; he considers additional possibilities, which I will not delve into here (see Roth 2020, 32ff).

In sum, Roth's reflections give argumentative substance to Mink's observation that "the determinateness of the past is part of common-sense ontology; it is not a theory but a presupposition of unreflective common experience," (Mink 1987, 195).

Roth, in effect, shows that "an eventful historical past" is unthinkable without our human interests and preferred categories and descriptions. There is no Universal History; there is no investigation-independent historical reality (Roth 2020, 33ff).

I find Roth's quarrel with historical realism straightforward and argumentatively transparent, so I now focus on his critique of anti-realism in historiography. Why is Roth not an anti-realist? Roth (2020, 47–49) acknowledges that anti-realists can, contra realists, tolerate what might be called "truth-value gaps" which appear due to the fact that our investigation of the past is typically "constrained" or "incomplete" with regard to our "access to" or "evidence of" the past. So, anti-realism, again contra realism, in this particular matter works with the investigation-*de*pendent nature of historical events/explananda—historical theorizing depends on our present options of investigation. But, for Roth, anti-realism is still too loyal to the realist picture of the past. "Absent a 'view from nowhere,' no good reason exists for asserting either verification independent truth-conditions or truth-value gaps," (Roth 2020, 48). Roth thus opts for irrealism:

> Irrealism, that is, does not require a semantic structure that assumes either a perspective out of time or one where meaning requires verification conditions dependent on what might be available now. Since there will be narrative sentences true of any particular time t but not knowable at *t*, these could not be verified at just that moment. Thus, [...] there can be true but in principle unverifiable sentences at a given time even absent the assumption of realism. Irrealism results by acknowledging that one's own history must play an important and ineliminable role "in determining our conception of what states of affairs may coherently be conceived as possible." (Roth 2020, 48; Roth quotes here Wright 1993, 187)

> Irrealism denies an assumption that both realism and antirealism turn out to share: an imagined view from nowhere, a past seen sub specie aeternitatis. (Roth 2020, 64)

If the realist belief is that "histories are found, not made," Roth's irrealist position postulates the exact opposite; histories are made by us—human theorizers (Roth 2020, 50). It goes without saying that such a position is deeply underwritten by Roth's holistic and naturalistic commitments, too. "*The real* – truth-makers for statements about the past – emerges from *within* a constituted past. Items appear as candidate truth-makers by virtue of their location within a constituted framework," (Roth 2020, 51, first emphasis mine; see also the whole discussion Roth offers on the views of Leon Goldstein, from 48ff).

The view that *histories are made, not found* deserves additional attention which will, I hope, dispel potential dissatisfaction with irrealism as a species of relativism. At the basic descriptive level, irrealism certainly has relativist implications; a historical past emerges from—hence is always relative to—our theorizing and narrativizing efforts. However, when understood naturalistically, this holds for the whole empirical inquiry. Without having socially established and enforced agreement on basic categories/concepts making sense of our causal interactions with the world (experiences), we would not have any way of constituting and communicating facts, nor subsequently of composing any kind of theories and explanations. For Roth, and all naturalists, there is no need to posit a sphere of "real reality" behind or above these linguistic frameworks (cf., Roth 2020, 57–59).

On the other hand, it is hard to imagine an argument that would once and for all discourage someone from simply presupposing the existence of such a sphere of "things-in-themselves." One is only offering reasons for a given pragmatic/conceptual choice here. Irrealism is, in this respect, possible to see as a choice to take seriously our entrenchment in history while doing historiography. Irrealism thus deliberately gives up on the idea of "The Past" marking or mirroring an "ontologically and theoretically well-defined world" (Roth 2020, 48, 35).

Put another way, since we cannot step out of space and time, our own natural and cultural histories are the only foundation for understanding our past. *A history is a past enough from the irrealist standpoint.* Consider in this setting the following comment made by Rorty:

> We can give what Locke called a "plain, historical, account" of how animals came to talk. But we cannot give a plain historical account of how they stopped coping with reality and began representing it, much less of how they stopped being merely phenomenal beings and began to constitute the phenomenal world. We can, of course, stick with Kant and insist that Darwin, like Newton, is merely a story about phenomena, and that transcendental stories have precedence over empirical stories. But the hundred-odd years spent absorbing and improving on Darwin's empirical story have, I suspect and hope, made us unable to take transcendental stories seriously. In the course of those years we have gradually substituted making a better future – a utopian, democratic, society – for ourselves, for the attempt to see ourselves from outside of time and history. (Rorty 2021, 102–103)

In this reading, naturalism positions irrealism among types of anti-representationalism advanced, among others, by American pragmatists (at a minimum, John Dewey and the aforementioned Rorty, cf., also with Zeleňák 2018, for an instructive attempt to introduce Roth's position via anti-representationalism).

From a normative perspective, Roth explicitly resists subscription to any form of "anything goes" relativism.[17] As I indicated in Sect. 4.2.2, Roth defends an essentially comparative way of evaluating individual historical accounts on several levels; from reasoning about the facts to juxtaposing several historical accounts (inclusive of historical accounts about "the same facts"). In any way, Roth qua naturalist does not offer any philosophical arguments here, the comparison he argues for is thoroughly empirical; it works with the Quinean web of belief, with the totality of our scientific practices and paradigms (cf., also Roth 2020, 48).

Thus, the suspicion that constituting historical events by narrativization equals to saying whatever we want about a past—or alternatively that whatever we say about a past makes sense, absent The Past as a metaphysical arbiter—is utterly unwarranted. There is, still above all, a powerful arbiter in the form of our epistemic, and itself historical, culture as a whole. That is, a culture full of human communities with their diverse and diverging commitments, conventions, and corrections. Del Mar, for instance, puts the matter in the following way:

> There are, within certain communities at certain times and places, conventions as to what are the units of historical knowledge, although these conventions are contingent and also

[17] See Roth (2020, 149, footnote 2) for his own Rortian disclaimer with regard to normative forms of relativism.

> subject to contestation and change. This already shows that historical knowledge is unlikely to be – or simply cannot be – entirely monadic, subjectivist, and relativist. As Ludwig Wittgenstein taught us, none of us, historians included, can speak an entirely private language. None of us is ever entirely in control of what we can see and tell about the past: we are social creatures through and through, and our modes of seeing and telling are shot through with more convention (and interested convention, with much at stake, including economically) than we can ever be reflexive about. (Del Mar 2020)

So, on the assumption that we author our own histories and that our idea of "the totality of the past" is in a certain sense an abstraction (i.e., an organization and selection) from this authorization, Roth thinks that it makes better sense to speak of the past*s* instead of the past. Once we reject the notion of Universal History, we get only individual, often incompatible, histories. After all, Roth co-opts the very term "irrealism" from Nelson Goodman (1978, 1979), and then especially Ian Hacking (e.g., 1990, 1992, 1999), who originally develops Goodman's well-known elaborations on "the new riddle of induction" in relation to historical theorizing. In particular, Hacking learns this general lesson from Goodman:

> It [Goodman's new riddle] shows that whenever we reach any general conclusion on the basis of evidence about its instances, we could by the same rules of inference, but with different preferences in classification, reach an opposite conclusion. [...] There is no general solution to his new riddle. Its scope goes far beyond induction and other trifling modes of reason. It confirms his doctrine [...] that we can and do inhabit many worlds. (Hacking 1992, 181, emphasis mine, this quote is chosen and reconstructed by Roth 2020, 36)

Hacking's application of this lesson, viz., that "different preferences in classification" give rise to different worlds, to historical theorizing, then consists of the following appreciation:

> We can well understand how new kinds create new possibilities for choice and action. But the past, of course, is fixed! Not so. As Goodman would put it, if new kinds are selected, then the past can occur in a new world. Events in a life can now be seen as events of a new kind, a kind that may not have been conceptualized when the event was experienced or the act performed. What we experienced becomes recollected anew, and thought in terms that could not have been thought at the time. Experiences are not only redescribed; they are refelt. This adds remarkable depth to Goodman's vision of world-making by kind-making. (Hacking 1999, 130, the quote is again chosen by Roth 2020, 37)

Now, I do not wish to rehearse all the thought-provoking points Roth finds in Hacking, as that would require a detailed exposition of Hacking's work (see Roth 2020, 36–47). Rather, I aim to underscore here that Roth's subsequent treatment of Hacking's views urges us to take his own idea of the pasts *literally*, i.e., in a non-metaphorical sense. Of course, to a significant degree, this issue hinges on our concept of metaphor.

To avoid examining numerous theories of metaphor, consider the following example. In "On the Very Idea of a Conceptual Scheme," Davidson (1973–74) claims: "According to Kuhn, scientists operating in different scientific traditions (within different 'paradigms') 'live in different worlds.' Since there is at most one world, these pluralities are metaphorical or merely imagined. [...] Kuhn [...] wants us to think of different observers of the same world who come to it with incommensurable systems of concepts." [...] "Kuhn's one world is seen from different points of view." To this account then Davidson says that: "Instead of living in different worlds, Kuhn's scientists may, like those who need Webster's dictionary, be only words apart," (Davidson 1973–74, 8–9, 11). Putting aside the question whether Davidson is being just to Kuhn (for sure, he is not a charitable reader here), he can be interpreted as outlining a distinct understanding of how different preferences in our use of words connect to our lived world(s). Davidson might be interpreted as dismissing the talk about "world" in plural, just on the basis of the available plurality of concepts.

Indeed, if one chooses to speak of the world, for example, in the spatiotemporal sense, no word or any vocabulary adds any layer to this world. It is the world where, e.g., a thing cannot be rectangular and round at the same time, nor occupy the same place in space at the same time as another thing. But even if one speaks of the world only in this sense, some parts of the world display unique patterns. We, the holders of different systems of concepts/points of view and certainly the denizens of the spatiotemporal world, are responsive to different vocabularies; we understand ourselves and our actions differently when described differently. Thus, by inheriting or purposely inventing certain vocabularies, we can navigate ourselves differently through the world. Our self-understanding is determinative for our identities and especially then actions.[18]

A similar line of reasoning leads Hacking to distinguish between "indifferent" and "interactive" kinds (Hacking 1999, 103ff). We humans are an interactive kind; unlike ants, genes, or metals, we are "classification-sensitive" things, responding in various ways to our own classification (Roth 2020, 38). On the other hand, there are various types of this classification into "kinds of people" with each having its unique history (Hacking 1986, 234).

As Roth neatly summarizes: "Some classifications of people into kinds prove relatively robust and stable (for better or worse), e.g., male and female as kinds. Some classifications, e.g., homosexual, Hacking suggests were 'socially constructed' but then became the basis for stabilizing and forming a distinct social identity. In this regard, some labels become a type of self-fulfilling prophecy. By creating a social and conceptual space for a type of behavior or person, people

[18] I actually think that this gives us good and sufficient reasons to speak of many world*s* in cases where our various linguistic frameworks endow us with different ways of life (historically viewed, but also in comparison with other non-human animals). Our linguistic practices create unique possibilities, therefore: unique goals; interests; or intentions. Language enables us to do unique things by endowing us with, in principle, an infinite number of conceptual toolkits. And in this mundane sense, language ultimately enables a unique way of life*s*. In the same spirit, Robert Brandom says: "It is not open to your dog to aim to reform the international monetary system, prevent global warming, or determine the mass of neutrinos. What makes something a language is that it makes this goal attainable. But that is not a goal that one could have in advance of speaking. [...] And there are in principle no other means possible for achieving that particular goal," (Brandom 2010, 307–308). A world with the international monetary system is a different world than the one without it, and by choosing to speak like this, one is not denying any "reality" of the spatiotemporal or causal world. *One is just acknowledging our specific—historical way—of life.*

invariably come to fill that space in a way that both defines and is defined by the label," (Roth 2020, 39). And as Roth also recognizes: "The stock of socially available ways of thinking about oneself shape and influence both [...] causal processes – acting because of reasons, changing reasons by reflecting on them," (ibid., emphasis mine).

In short, these social and conceptual spaces ultimately transform the very mode in which we causally interact in and with the world. Viewed in this way, by social pressures enforcing given conceptual frameworks, we "make up people;" we determine what kinds of actions are possible to perform, i.e., we in fact determine both what is done and what was done (Hacking 1986, 231). One of the examples Hacking develops and Roth discusses concerns the category of suicide.

> Any number of people have over the course of history had a hand in actively bringing about their own demise. But what interests and concerns Hacking [...] is the emergence of a special notion of suicide, one that classifies a suicide, any suicide, as a type of insanity. Suicide thus becomes an index of mental health for individuals, and the rate of suicide becomes a corresponding index for that of national groups. This connects, Hacking suggests, Foucault's "two poles of development," one centered on how to classify individuals – his "anatomo-politics of the human body" – and the second that characterizes the "biopolitics of the population." Durkheim's (1951) Suicide virtually creates a discipline, or at least a paradigm for one, by forging one of the first suggested links between these two poles. This link would have been impossible but for the statistics collected over the previous years, and the statistics required a prior commitment to counting and classifying the intention leading to death. But the intention here can be for Hacking nothing other than a piece of behavior we choose to describe in a particular way. "Even the unmaking of people has been made up." (Roth 2020, 41–42, Roth refers here to Hacking 1986, 234–235).

According to Roth, this example shows how re-classifying suicide with the aid of medical and statistical vocabulary (in fact also temporarily asymmetrical conceptual frameworks) "literally changes what someone previously did;" it "changes the past," (Roth 2020, 42, for further examples see 42ff). Roth then pointedly concludes: "We make, unmake, and remake the past not out of ignorance but because, regarding human actions, a past consists of just such makings and remakings. Talk of changing "the past" rings odd only because we (wrongly) attribute to interactive and intentional kinds the robustness possessed by some indifferent kinds." [...] "We project some predicates rather than others, but what we take to be stable features turn out to be socially entrenched practices which have become, in the fullness of time, taken for metaphysical verities," (Roth 2020, 44, 45). We can simply make up "the past" in ways not possible in the case of "the world" conceived as, for example, the spatiotemporal world.

Absent any form of standardizing and stabilizing the events of human history, "the past" is essentially indeterminate. Put another way, as the discussion on the plausibility of something akin to I.C. at the beginning of this section revealed, it is impossible to coherently flesh out the idea of "the past" apart from favoring certain kinds of descriptions driven by certain kinds of interests. Therefore, *apart from choosing a past*. And, as was emphasized generally in this whole chapter, that is an important aspect differentiating historical discourse from other scientific discourses, for example, physicalist discourse. As I have attempted to show here in this section, for Roth, this epistemic difference translates also into our understanding of historical reality; into our metaphysics of the past(s).

> Given alternative modes for structuring what happens, changes in descriptions can alter relations among events imputed to a past, and so how a past thus structured impacts what becomes possible going forward. A plurality of pasts results because constituting a past depends to some degree on socially mediated negotiations of a fit between descriptions and experience. Even what we take to distinguish what can change and what cannot itself depends on the descriptions deployed. Unless for reasons now unknown there ceases to be a possibility of descriptive change or reclassification, human histories will continue to reveal a multiplicity of pasts. (Roth 2020, 64)

In the next chapter, Chap. 5, I deliberately return to the category of "suicide" with my own case study concerning the death of one of the most analyzed yet mysterious artists of the last two centuries, Vincent van Gogh. This case study further, directly and indirectly, elaborates on the irrealist and naturalist outlooks vis-à-vis existing historiographical literature on Van Gogh's life, which exhibits not only a plurality of "facts," or "theories," but also a range of explanatory strategies oscillating between diachronic and synchronic approaches.

4.4 The Significances of Historical Knowledge

The aim of this chapter was to introduce Roth's views on historical explanation and the significance of historical knowledge. I started with Roth's attempts to give back a substantial content to historical narratives, which ends in identifying a special category of explanation he labels as essentially narrative explanation. This explanation is constructed via our retrospective insight creating the very explanandum and theoretical posits comprising respective explanans. Historical events are then, according to Roth, non-detachable from our narrativizing and theoretical practices. Historical reality is thus of our own creation; it is the result of our "disciplined disciplinary imagination." That is the gist of Roth's historical irrealism. I also stressed that the unwanted, and ultimately untenable, "constructivist" and "relativist" readings of irrealism can be avoided, or at least alleviated, by understanding irrealism from within naturalism.

There is no better testament of Roth's naturalist commitments than the fact that he is consistently drawing upon existing historical and generally scientific theories and literature. He openly states in PSHE: "Nothing in this book represents a proposal to reform historical practice. Ironically, at least relative to where this debate begins, it is philosophical attitudes and standards that need to change regarding what to count as an explanation," (Roth 2020, xv). And, as I have shown, this applies also to (folk) philosophical attitudes towards understanding the idea of the past(s).

As always, such a rather unorthodox approach will be digested very slowly. Apart from being unorthodox, the reception of Roth's position, at least in my view, complicates the fact that, within our contemporary intellectual and philosophical climate, we still do not have, and for better or worse perhaps never will have, a homogenous understanding of post-positivistic science, naturalism (in philosophy), and their mutual relationship. Zammito (among other things, a practicing historian) for instance worries that:

> *Roth's historicizing proposes to engulf epistemic goodness in the contingency of history and its transient communities of inquiry, all seeking retrospective footing for their ongoing endeavors.* [...]
>
> Empirical knowledge requires general beliefs about the world to salvage any order from the flux of experience. And yet Roth's brave new world seems to offer little purchase on what would make historical inquiry empirical. On his reading of that term, the empirical in fact *disappears*. He explains that, in methodological naturalism, empiricism makes sense only from *within* science. [...] From *within*, empiricism retains unquestioned importance, but this is only because theories within a science accept a standardized sense of what should be taken as a fact. But the empirical disappears when competing theories determine what counts as evidence or can explain away any discordant evidence, which he takes to be pre-eminently the case in history. Accordingly, appeal to facts in contestations of historical accounts turns out to be nothing but historians (including myself) "thumping on the lectern or praising the standards practitioners embrace," [Roth 2020, 137]. [...] *I find it difficult to understand how historical practice can remain an empirical inquiry once Roth's irrealism goes through.* (Zammito 2020, long emphasis mine, rewritten to respect my style of referencing)

However, this is exactly a response demanding ahistorical philosophical solutions to empirical problems. Roth qua naturalist gives us no philosophical solutions. Never intended to. Working scientific solutions will be only empirical, meaning that they will emerge and evolve from within the historiographical and other scientific works and theories themselves. In this sense, Roth's analysis of historical reasoning complements the naturalistic account of empirical inquiry developed by Quine as I unpack it in Chap. 2. It is as if Zammito does not appreciate the actual depth in which Roth's naturalism relies on *scientific pluralism and particularism* (however, he is not alone in this, see Currie and Swaim 2021; Swaim and Currie 2022; cf., Dewulf and Roth 2022).

Also, Branko Mitrović (2024, cf., 2022) thinks of Roth's work as exemplifying the tendentious post-modern/anti-realist positions in the theory and philosophy of history. However, it is symptomatic of Mitrović's abbreviated conclusions that he completely ignores Roth's naturalistic grounding of irrealism. So, instead of criticizing Roth, e.g., for having too liberal a view of science or too theory-laden a view of historical events themselves, he chooses to regard Roth as a Platonist.

Mitrović (2022) at one point asks Roth where the theories and particular descriptions of historical events are in fact; what is their ontology? With respect to the naturalistic rebuttal of idealism (see Chap. 3 again), and Mitrović's overtly and overly antagonistic attitude, I cannot think of a better reply than to answer in Samuel Johnson's spirit that the descriptions of historical events—as well as, for instance, the physicalist descriptions—exist in books, are written on papers, are stored in various digital forms nowadays, are taught at universities, and are shared and talked about at different other (spatiotemporal) places by us (among other things, biological) beings. Where else would they be?

Tsilipakos (2020) offers much more helpful and still rigorous critique, centering on Roth's (as well as Danto's and Mink's) assumption that the past reality is indeterminate until retrospectively re-constructed. In short, Tsilipakos tries to show that Roth's narrative explanations might lead us to a form of historiographical presentism and/or anachronism neglecting the factual and conceptual configuration (inclusive of the conceptual self-understanding) of a chosen past society/epoch. Recall that Roth follows Mink in emphasizing that "the past" is open to redescriptions, inter alia, because one can always grasp the past events via chronologically newer conceptual frameworks. As I have already specified, Mink himself speaks of temporarily asymmetrical concepts, where one of his examples is "the unpropertied citizens of Rome constituted the first urban proletariat." Tsilipakos writes:

> Roth [...] approvingly alludes to this example as not only made possible by, but as reinforcing the "nonstandardization" thesis: it is thought to constitute 'a further principled barrier to any hope of normalizing descriptions of historical events' and, thus, to reaffirm the distance between history and normal science. [...] *This brushes aside the issue of accuracy and the fact that the historian might need to exercise conceptual sensitivity*. (Tsilipakos 2020, 301, emphasis mine).

Tsilipakos (ibid., 301ff) then moves on to scrutinize the example from the historiographical angle and concludes that to apply the concept of "urban proletariat" to Antiquity is at least misleading; meaning primarily that it is (conceptually) insensitive to the factual/conceptual matters known about the Ancient Roman society. I actually think that Tsilipakos offers sound (although sketchy) arguments for such a conclusion. However, I do not think that Tsipilakos is justified in inferring that since "the social life of any period is populated by those who lived through it and the conceptual forms through which their conduct took shape," the past is therefore *not* "a conceptual *terra nullius*" (Tsilipakos 2020, 305).

The trouble is that there is no algorithmic conversion from the available knowledge about a chosen ancient society/epoch to our present/future account of it, let alone our account of "the totality of the past." We, scientists and scholars, do the accounting. Again, there is no Universal History, just our histories. On my reading, Roth's irrealism counts on both symmetrical and asymmetrical interventions with and interpretations of our past. On the other hand, the asymmetrical cases are often not made explicit by historians themselves, and, more importantly, express the unique features of historical explanation qua a form of scientific explanation. That is why Roth puts more weight on them while developing his irrealist tenets.

When Roth directly addresses the issue of symmetric conceptualizations, his views are not that dissimilar to Tsilipakos'. Commenting on a different example, Roth together with Dewulf pinpoint the following: "Statements about Pericles' career have truth-value and become an object of rational debate from within modern biographical concepts like 'career' and an entire philological tradition capable of reconstructing all the important testimonies about Pericles, etc. *Absent an epistemic culture necessary to study the history of ancient Greece, including its theories and practices of inquiry, statements about Pericles' career could not have a truth value. Then, those statements would be akin to fiction*," (Dewulf and Roth 2022, 214, emphasis mine). So, on the one hand, Tsilipakos and Roth are not, at least in general terms, in disagreement about the necessity to stay factually and conceptually sensitive to respective historical periods. On the other hand, our historical sensitivity in practice hinges on the criteria we ourselves choose to decide whether, e.g., applying a given asymmetrical concept truly expands our understanding of history.

This brings us back, in a fitting way, to Zammito's complaint above. That is, to the question of to what extent historiography can remain empirical, *and what it actually means within the historiographical practice* (cf., Tsilipakos 2020, 305).

My own strategy in this book remains to address this question through individual case studies in Chaps. 5 and 6, but *a broadly pragmatic strategy* is readily at hand, I think. I find an acknowledgment of this broadly pragmatic strategy even in Tsilipakos' own text. When Tsilipakos contemplates the need to authentically capture the conceptual strata of individual historical epochs, he remarks that the concepts central to past societies ("slavery," "class," and other sociopolitical concepts) "are not dispensable or merely secondary features […]; they are the concepts that enter into social life of the period and give it its distinctive identity. If these concepts are not preserved by the historian or social scientists in their descriptions, then we might as well describe the entire history of the world as the history of one type of society," (ibid., 304, emphasis mine). Here, Tsilipakos rightly stresses that the symmetrical conceptual approach matters as well, and that, for obvious reasons, it critically contributes to what will ultimately be considered as of historical significance. And right after that, he nicely articulates what I see as the broadly pragmatic strategy of applying asymmetrical concepts/descriptions:

> I do not mean to say that consideration of the identity of a historical period absolutely precludes the description of a class as an urban proletariat but, rather, that the specific character of the society as expressed in institutions and concepts is presupposed and, crucially, serves as a check on the applicability of any such description even at the level of a loose analogy – which, let it be noted, might be perfectly appropriate for some purposes, which may work themselves in various ways into the many different kinds of historical inquiry. (ibid., emphasis mine)

This is perfectly compatible with Roth's scientific pluralism (which is a naturalistic commitment, for that matter). Also, as I indicated, Roth's historical irrealism does not imply that all historically significant happenings will be narratively structured. *There is more history than narrative history*. Certainly, often "only" capturing the character of a given historical society/epoch will be of historical significance, too. Roth's disconnection of historical *explanation* from any form of predictive reasoning is balanced by his naturalization of historical *inquiry* as such.

Indeed, in Roth's picture, history is a science in the disciplinary sense, but it must inevitably borrow, exploit, and experiment with various facts, methods, styles, and standards present-day empirical practices and paradigms offer. History must also be written from within science. So, even historical/narrative explanations are, in principle, still in a connection to empirical practices and paradigms. With Quine, one could say that every "historical proposition" connects to our experience indirectly "through participating in a large, cohesive section of scholarly and scientific lore and common sense which has its ultimate links somewhere outside the historical quarter," (Quine 1987, 162). That is just about the most one can ask of a philosopher. Notwithstanding, it would be unjust to conclude that Roth's irrealism, by battling philosophical biases and dogmas, does not in any way contribute to formulating certain solutions to problems related to our understanding of historical significance and science in general.

In my book, Roth's (historical) irrealism first and foremost amplifies the view that our human relation to "past reality" is itself essentially historical. In this sense, irrealism also philosophically contributes to our proper self-understanding, because it substantially expands the idea of us as historical beings. Also, it should be obvious at this point that it is possible to say the same about "present reality." *The very idea of reality is naturalized hence historicized. Put another way, both present and past reality emerge from within the same holist and historical constrains.*

Whether this will be understood as a limitation rather than a liberalization of our epistemic condition will always depend on one's background beliefs and especially further purposes and interests. However, as I have repeatedly pointed out, naturalists can always find some liberalizing power precisely in learning to live with and exploit our epistemic and other human limitations. We cannot escape our past, true. We cannot even access "the past," granted. But if we can choose a past, as Roth suggests, we can choose how to move on from there going forward. We can choose a future.[19]

References

Beatty, John, and Isabel Carrera. 2011. When what had to happen was not bound to happen: History, chance, narrative, evolution. *Journal of the Philosophy of History* 5 (3): 471–495.

Bird, Alexander. 2022. Thomas Kuhn. In *The Stanford encyclopedia of philosophy*, ed. Edward N. Zalta. https://plato.stanford.edu/entries/thomas-kuhn/.

Borradori, Giovanna. 1994. *The American philosopher: Conversations with Quine, Davidson, Putnam, Nozick, Danto, Rorty, Cavell, MacIntyre, Kuhn*. Trans. Robert Crocitto. Chicago: University of Chicago Press.

Brandom, Robert. 2010. Reply to Daniel Dennett's "The evolution of 'why?'". In *Reading Brandom: On making it explicit*, ed. Barbara Weiss and Jens Wanderer, 305–308. Abingdon: Routledge.

Currie, Adrian, and Daniel G. Swaim. 2021. Past facts and the nature of history. *Journal of the Philosophy of History* 16 (2): 179–206.

[19] This is also certainly a philosophically significant insight. Here I paraphrase Roth himself, compare with Roth (2020, 147–148). And also here: "Although Hayden White will be the person least directly discussed in the ensuing pages, his work also has been determinative of my own thinking on this topic. As I told him, for many years I had taped to my computer screen a quote from him that also ultimately served as an epigraph for my paper 'The pasts:' 'We choose our past in the same way that we choose our future.' What I often imagine myself to be doing is offering an account of why I take this to be correct by reformulating his insights as a conclusion of a type of philosophical argument," (Roth 2020, vii–viii; White's quote is from White 1966; "The pasts" was originally published in 2012; see Roth 2012).

Danto, Arthur C. 1962. Narrative sentences. *History and Theory* 2: 146–179.
Danto, Arthur C. 1965. *Analytical philosophy of history*. Cambridge, UK: Cambridge University Press.
Danto, Arthur C. 1985. *Narration and knowledge*. New York: Columbia University Press.
Danto, Arthur C. 1995. The decline and fall of the analytical philosophy of history. In *A new philosophy of history*, ed. Frank Ankersmit and Hans Kellner, 70–85. Chicago: Northwestern University Press.
Davidson, Donald. 1973–74. On the very idea of a conceptual scheme. *Proceedings and Addresses of the APA* 47:5–20.
Del Mar, Marcus. 2020. Is historical knowledge philosophically interesting? [Review]: The philosophical structure of historical explanation. *The New Rambler*.
Dewulf, Fons, and Paul A. Roth. 2022. Real true facts: A reply to Currie and Swaim. *Journal of the Philosophy of History* 16 (2): 207–225.
Douven, Igor. 2011. "Peirce on abduction" (supplement to Abduction). In The Stanford Encyclopedia of Philosophy, ed. Edward N. Zalta. https://plato.stanford.edu/entries/abduction/peirce.html
Durkheim, Emile. 1951. *Suicide: A study in sociology*. New York: The Free Press.
Goodman, Nelson. 1978. *Ways of worldmaking*. Indianapolis, IN: Hackett.
Goodman, Nelson. 1979. *Fact, fiction, and forecast*. 3rd ed. Indianapolis, IN: Hackett.
Hacking, Ian. 1986. Reconstructing individualism. In *Reconstructing individualism*, ed. Thomas C. Heller, Morton Sosna, and David E. Wellbery, 222–236. Stanford, CA: Stanford University Press.
Hacking, Ian. 1990. Natural kinds. In *Perspectives on Quine*, ed. Richard Barrett and Robert F. Gibson, 129–141. Cambridge, MA: Blackwell.
Hacking, Ian. 1992. World-making by kind-making: Child abuse for example. In *How classification works*, ed. Mary Douglas and David Hull, 180–238. Edinburgh: Edinburgh University Press.
Hacking, Ian. 1999. *The social construction of what?* Cambridge, MA: Harvard University Press.
Hempel, Carl G. 1942. The function of general laws in history. *Journal of Philosophy* 39:35–48.
Hilberg, Raul. 1985. *The Destruction of the european jews*. Revised and definitive ed., 3 vols. New York: Holmes & Meier.
Hull, David L. 1988. *Science as a process: An evolutionary account of the social and conceptual development of science*. Chicago: University of Chicago Press.
Kuhn, Thomas S. 1977. *The essential tension*. Chicago: University of Chicago Press.
Kuhn, Thomas S. 2012. *The structure of scientific revolutions*. 4th ed. Chicago: University of Chicago Press. (Originally published 1962).
Little, Daniel. 2010. *New contributions to the philosophy of history*. New York: Springer.
Mink, Louis O. 1987. In *Historical understanding*, ed. Brian Fay, Eugene O. Golob, and Richard T. Vann. Ithaca, NY: Cornell University Press.
Mitrović, Branko. 2022. A naïve realist rumination on the Roth-and-Dewulf versus Currie-and-Swaim exchange. *Journal of the Philosophy of History* 16 (3): 295–307.
Mitrović, Branko. 2024. What is historical anti-realism and how to define it? *Journal of the Philosophy of History* 18 (2): 113–124.
Popper, Karl R. 1972. *Objective knowledge*. Oxford: Oxford University Press.
Quine, Willard Van Orman. 1987. *Quiddities: An intermittently philosophical dictionary*. Cambridge, MA: Harvard University Press.
Rorty, Richard. 2021. *Pragmatism as anti-authoritarianism*. Cambridge, MA: Harvard University Press.
Roth, Paul A. 1998. The epistemology of 'Epistemology Naturalized'. *Dialectica* 53 (2): 87–110.
Roth, Paul A. 2004. Hearts of darkness. *History of the Human Sciences* 17:211–251.
Roth, Paul A. 2012. The pasts. *History and Theory* 51 (3): 313–339.
Roth, Paul A. 2020. *The philosophical structure of historical explanation*. Evanston: Northwestern University Press.

Swaim, Daniel, and Adrian Currie. 2022. Minimal metaphysics vs. maximal semantics: A response to Paul Roth and Fons Dewulf. *Journal of the Philosophy of History* 16 (2): 226–236.
Tsilipakos, Lazaros. 2020. Descriptive accuracy in history: The case of narrative explanations. *Philosophy of the Social Sciences* 50 (4): 283–312.
Uebel, Thomas. 2019. Acts, events, and stories: On the history of Danto's compatibilist narrativism. *Journal of the Philosophy of History* 14 (1): 47–79.
White, Hayden. 1966. The burden of history. *History and Theory* 5 (2): 111–134.
Wright, Crispin. 1993. *Realism, meaning, and truth.* 2nd ed. Cambridge, MA: Blackwell.
Zammito, John H. 2020. The philosophical structure of historical explanation [Review]. *Notre Dame Philosophical Reviews.*
Zeleňák, Eugen. 2018. Non-representationalism in philosophy of history: A case study. In *Towards a revival of analytical philosophy of history*, ed. Krzysztof Brzechczyn, 116–129. Leiden: Brill.

Chapter 5
First Case Study: Lust for Knowing: The Death of Vincent van Gogh

Abstract Borrowing from the cybersecurity lingo, what I do in this chapter could be compared to constructing a sandbox environment. It is a restrained playground where I experiment with the application of the central distinctions and conclusions presented in Chaps. 2 and 4. To be specific, I play here with the general naturalistic approach, as well as the basics of Roth's historical irrealism, by applying them to an actual historiographical discussion. I take my whole approach to the issues discussed in this chapter as exemplifying the naturalistic approach I defend here in this book, inclusive of the literature and sources I select. The particular discussion I choose concerns the life and death of the painter Vincent van Gogh. It must be noted that the literature on Van Gogh is a massive and multi-genre affair. I am thus deliberately limiting myself to accounts of his (alleged) suicide death. Apart from engaging with the commonly accepted facts, contexts, and debates, my controlled environment is historiographically shaped predominantly by the theory/explanation advanced by Steven Naifeh and Gregory White Smith in their *Van Gogh: The Life* (2011). Naifeh and Smith challenge the widely accepted theory that Van Gogh committed suicide, by offering an alternative explanation respecting the majority of known facts about Van Gogh.

Keywords Van Gogh · Narrative · Historical explanation · Irrealism · Naturalism

5.1 Introduction: Beyond the Myths (or At Least Some of Them)

> What am I in the eyes of most people? A nonentity or an oddity or a disagreeable person – someone who has and will have no position in society, in short a little lower than the lowest. Very well – assuming that everything is indeed like that, then through my work I'd like to show what there is in the heart of such an oddity, such a nobody.
>
> —Vincent van Gogh, 1882[1]

---

[1] (van Gogh 2024, 249).

M. Hubálek, *Dissolving Naturalism and Historicism Into Each Other*, Studies in History and Philosophy of Science 67,
https://doi.org/10.1007/978-3-032-07535-2_5

> Now I have a horror of success; I'm afraid of the morning after following a success by the Impressionists; even the present difficult days will later seem like "the good times" to us.
> —Vincent van Gogh, 1888[2]

Vincent van Gogh (hereafter Van Gogh), born in March 1853 and died in July 1890, is currently one of the best known and most celebrated painters. Some people think of him as Impressionist, while others think that the more apt classification is Post-Impressionist, or the first Expressionist. He is sometimes even perceived as the father of modern art, who profoundly changed the trajectory of art history. However, during his lifetime and shortly after, he was unknown to the wider public and his paintings were often judged poorly by the critics. Also, he sold only a few paintings during his lifetime; rumor has it that he sold in fact only one painting—*The Red Vineyard* (cf., VGMA 2024a).

Two telling examples of the negative as well as *prophetic* criticism: in 1891, a year after Van Gogh's death, Johan de Meester doubted "that people will ever hear much about the art of Vincent van Gogh (de Meester in Turnbull 2022, 17).[3] Even Julius Meier-Graefe thought, back in 1904, that "it is unlikely that the time will come when [Van Gogh] will be appreciated by the layman," (Meier-Graefe in Turnbull 2022, 17). Van Gogh also thought negatively of himself as an artist. In May 1890, a couple of months before his death, Van Gogh wrote a letter addressed to his brother Theo and his wife Johanna (often addressed simply as "Jo"), in which he declares: "I feel – a failure – that's it as regards me – I feel that that's the fate I'm accepting. And which won't change any more," (van Gogh 2024, RM20). Van Gogh never sent this letter, and Theo discovered it only after his brother's death (Turnbull 2022, 17).

Ironically enough, one of the first and few reviews of Van Gogh's paintings published during his life was a rather positive one (cf., Mathews 1986, 94). I am referring to the now well-known review written by G.-Albert Aurier (among other things, a poet), which appeared as an article in the *Mercure de France* in January 1890. Consider this excerpt:

> Beneath skies that sometimes dazzle like faceted sapphires or turquoises, that sometimes are molded of infernal, hot, noxious, and blinding sulfurs; beneath skies like streams of molten metals and crystals, which, at times, expose radiating, torrid solar disks; beneath the incessant and formidable streaming of every conceivable effect of light, in heavy, flaming, burning atmospheres that seem to be exhaled from fantastic furnaces where gold and diamonds and singular gems are volatilized – there is the disquieting and disturbing display of a strange nature, that is at once entirely realistic, and yet almost supernatural, of an excessive nature where everything – beings and things, shadows and lights, forms and colours – rears and rises up with a raging will to howl its own essential song in the most intense and fiercely high-pitched timbre [...].

[2] (van Gogh 2024, 660).

[3] I dedicate this chapter to Alan Turnbull, who wrote the magnificent book *Who Shot Van Vogh?: Facts and Counterfacts About the World's Most Famous Artist* (Turnbull 2022), which ultimately inspired me to put together this text. I rely heavily on this book in this chapter.

> Such … is the impression left upon the retina when it first views the strange, intense, and feverish work of Vincent van Gogh, that compatriot, and unworthy descendent of the old Dutch masters. […]
>
> *Nevertheless, in the case of Vincent van Gogh, in my opinion, despite the sometimes misleading strangeness of his works, it is difficult for an unprejudiced and knowledgeable viewer to deny or question the naive truthfulness of his art, the ingeniousness of his vision.* Indeed, independent of this indefinable aroma of good faith and of the truly seen that all his paintings exude, the choice of subjects, *the constant harmony between the most excessive colour notes, the conscientious study of character, the continual search for the essential sign of each thing, a thousand significant details undeniably assert his profound and almost childlike sincerity, his great love for nature and for truth – his own personal truth.* (Aurier in Bónafoux 1992, 152–154, emphasis mine)

Van Gogh himself valued not only the praise itself, but the literary style of the review as well. He wrote Aurier a letter in response to the article, in which he notes: "I like it very much as a work of art in itself, I feel that you create colors with your words," (van Gogh 2024, 853). On the other hand, Van Gogh does not hesitate to diminish Aurier's compliments and immediately adds: "Anyway I rediscover my canvases in your article, but better than they really are – richer, more significant. However, I feel ill at ease when I reflect that what you say should be applied to others rather than to me," (ibid.).[4] Be that as it may, Aurier's review article has contributed its share to a particular perception of Van Gogh as a person and/or to a particular reception of his art works. The article is titled "The Isolated Ones: Vincent van Gogh," after all.

At this point, i.e., the beginning of 1890, people who knew Van Gogh were obviously aware of his fragile mental condition. He had already cut off, better said, self-mutilated, his left ear (in December 1888), was committed voluntarily to the Saint-Paul-de-Mausole psychiatric hospital in Saint-Rémy (southern France), where he was able to create circa 150 paintings (he stayed there for 1 year, from May 1888 until May 1890, cf., VGMA 2024b). And it was already reported that he had attempted to commit suicide (cf., Naifeh and Smith 2011, Chapter 40) To come to the point, various conclusions about Van Gogh's unstable "mental health" have always been used as manuals for interpreting his art and personality (see Turnbull 2022, 157ff).

Let us return to Aurier's review article for a moment. This passage directly follows the last paragraph of the excerpt I quoted above:

> *Given this, we are thus able to infer legitimately from Vincent van Gogh's works themselves his temperament as a man, or rather, as an artist – an inference that I could, if I wished,*

[4] It was actually Theo who mailed his brother Aurier's article, and this is how Van Gogh reacted: "I was extremely surprised by the article on my paintings that you sent me, no need to tell you that I hope to go on thinking that I don't paint like that, but rather I do see from it how I ought to paint. For the article is quite right in the sense that it indicates the gap to be filled, and I think that basically the writer writes it rather to guide not only me but also the other Impressionists, and even rather to make the breach in the right place. So he proposes a collective self, as ideal for the others as it is to me. He tells me simply that there's something good here and there, if you like, in my very imperfect work as well, and there's the consolatory side which I appreciate and which I hope I'm grateful for," (van Gogh 2024, 850; cf., also the letter number 854).

> *corroborate with biographical facts*. What characterizes his works as a whole is its excess ... of strength, of nervousness, its violence of expression. In his categorical affirmation of character of things, in his often daring simplification of forms, in his insolence in confronting the sun head-on, in the vehement passion of his drawing and colour, even to the smallest details of his technique, a powerful figure is revealed [...] masculine, daring, very often brutal [...] yet sometimes ingeniously delicate. (Aurier in Bónafoux 1992, 154, emphasis mine)

Aurier's review, according to an art historian, Patricia Mathews, manifests the paradigmatic nineteenth-century assumption that "suffering" has certain "beneficial effects" for creative people. So, by Aurier and undoubtedly by many other contemporaries, Van Gogh's "excessive temperament was seen as a sign of his superior ability to experience and to reveal," (Mathews 1986, 99). Mathews then also continues: "For this reason, Aurier connected Vincent's genius directly to madness as well as to pain and suffering. He called him "a terrible, maddened genius, often sublime, sometimes grotesque, always near the brink of the pathological," (ibid., 100; cf., Aurier in Turnbull 2022, 158).

I put aside the complex question of whether or not today we still hold this pain-makes-the-artist assumption, but I feel safe to say that it is still the defining assumption for the attempts to interpret Van Gogh's artistic identity. It is the intriguing dichotomy between his antemortem *struggle* and posthumous *success* that makes this interpretative approach almost irresistible.[5] And, what is more, "the Van Gogh madness narrative," as Turnbull at one point puts it (2022, 160), has become the lens through which we understand significant portions of Van Gogh's personal life and even death, which will be the main theme of this chapter.

Here and now, I just want to mention the basic facts generally reported about the circumstances of Van Gogh's death.

It is believed that Van Gogh died of suicide. That, on July 27, 1890, for one reason or another, he shot himself with a (borrowed) gun in his upper abdominal area in (the fields near) Auvers-sur-Oise, with the intention of ending his life. He failed to kill himself, so he returned to the inn in which he was accommodated, and it took approximately another 30 h before he passed away in his bed on July 29. Van Gogh came to Auvers shortly after leaving the asylum in Saint-Rémy to stay near his brother Theo, who resided in Paris, and to receive specialized medical care from a homeopath, Dr. Paul Gachet.

However, an alternative theory about Van Gogh's death appeared not that long ago. As Martin Bailey, a renowned expert on Van Gogh, observes:

> The theory that Van Gogh's death was murder (or manslaughter) first surfaced seriously in 2011, in a comprehensive biography [*Van Gogh: The Life*] by two American writers, Steven Naifeh and Gregory White Smith. They wrote that Van Gogh was shot in the abdomen on 27 July 1890 by 16-year-old René Secrétan, a summer visitor in Auvers-sur-Oise, who taunted the artist. Van Gogh managed to stagger back to his inn, dying two days later from his wounds. (Bailey 2019).

---

[5] Here I deliberately allude to the great (audio-)book entitled *Van Gogh: Struggle & Success* (Leeman 2013). However, the "success" the book deals with is rather Van Gogh's treatment of the very idea of "success."

I state here for the record that Bailey actually considers this "murder story" a myth, even though he thinks that Naifeh and Smith's (hereafter N&S) book is otherwise "meticulously researched and adds considerably to our knowledge," (Bailey 2018). Also, I do not want to suggest that the circumstances of Van Gogh's death have always been considered as being unequivocal and have not attracted the attention of (art) historians, the opposite is in fact true. I rather want to say that the account of N&S was quite impactful and successfully penetrated the popular culture. Bailey indirectly reflects this fact when he admits: "A few years ago, the question I was most often asked about Vincent van Gogh was, 'Why did he cut off his ear?' Now it's, 'Did he commit suicide, or was he actually murdered?'" (Bailey 2019).

N&S presented their "new theory" in "a legal-brief-like appendix" at the end of their 953-page-long book *Van Gogh: The Life* (2011), and perceive it structurally as "*an alternate circumstantial explanation*," which "better fit[s] the few firmly known facts about the shooting, and the even fewer reliable accounts, than the traditional suicide story," (Naifeh and Smith 2014, emphasis mine). Also, instructively to my aims in this chapter, N&S are convinced that the traditional suicide story gains its credibility (at least partially) from the fact that it is a convenient story for portraying Van Gogh as an isolated and struggling yet genial artist. "Vincent's suicide has become the grand finale of the story of the martyr for art, it's his crown of thorns," (Naifeh and Smith 2014 quote an unnamed curator here). Let me now finally quote from N&S's book here as well. This is how they comment on the suicide story in the Appendix of their aforementioned book:

> *It was, and is, a satisfying narrative. It provides a suitably tragic end to an undeniably tragic life: a troubled, unappreciated artist seeks escape from the neglect of the world by taking his own life.* The story not only appeared early, it caught on quickly, and it played an important role in the meteoric ascent of Vincent's celebrity in the decades immediately after his death. By 1934, when it was immortalized in Irving Stone's bestselling novelization, Lust for Life, the story of Vincent's suicide in the wheat field had become firmly lodged in the artist's legend. Two decades later, in the 1950s, when the celebrity of Vincent van Gogh reached new heights with the centenary of his birth in 1953, it was sealed in the mythology permanently with the release three years later of the Academy Award-winning movie adaptation of *Lust for Life*. (Naifeh and Smith 2011, 792, emphasis mine)

It is not without interest that N&S's new theory has already also materialized in two different movie adaptations; in *Loving Vincent* (2017) and *At Eternity's Gate* (2018). See an illuminating comparative study of these two biographical movies by Paula Arantzazu Ruiz (2021). A little teaser of Riuz's study approach: "Nuances aside [...] functions and narrative characteristics of the biopic conceive of the genre in terms of its "historical obligation," that is, the description of the meritorious role in history of the character portrayed, a supposedly important figure whose actions in life are deemed worthy of committing to posterity in art—the same rationale behind pictorial and literary portraits," (Ruiz 2021, 94). Van Gogh's case nicely shows how the personal history of an artist can be heavily molded by other art forms (nowadays, especially by movies, but literature is significant, too).

In this chapter, I want to show how N&S's attempts to push for their new theory of Van Gogh's death, as well as their attempts to demonstrate why the widely accepted suicide story is a worse explanation of the available facts, relate to the naturalistic tenets as well as Roth's historical irrealism (see again Chaps. 2 and 4).

In Sect. 5.2, I enlarge the list of known basic facts and contexts about Van Gogh, to appropriately set the stage for discussing his (suicide) death. Section 5.3 then directly surveys the various accounts of Van Gogh's death, principally by following N&S's alternative theory, proposing that Van Gogh was killed, as well as its tension with the widely accepted suicide theory. These two sections together function as an indicator of how much the discussion is, in practice, about questioning facts (answering what-questions) and how much about theorizing these facts (answering how- and why-questions). Section 5.4 then most explicitly connects the content of this chapter with the conclusions and distinctions put forward in this book.

5.2 Who Was Van Gogh?

> In character I'm quite different from the various members of the family, and I'm actually *not* a "Van Gogh."
>
> —Vincent van Gogh, 1883[6]

When Van Gogh died, he was only 37 years old. I write "only" not because that age would significantly deviate from the life expectancy in nineteenth-century Europe. I write "only" because it was enough time for him to create over 2000 artworks (see Wikipedia Contributors 2024). And he made the choice to pursue a path in the field of art only at the age of 27! It was his brother Theo (working as an art dealer) who encouraged him to try to be a professional artist, and who supported Van Gogh morally and financially ever thereafter. Before that, Van Gogh was also an art dealer, taught children in a school for a while, and worked as a lay preacher in a poor mining area (as a "lay" preacher, since he did not finish his theology studies, cf., VGMA 2024c). In addition to Dutch, his mother tongue, he was fluent in French and English. Without a doubt, Van Gogh was a persistent and passionate individual, not only an artist. I choose this piece of information as a corroboration of such a claim: "Vincent lived among the miners and their families in the Borinage [Belgium], sharing their poverty. He slept on the floor and gave away his possessions. His dedication was such that he was nicknamed 'The Christ of the Coal Mine.' Vincent nevertheless failed to establish a close-knit community of worshippers and his contract was not renewed," (ibid.). And since Van Gogh's personal life became widely publicized (among other things, by publishing his detailed and sizeable correspondence, cf., VGMA 2024d; Turnbull 2022, 159), it has become wholly open for our retrospective insights and re-interpretations. So, as Turnbull with a touch of sarcasm highlights:

> He may have died a painter, but his status was rapidly elevated to become emblematic of the individual struggling against conventional society to express a personal visionary truth.

[6](van Gogh 2024, 411).

> Van Gogh has now become a representation of whatever the viewer wishes to see: a suffering tortured genius; a man of violence and self-harm; an embodiment of peace and compassion; a religious fanatic; a Marxist; a madman; a cultural commodity; an investment opportunity. (Turnbull 2022, 8)

Of course, as I already stated, one of the main concerns of our retrodictive reasoning with respect to Van Gogh's personal history has always been the nature of his "illness" (that is the word Van Gogh himself used to describe his condition, see VGMA 2024e). Turnbull crisply consolidates opinions on/diagnoses of Van Gogh's illness into three distinct groups:

1. Psychiatric: that he suffered from schizophrenia, manic depression or a psychogenic condition.
2. Neurological: that he suffered from epilepsy.
3. Somatic (i.e. related to the body, not the mind): that he suffered from one or more of the following—syphilis, gonorrhoea, alcohol poisoning, delirium, eye disease, Ménière's disease, sunstroke. (ibid., 161, Turnbull moreover offers many other references on this issue)

In any case, and to stick to the symmetrical conceptual description here as well, when Van Gogh entered the asylum in Saint-Rémy in May 1889, it was with the following diagnosis: "Dr Peyron determined that Van Gogh was suffering from a form of epilepsy, accompanied by acute insanity and hallucinations. He based this on the diagnosis of the doctors in Arles [the place where he self-mutilated his ear] and Vincent's own description of his attacks," (VGMA 2024e).

Let us now zoom in on the last 2 months of Van Gogh's life. In May 1890, he checked himself out of the mental health facility at Saint-Rémy. "On the day of his release the hospital authorities wrote a one-word summary of his condition in the final column of the hospital report: 'Cured.' 'Damaged by boredom and grief,' was the artist's own assessment as expressed in a letter to his brother," (Turnbull 2022, 158). Van Gogh afterwards settles in Auvers. The website of the Van Gogh Museum clearly puts together the basic info on this period:

> Auvers offered Vincent the peace and quiet he needed, while being close enough to Paris for him to visit his brother Theo. There was a doctor there too, Paul Gachet, who could keep an eye on him. Vincent quickly befriended Gachet, himself an amateur painter, who advised Van Gogh to devote himself completely to his art. He did precisely that, painting the gardens and wheat fields around the village at a feverish rate.
>
> Vincent threw himself entirely into his painting in this period, completing virtually a work a day. His health seemed to be improving, too. (VGMA 2024f)

Recollect that Van Gogh is financially dependent on the generous support of his brother Theo (now already for a decade). The typically narrated twist the Van Gogh story has at this point, are the changes in his brother's personal and professional life. Theo was a married man now, and in January 1890 had a baby (named after his brother—Vincent). Bailey, for instance, soberly concludes that: "[Van Gogh] was therefore concerned that a wife and an infant would mean he might lose much of his brother's emotional and financial support. He was also worried about Theo's recent problems at work with the owners of the gallery. Although most of the time Vincent could cope with the challenges of his life, he was upset," (Bailey 2019). N&S also realize that this must have been a humanly very difficult period for Van Gogh. Below, I reproduce N&S's elucidatory reconstruction of thoughts behind a letter posted by Theo and Van Gogh's reaction to it. The letter was delivered at the beginning of July 1890; Van Gogh would die a few weeks after receiving it (see the letter in question here, van Gogh 2024, 894). Firstly, Theo's thoughts:

> In a cri de coeur unlike anything else in their long correspondence, Theo described the hell his life had become. The baby was sick – "continuously crying, day and night." "We don't know what to do," he wrote, "and everything we do seems to aggravate his sufferings." Jo was sick, too – sick with worry: so terrified that her child might die that she "moaned in her sleep."
>
> In all these woes, Theo saw one single cause: not enough money. "I work all day long but do not earn enough to protect that good Jo from worries over money matters," he confessed. He blamed his employers of seventeen years – "those rats" – for paying him so little and treating him "as though I just entered their business." But mostly he blamed himself. In the ultimate test of manhood – providing for his wife and child –he had failed miserably. The shame of it drove him, yet again, to the thought of quitting: of "taking the plunge" by setting up as an independent dealer. For the risk-averse Theo, it was tantamount to a threat of suicide. (Naifeh and Smith 2011, 768)

And here are Van Gogh's thoughts:

> In every cry of anguish, Vincent heard an accusation. When Theo listed Vincent among the mouths he had to feed, or portrayed himself as an exhausted draft horse pulling a heavy cart in which Vincent rode, or predicted "going through the world like a down-and-out beggar," the Furies of guilt swept back into Vincent's life with a vengeance. [...] Theo even touched the most sensitive subject of all, his deteriorating health, with a teary wish that he might live to see his son "grow up to be Somebody" – unlike his failed father and dissolute uncle. (ibid.)

However, after a closer look at the letter, one also sees many words of encouragement and calls for endurance from Theo. I choose the following:

> What do you say to this old chap? Don't bother your head about me or about us, old chap, be aware that what gives me the greatest pleasure is when you're well and when you're at your work, which is admirable. You already have too much fire, and we must still be ready for battle a long time from now, for we'll battle all our lives without taking the oats of charity they give to the old horses in grand houses. We'll pull the plough until it moves no longer, and we'll still gaze with admiration at the sun or the moon, according to the time of day. [...]
>
> Your portrait of Miss Gachet must be admirable, and I'll be pleased to see it, oh those little patches of orange in the background. (van Gogh 2024, 894)

It is thus worth underscoring that, without knowing how unusual the letter is in the context of Van Gogh and Theo's correspondence numbering hundreds of letters, one would not be able to instantly detect the seriousness of the given situation. One trivial but telling detail: Theo always addresses his brother as "My dear Vincent," except for this particular letter, which starts with the salutation "My very dear brother" (see ibid.).

To return to Van Gogh, he hesitated for a few days after reading the letter, but eventually decided to take his brother and his family by surprise by visiting them in person in Paris. And as N&S (Naifeh and Smith 2011, 769, cf., 769ff) note and further elaborate: "In his effort to undo a disaster, Vincent triggered a catastrophe." So it really seems like Van Gogh subsequently came back to Auvers as "a worried man," (VGMA 2024f, cf., Naifeh and Smith 2011, 770ff). On approximately July 10, Van Gogh laments: "Once back here I too still felt very saddened, and had continued to feel the storm that threatens you also weighing upon me. What can be done – you see I usually try to be quite good-humoured, but my life, too, is attacked at the very root, my step also is faltering," (van Gogh 2024, 898).

Van Gogh might indeed have been worried, but he still tried both to channel his worries through his art and to orient himself towards a (at least near) future: "I made a point of trying to express sadness, extreme loneliness. You'll see this soon, I hope – for I hope to bring [the paintings] to you in Paris as soon as possible, since I'd almost believe that these canvases will tell you what I can't say in words, what I consider healthy and fortifying about the countryside," (ibid.). And, in his very last letter to Theo, written on July 23, Van Gogh, among other things, casually asks his brother to order some paints for him as well as for a fellow painter Anton Hirschig, and ends again with the prospect of a future (of a future letter, at the very least): "More soon. Look after yourself, and good luck in business &c. Warm regards to Jo, and handshakes in thought. Yours truly, Vincent," (van Gogh 2024, 902).

Then comes the fateful July 27. It was a Sunday and Van Gogh went about his daily routine in the Auvers countryside: he painted in the morning, returned to his inn (the Ravoux Inn) for lunch and finally (fully equipped with his painting tools) left to continue his work. Where exactly Van Gogh went is not known, and we will probably never know for sure (see, e.g., Turnbull 2022, 191). N&S depict the aftermath of events as follows "Hours later, after the sun had set, he staggered back to the Ravoux Inn without his bag or his easel or his canvas. The Ravouxs and their other boarders, who had taken dinner outdoors on a hot summer evening, were lingering on the café terrace. [...] He passed them without a word and went straight to his room. Gustave Ravoux [the owner of the inn], concerned about his guest's strange behavior, listened from the bottom of the stairs. When he heard moaning, he climbed to Vincent's attic room. He found Vincent lying on his bed, curled up in pain. He asked what was wrong. "Je me suis blessé," Vincent replied as he lifted his shirt and showed Ravoux the small hole under his ribs—"I wounded myself," (Naifeh and Smith 2011, 775; cf., VGMA 2024g).

5.3 What Happened to Van Gogh?

> Wings, wings above life!
>   Wings over the grave and death!
>
> That's what we need, and I'm beginning to see that we can acquire them.

—Vincent van Gogh, 1875[7]

What happened on July 27, 1890, before the "wounded" artist "staggered back" to his inn in Auvers? Turnbull again transparently merges the relevant theories into three what he calls "scenarios." N&S's "new theory" is also cited—as the third scenario.

Let me start by introducing the chosen key points of each of these scenarios, so that I can move on to scrutinize N&S's theory in greater detail, and also with respect to the general aims of this book. "Scenario One" is the paradigmatic "madman" suicide story, dominant for the whole of the twentieth century, which springs from numerous non-fictional biographies, popularizing texts, as well as from Stone's well-received novel *Lust for Life*; "a multimillion bestseller, never out of print since its first appearance," (Turnbull 2022, 18, cf., 195). Turnbull summarizes the first scenario in two claims:

Scenario One:

- That Van Gogh shot himself in the wheat fields above Auvers.
- That the act was committed in a state of madness and occurred at the site of his final painting *Wheatfield with Crows*,[8] a canvas that reflected the instability of his mind. (Turnbull 2022, 195)

"Scenario Two" is a slight variation, and further development, of the paradigmatic suicide story; the place of the suicide changes, a different painting is taken to be the final one, the existence of a suicide letter is postulated, and Van Gogh's action is considered to be "lucid." Turnbull breaks down this scenario into ten claims, as he mentions another sub-variation/actualization of this second scenario.

I largely omit this sub-variation and choose six representative claims Turnbull creates (ibid., 196–197, long emphasis mine). Here are the claims:

Scenario Two:

- That on Sunday, 27 July, after a morning spent painting the picture Tree Roots, Van Gogh returned to the inn, had lunch as usual and went out again—this time without his paints, easel or canvas.
- That he walked about three-quarters of a mile to the wheat fields beyond the château in Auvers—back to the site of his morning painting session—and, at an unspecified point in that vicinity, in a deliberate and lucid act, shot himself in the chest.
- That at some time in the evening he returned, injured, stumbled past Ravoux's clients and went up to his room, and that he died of his wounds in the early hours of 29 July.

---

[7] (van Gogh 2024, 047).

[8] However, Turnbull rightly adds that: "the claim made in the popular literature that *Wheatfield with Crows* is Van Gogh's final painting cannot be verified by any evidence in the letters or from any other sources," (ibid.).

- That the unfinished letter later found in his jacket pocket constituted a suicide note, in that it expressed feelings of failure and hopelessness and had all the characteristics of a farewell letter.
- That Tree Roots was his final canvas, painted on the morning of the shooting and which remained unfinished.[9]
- That Tree Roots, both in its theme ("a relentless struggle between life and death") and in its proximity to the suicide, is "a painted farewell note." (van der Veen in Turnbull)

Especially with regard to N&S's new theory, the most attention-grabbing is, of course, the claim about the alleged suicide note. Turnbull himself immediately tries to balance this claim: "The suggestion that the letter found on Van Gogh's body bore the characteristics of a suicide note is not a view shared by all. A psychiatric assessment published in the 1960s said: 'An unfinished letter to his brother found in his pocket gave no indication of his intentions or of his reasons,'" (Turnbull 2022, 197; and Nagera in ibid., emphasis mine.). The letter in question is in fact a draft (or simply a different version) of a letter Van Gogh had already sent to his brother. It is the letter I also quoted above, which Van Gogh ends, inter alia, by the future-oriented phrase "more soon." (van Gogh 2024, 902).

The different version of this letter, or the alleged suicide note, indeed contains this quite emotionally loaded passage, which to some might give the impression of a kind of farewell:

> Ah well, really we [the painters] can only make our paintings speak. My dear brother, there's this that I've always told you, and I tell you again once more with all the gravity that can be imparted by the efforts of thought assiduously fixed on trying to do as well as one can – I tell you again that I'll always consider that you're something other than a simple dealer in Corots, that through my intermediacy you have your part in the very production of certain canvases, which even in calamity retain their calm. For that's where we are, and that's all, or at least the main thing I can have to tell you in a moment of relative crisis. In a moment when things are very tense between dealers in paintings – by dead artists – and living artists.
>
> Ah well, I risk my life for my own work and my reason has half foundered in it – very well – but you're not one of the dealers in men; as far as I know and can judge I think you really act with humanity, but what can you do [que veux-tu]? (van Gogh 2024, RM25)

However, this is all written in the context Van Gogh himself depicts as "the storm that threatens you also weighing upon me," i.e., in the context of the turbulences in Theo's personal, and especially professional life (see again the end of previous section). Moreover, Van Gogh was a prolific and pedantic letter-writer; he often drafted and then discarded his letters. Nothing unusual here. It is perhaps the fact that he still carried the letter with him that seems to be indicative of some kind of significance (for him at least). But it could equally be argued that Van Gogh, the letter-writer, would have written a (more specific) suicide note/letter. The absence of such

[9] Turnbull also notes that: "The centenary exhibition of 1990 was the first to place *Tree Roots* as Van Gogh's final work. Although this cannot be verified, most researchers cite it or *Farms near Auvers* as his last painting," (Turnbull 2022, 197).

a letter thus might be significant, too. Let me suppose for now that these two details cancel each other out.

The last scenario, "Scenario Three," is finally N&S's alternative explanation in a nutshell. Turnbull squeezes it into ten claims while referring to N&S's Appendix (Naifeh and Smith 2011, 791ff).

Scenario Three:

- That at midday on Sunday, 27 July 1890, Van Gogh returned from a morning of painting and, following his usual meal at Ravoux's inn, set off again with a canvas, easel and painting gear.
- That two separate witnesses report seeing him walking along the road towards Chaponval, half a mile from Auvers.
- That the same witnesses report seeing him enter a small farmyard in the area of Chaponval.
- That one of the witnesses heard a gunshot, but when they investigated a little later in the day, the farmyard was empty.
- That René Secrétan, a 16-year-old youth known to Van Gogh, carried around an antiquated pistol which had an unfortunate tendency to misfire. That intentionally or accidently, the gun was fired and Van Gogh was fatally injured.
- That the bullet which entered the body did so at an unusually oblique angle and was in fact fired from a position "too far out" for Van Gogh to have pulled the trigger.
- That in the evening, about 6 h after last being sighted, Van Gogh returned injured to the inn and stumbled past Ravoux's clients, and that he died of his wounds in the early hours of 29 July.
- That his final canvas, easel and painting gear disappeared and were never recovered.
- That no gun was ever found.
- That he wrote no suicide note. (Turnbull 2022, 197–198)

Just to be clear, Turnbull does not take sides here, he does not openly prefer any of these scenarios, as his approach in the book is generally expository, not argumentative. He even mentions facts and refers to sources that contradict several claims of this "Scenario Three" (cf., ibid., 198ff). I overlook that now, since I will mention some of these contradictions indirectly while analyzing N&S's account.

So, let me concentrate on N&S's scenario in depth now, as it should be more than clear that N&S offer a "new" or "alternative" explanation with respect to the dominant "Van Gogh suicide narrative," to put it simply. That is the driving impulse of their theorizing. And it seems fair to say that N&S think that this narrative has its origin not only in the preconceived portrait of Van Gogh as a "madman" or "struggling genial artist," but also in the immediate present-day speculations of the residents of Auvers. Here is once again the Van Gogh suicide narrative in N&S's own words:

> Within hours of Vincent's return to the Ravoux Inn, rumors had begun to circulate about the circumstances that had led to his fatal wounding. Those rumors quickly coalesced into a narrative about the events of July 27. According to this narrative, which was taken up in

> virtually all subsequent accounts, Vincent borrowed a revolver from Gustave Ravoux, the owner of the inn where he was staying, and took it with him on his regular afternoon painting expedition that day. He then climbed up the riverbank and walked some distance into the wheat fields that lay above and outside the town. There he put down his load and shot himself. The blow did not kill him (it missed his heart), but it did knock him out. By the time he regained consciousness, darkness had fallen and he could not find the gun. So he staggered back down the steep riverbank and returned to the Ravoux Inn seeking medical attention. (Naifeh and Smith 2011, 791)

However, N&S argue that this narrative does not harmonize with the available facts. And the goals of their "note" (the appendix in their book is actually titled "Appendix: A Note on Vincent's Fatal Wounding") are "to put forward an account of the events of July 27 that better fits the known facts about the incident and about the man; to examine the origins of the traditional account; and to explain why, in our opinion, that account falls short," (ibid., 792). On the next few pages, I focus predominantly on the first goal.[10] N&S's hypothetical reconstruction of what happened on July 27 of 1890 centers around René Secrétan—at that time a 16-year-old boy—who liked to play at being a cowboy and so also acquired a real gun. Secrétan came from a wealthy Parisian family, who regularly spent their summers in Auvres.

A few notes are in order before I continue; I paraphrase and quote here from N&S's Appendix, and, since N&S prefer to speak simply of "René," I follow them on that for a matter of clarity. Also, N&S's reconstruction is language- and content-wise much richer; I confess that I often strip it down to the absolute minimum in both respects. On the other hand, I count on the fact that readers of this chapter have also read Turnbull's ten-point summarization of N&S's account (Turnbull's "Scenario Three"). Finally, René gave interviews in 1956 and 1957 (!) and so N&S built their case on René's own recollections. Interestingly, 1956 is the year when the movie adaptation of Lust for Life was released, and 1957 marks the year when René died (ibid., 792, 798).[11]

Back to René and his gun: "René described the gun as an old .380-caliber pistol that was falling apart and worked only erratically," (ibid., 793). And, "according to René, the gun was sold (or lent) to him by Gustave Ravoux, the innkeeper," (ibid.). René was also the leader of a group of boys who constantly bullied Van Gogh:

> They put salt in his coffee and watched from a distance as he spat it out and cursed with anger. They put a grass snake in his paint box; when he discovered it, he almost blacked out, René recalled. René noticed that Vincent would sometimes suck on a dry paintbrush when he was thinking, so they rubbed the brush with chili pepper when he wasn't looking. It was all part of a campaign to "drive [Vincent] wild," René admitted. (ibid.; and Secrétan in ibid.)

---

[10] Later, N&S also write that their "hypothetical reconstruction of the events of July 27, 1890, resolves many of the contradictions, fills many of the gaps, and fits together many of the misshapen pieces of the traditional narrative of suicide that has dominated Van Gogh mythology since the day of the shooting," (ibid., 796).

[11] N&S at one point characterize René's statements given in these interviews as "guilt-ridden" (ibid., 803). The reasons N&S have for this characterization will become clear on the few next pages.

Surprisingly, as N&S (ibid.) also observe, "Vincent suffered René's abuses without complaint, even in good humor (he never mentioned them in letters to Theo)." N&S offer several reasons for Vincent's tolerance; Van Gogh liked to meet René's brother, Gaston, who shared an interest in the arts and frequently accompanied René. Both brothers Secrétan were also happy to pay the bill for Vincent when they all drank together at Ravoux Inn (ibid., 794).

Then, "The Secrétan brothers were also the right kind of company – sons of a respectable bourgeois family that could play an important role in Vincent's delusive plan to lure Theo and his family to Auvers. But René Secrétan also supplied Vincent with something he could not get any other way: women. (There was no brothel in Auvers.)," (ibid.). Last but not least, "René discovered the painter masturbating in the woods. This humiliating encounter offered new opportunities for even crueler forms of mockery and torment," (ibid.).

The relationship of René and Van Gogh was thus undoubtably a toxic one (cf. ibid.). René also admitted that, after some time, Van Gogh started to react more sensitively to the bullying: "[he] took it very badly;" "One day he became red with anger and wanted to kill everyone," (Secrétan in ibid.). N&S thus hypothesize that:

> The shot that killed Vincent van Gogh was probably fired not in a wheat field, but in or near a farmyard on the road to Chaponval [...]. Moreover, the gun that delivered the fatal blow was probably not brought into that farmyard by Vincent van Gogh, who knew nothing about guns and had no need of one, but by René Secrétan, who rarely went anywhere without his .380-caliber peashooter. The two may have encountered each other by accident on the Chaponval road, or they may have been returning from their favorite watering hole together [meaning from Ravoux Inn]. Gaston was almost certainly with them, as Vincent would have avoided René, whether alone or in the hostile company of his followers.
>
> René had a history of teasing Vincent in a way intended to provoke him to anger. Vincent had a history of violent outbursts, especially when under the influence of alcohol. Once the gun in René's rucksack was produced, anything could have happened – intentional or accidental – between a reckless teenager with fantasies of the Wild West, an inebriated artist who knew nothing about guns, and an antiquated pistol with a tendency to malfunction. (ibid., 795)

Just a reminder: N&S themselves see the explanatory value of this alternative reconstruction as the possibility of explaining many other details. Put another way, by answering a what-question, N&S give answers to many other why-questions. I choose and quote the most representative details N&S think that their reconstruction ultimately explains (ibid., 796–798). And, for reasons of space, I try to shorten them as much as possible:

- It explains the immediate and permanent disappearance of all the evidence surrounding the incident [...]. In his wounded state, Vincent would never have been able to clean up so thoroughly afterward [...]. In their hasty cover-up of the deed, with night falling, the Secrétan brothers may well have missed something—a spot of blood or a spent cartridge—but the police would never have found it, because they were searching the wheat fields far away, not the farmyards on the Chaponval road.

- It explains the oddities of Vincent's wound as reported by the doctors who examined it: that the shot was to the body, not to the head; that the bullet entered from an unusual, oblique angle—not straight on as one would expect in a suicide; and that the shot appeared to have been fired from "too far out" for Vincent to have pulled the trigger. (Hirschig in ibid., 796)
- It explains why Vincent left no suicide note, and why Theo could find no trace of a "farewell" when he searched Vincent's room and his studio in the days after the shooting. It explains why Vincent bothered to take a load of canvas and paint and other supplies on his expedition that day—something he would have been unlikely to do if he did not intend to return.
- It explains why he failed to "finish himself off" when his first (and only) shot went wrong, and chose instead to take the far more painful and embarrassing path back to his attic room at the Ravoux Inn.
- It explains why the origin of the gun that inflicted the fatal wound was not revealed until 70 years later, even though it must have been known by a number of people at the time of the shooting. Auvers being a small town and revolvers being a rarity in rural France, many friends of both Gustave Ravoux, who lived a very public life, and René Secrétan, who brandished the weapon openly, surely were familiar with the innkeeper's exotic firearm. [...]
- Finally, this reconstruction explains why Vincent's "confessions" to a suicide attempt, as reported by witnesses at the time, were so hesitant, halfhearted, and oddly hedged. When the police asked Vincent directly, "Did you want to commit suicide?" he answered with an indecisive "Yes, I believe so." When they told him it was a crime to attempt suicide, he appeared more concerned that no one else should be blamed than that he himself might be charged with a crime. "Do not accuse anyone," he responded; "it is I who wanted to kill myself." [...] (Van Gogh in ibid., 798, cf., also footnote 55 on page 806)

To put it crudely, N&S consider Van Gogh's hesitancy as revealing his effort to cover for the boys, the Secrétan brothers (ibid.). Moreover, N&S's reconstruction is, by design, compatible with both versions; the accidental killing of Van Gogh, "an adolescent prank gone terribly wrong" (ibid., 797), as well as with Van Gogh being murdered. My sense, however, is that N&S obliquely favor the first version. At least in the sense that the first version corresponds to the toxic relationship René and Van Gogh had, and nothing is directly indicative of the second version. What would be René's motive for intentionally killing Van Gogh?

So, N&S search for a different motive, i.e., why would Van Gogh obstinately lie about and cover the actions of his tormenter René, and his brother Gaston, by claiming that he had tried to commit suicide? N&S are of the view that Van Gogh "welcomed death" (ibid., 798). "In fact, whether by accident or negligence or malice, René Secrétan may have provided Vincent the escape that he longed for, but was unable or unwilling to bring upon himself, after a lifetime spent disavowing suicide as 'moral cowardice' and 'the deed of a dishonest man,'" speculate N&S (ibid., and Van Gogh in ibid.). Therefore, even for N&S's scenario challenging the suicide narrative preferred by the official history, the determining explanatory factor is still Van

Gogh's worry that his plight in life was to be a burden on his brother Theo and his family. Only with this detail in the picture, does Van Gogh's secrecy/martyrdom seem intelligible at this particular point of his generally troubled life (cf. ibid.).

All in all, N&S and their "accident story" of Van Gogh's death do not proceed by "discovering" any new facts (at a minimum, in the sense of previously non-discussed facts). Indeed, at first glance, their approach may appear to be "only" imaginative gap-filling. So, in the next concluding section, let us go over N&S's explanatory approach with a fine-tooth comb.

5.4 Why Did Van Gogh Die?

With respect to inquiring into Van Gogh's death, N&S's principal question is "what happened to Van Gogh?" N&S's principal question is not a why-question, at least on my reading, since they primarily try to challenge the already existing paradigmatic explanation of Van Gogh's death (that is, the already existing answer to a what-question), and, secondarily, since they do not intend to explain, at least not directly, any significant implicated why-question of their alternative reconstruction of what happened on July 27, 1890. They are not interested in answering, e.g., the question "why would René want to murder Van Gogh?"

That being said, I feel obliged to note here that the distinction I introduce in Chap. 2 (Sect. 2.3) between what- and why-questions is not always as sharp as one would probably want it to be. It seems to be the case that every significant what-question can, in principle, be re-articulated in the form of a why-question. It seems to be only a pragmatic/contingent distinction one often makes ad hoc. One can simply ask: "why did Van Gogh die?"

What I see as the main benefit of the distinction between the what-question/why-question is that it is an analogous distinction to a popular distinction often exploited in historiography, viz., the distinction between chronology (the selection of events and structuring them into an order) and history (explaining the selected events and their order). On the other hand, when formulated in the question format, the distinction loses its sharp margins; it is suddenly "only" a matter of our perspective/language use. I have no objections to that. However, I do not see these mostly linguistic modalities as changing my arguments in any substantial way. It could even be argued that this linguistic modality once again accentuates the continuity between our everyday reasoning and "scientific" reasoning. The continuity much welcomed by naturalists of the Quinean kind.

In order to plausibly answer the question "why did Van Gogh die?," one needs to know many details (facts and contexts, if you will). Put another way, many what-questions need to be answered within the process of answering this why-question, too. N&S know this very well and frame their account accordingly (cf., also ibid., 552ff).[12]

So, in this sense, the explanatory structure of N&S's hypothetical reconstruction is simultaneously abductive and comparative. It is an abductive explanation, as when one accepts the murder/manslaughter scenario, it explains many other known facts/events. And it is also comparative, as it explains the known facts in a different way. One example: the immediate absence of any evidence related to the happenings in the afternoon of July 27, 1890 (at that time, no gun was found and Van Gogh's gear was not found either) is explained by N&S as the purposeful destruction of the evidence by the Secrétan brothers. The traditional suicide scenario presupposes that either Van Gogh managed, despite his injury, to discard the evidence by himself, or that he, e.g., did not take the painting gear along with him for the afternoon painting session (see again the "Second Scenario" above). Also, N&S's explanatory approach nicely demonstrates that abductive inferences work smoothly with mentalistic and mentalistic/intentional concepts, whatever their metaphysical status is.

From the naturalistic perspective, abductive and comparative inferences are perfectly acceptable, in fact quite common, instruments of our scientific theorizing. Thus, the charge against N&S's alternative reconstruction put forward, for instance, by William Arnold, namely that: "The scenario they constructed for the deed was even more of a leap of faith and based on nothing but their own imaginations" (Arnold 2024), simply ignores the explanatory structure of N&S's reconstruction. Or, for some unspoken reason, the charge possibly beforehand dismisses abductive and comparative inferences as proper scientific and/or historiographical inferences. But that is a non-starter for naturalism defended here in this book.

As naturalists can with Quine happily accept that, quite often, "hypothesis is guesswork; but it can be enlightened guesswork," (Quine and Ullian 1978, 65, cf., again Sect. 2.3). Recall that Quine also thinks that historical hypotheses also gain their scientific credibility indirectly by "participating in a large, cohesive section of scholarly and scientific lore and common sense which has its ultimate links with observation categoricals and prediction somewhere outside the historical quarter," (Quine 1987, 162, emphasis mine). And, in my interpretation, the slightly cryptic wording "outside the historical quarter" predominantly refers to various results of synchronic scientific inquiries. Put another way, we often write and re-write history from outside that particular history; for example, by using diachronically asymmetrical concepts and/or relying on diachronically posterior theories (cf., again Sect. 2.2.1).

N&S also believe that "the Van Gogh madness narrative" is to a large extent supported by several synchronic assumptions. The most magnetic one is, of course, the assumption that Van Gogh's suicide is in some way interconnected with his mental condition and/or self-harming behavior. "People who knew of Vincent's stay in an

[12] It could be said, I argue, that N&S are implicitly answering the question "why did Van Gogh *not* commit suicide?" And that they gather enough answers to related what-questions (facts, contexts) to answer this negative question.

asylum and had seen his deformed ear were quick to assume a connection between self-mutilation and suicide – a connection debunked by later research," writes N&S (2011, 800). N&S then cite a couple of studies from the 1970s and 1980s dealing with the phenomenon of self-mutilation (see the footnote 89 in Naifeh and Smith 2011).

One of these studies is *Self-Mutilation: Theory, Research, and Treatment* (1988) by Barent W. Walsh and Paul M. Rosen. Walsh and Rosen conclude that: "The similarities between suicide and self-mutilation can be deceiving. Both forms of behavior are self-directed, and both result in concrete physical harm. Nonetheless, when the behaviors are analyzed using the framework of Shneidman's commonalities, it is evident that the behaviors are different in many facets, and in several ways are even opposites," (Walsh and Rosen 1988, 51, this quote is chosen by N&S).

So, this kind of (broadly conceived) psychological research can be exploited to enlighten our (hypothetical) historical reconstructions. At a minimum, and setting aside the difficulties with classifying human action, if a strong correlation were shown between "self-mutilation" and "suicide," it would be a reason for supporting a Van Gogh suicide scenario. The same kind of retrodictive reasoning is available, e.g., in the case of Van Gogh's "epilepsy" diagnosis. Still, supposing for the sake of argument that we would be able to unambiguously diagnose Van Gogh's illness, even the best diagnosis would obviously not be a direct answer to the question "why did Van Gogh commit suicide?" "Pathology explains nothing" directly and in itself (Adamov in Perruchot 1969, 302).

In addition, the conventional suicide story was defended against N&S's alternative version by interpreting anew the preserved forensic evidence about Van Gogh's bullet wound (see van Tilborgh and Meedendorp 2013). N&S then, in response, requested an expert opinion from Dr. Vincent Di Maio; "one of the world's leading handgun forensic experts," (Naifeh and Smith 2014). N&S write: "Di Maio's conclusion? 'It is my opinion that, in all medical probability, the wound incurred by Van Gogh was not self-inflicted. In other words, he did not shoot himself.' Which brings us back to the question: If Van Gogh didn't shoot himself, who did shoot him?" (ibid.). However, to their credit, N&S knew that their original case in their 2011 book cannot be based only on this one particular piece of evidence. The reason for that is partially, of course, that this evidence underdetermines given theories about it (after all, Tilborgh and Meedendorp 2013 also consulted a forensic expert), but partially also the fact that this piece of evidence is too fragile to resist the force of Van Gogh's (madman) suicide narrative on its own.

I think that N&S knew very well that they had to fight one story with another story. And if one takes on board the working assumptions of the naturalism I defend in this book, there is nothing unscientific or irrational about their alternative hypothesized story. N&S gave us a plausible answer to the question "why did Van Gogh die?"

References

Arnold, William. 2024. (2020). *The companion book: A peep inside* [Introduction]. Van Gogh Death. https://www.vangoghdeath.com/death-in-auvers-1

Bailey, Martin. 2018. Van Gogh: It was suicide, not murder. *The Art Newspaper.* https://www.theartnewspaper.com/2018/09/07/van-gogh-it-was-suicide-not-murder

Bailey, Martin. 2019. Van Gogh's suicide: Ten reasons why the murder story is a myth. *The Art Newspaper.* https://www.theartnewspaper.com/2019/09/06/van-goghs-suicide-ten-reasons-why-the-murder-story-is-a-myth

Bónafoux, Pierre. 1992. *Van Gogh: The passionate eye*. New York: Times Mirror Company.

van Gogh, Vincent. 2024. *Vincent van Gogh: The letters* [online]. Van Gogh Museum and Huygens ING. https://vangoghletters.org/vg/. (the numbers I use when referring correspond to the numbers associated with each letter on this web page).

Leeman, Fred. 2013. *Van Gogh: Struggle & success.* Narrated by Helen Mirren. Tribute International.

Mathews, Patricia. 1986. Aurier and Van Gogh: Criticism and response. *The Art Bulletin* 68 (1): 94–104.

Naifeh, Steven, and Gregory White Smith. 2011. *Van Gogh: The life* [eBook]. New York: Random House.

Naifeh, Steven, and Gregory White Smith. 2014. NCIS: Provence: The Van Gogh mystery. *Vanity Fair*. https://www.vanityfair.com/culture/2014/12/vincent-van-gogh-murder-mystery.

Perruchot, Henri. 1969. *Život Vincenta van Gogha*. Praha: Orbis.

Quine, Willard Van Orman. 1987. *Quiddities: An intermittently philosophical dictionary*. Cambridge, MA: Harvard University Press.

Quine, Willard Van Orman, and Joseph S. Ullian. 1978. *The web of belief.* 2nd ed. New York: Random House. (Originally published 1970).

Ruiz, Pablo A. 2021. The Van Gogh enigma: Re-mythifications of the artist's genius in "Loving Vincent" and "At Eternity's Gate". *Comparative Cinema* 9 (16): 93–162.

van Tilborgh, Louis, and Teio Meedendorp. 2013. The life and death of Vincent van Gogh. *The Burlington Magazine* 155 (1324): 456–462.

Turnbull, Andrew. 2022. *Who shot Van Gogh?: Facts and counterfacts about the world's most famous artist*. London: Thames & Hudson.

VGMA. 2024a. *How many paintings did Vincent sell during his lifetime?* Van Gogh Museum Amsterdam. https://www.vangoghmuseum.nl/en/art-and-stories/vincent-van-gogh-faq/how-many paintings-did-vincent-sell-during-his-lifetime.

VGMA. 2024b. *Hospitalization*. Van Gogh Museum Amsterdam. https://www.vangoghmuseum.nl/en/art-and-stories/vincents-life-1853-1890/hospitalization.

VGMA. 2024c. *Looking for a direction*. Van Gogh Museum Amsterdam. https://www.vangoghmuseum.nl/en/art-and-stories/vincents-life-1853-1890/looking-for-a-direction.

VGMA. 2024d. *Van Gogh's letters*. Van Gogh Museum Amsterdam. https://www.vangoghmuseum.nl/en/art-and-stories/stories/all-stories/van-goghs-letters.

VGMA. 2024e. *Vincent's illness and the healing power of art: About Van Gogh's psychiatric disorder*. Van Gogh Museum Amsterdam. https://www.vangoghmuseum.nl/en/art-and-stories/stories/vincents-illness-and-the-healing-power-of-art.

VGMA. 2024f. *Vincent's final months*. Van Gogh Museum Amsterdam. https://www.vangoghmuseum.nl/en/art-and-stories/vincents-life-1853-1890/vincents-final-months.

VGMA. 2024g. *The last days of Vincent van Gogh*. Van Gogh Museum Amsterdam. https://www.vangoghmuseum.nl/en/art-and-stories/vincents-life-1853-1890/vincents-final-months.

Walsh, B. W., and P. M. Rosen. 1988. *Self-mutilation: Theory, research, and treatment*. New York: The Guilford Press.

Wikipedia Contributors. 2024. *List of works by Vincent van Gogh*. Wikipedia: The Free Encyclopedia. https://en.wikipedia.org/w/index.php?title=List_of_works_by_Vincent_van_Gogh&oldid=1216917052

Chapter 6
Second Case Study: A Brief (Hi)Story of Just-So Stories in Evolutionary Science

Abstract This chapter is construed as the conceptual analysis of the catchphrase "just-so story" that, from the end of 1970s, has become a very popular sticker for labeling weak or just unwarranted evolutionary hypotheses. First, I track the current ways in which the phrase "just-so story" is used and conclude that it is possible to categorize two mutually exclusive usages. I name these two as "positive strategy" and "negative strategy." Then, second, I underscore the fact that within these two strategies themselves there are additional and very instructive differences with respect to their usage. One could split them again into two basic differences; the usage either targets the description of evolutionary ontology (it targets the problem of adaptationism), or the epistemic difficulties associated with evolutionary explanations (it targets their historical and/or speculative features). However, third, I argue that these strategies do not satisfactorily capture the core of the "original" denotation of the phrase as advanced by Stephen Jay Gould and Richard Lewontin. I revisit the foundation(s) of their anti-adaptationist critique in order to reframe it as predominantly a critique of distinct methodological manners and epistemic maxims related to evolutionary inquiry qua historical inquiry.

Keywords Just-so story · Narrative · Adaptationism · History · Gould

6.1 Introduction: Bounding Speculations

> The meaning of a question is the method of answering it. Tell me how you are searching, and I will tell you what you are searching for.
>
> —Ludwig Wittgenstein[1]

[1] (Wittgenstein 1975, 12).

M. Hubálek, *Dissolving Naturalism and Historicism Into Each Other*, Studies in History and Philosophy of Science 67,
https://doi.org/10.1007/978-3-032-07535-2_6

> Rhetoric is never historically so isolable from the substance of theories that we cannot seriously ask whether the theory might have occurred to its discoverer without the rhetorical premises that recommended it.
>
> —Arthur C. Danto[2]

The main goal of this case study is to demonstrate, on an actual theoretical discussion, that evolutionists/anthropologists themselves recognize that they have to deal with historical explanations, i.e., that they are theorizing historical entities and events.

Their respective disagreements about the epistemic utility of what in this chapter I call "just-so storytelling" are informative not only with respect to their different starting points, viz., different notions of science and scientific explanation, but also with respect to how these differences influence their views on evolutionary ontology. On the whole, it once again shows that, though often unreflectively, when one tackles some obstacles related to explaining, ontological pictures follow epistemological commitments.

I thus present evolutionists/anthropologists not only as historical scientists, but also as scholars facing the history of their own field.

It is currently becoming increasingly common to admit openly that some parts of our science, even the parts traditionally perceived as "hard," are highly speculative. In other words, that the difference between (what might be called) "speculation" and (what might be called) "enlightened guesswork" is a matter of degree and not of a kind. In the case of biological evolution, the situation is even more complicated, as the theory of evolution supplements our world view with the origin story of our own species. Some conclusions regarding the speculative aspects of evolutionary science are thus often omitted, or simply overlooked, due to the associated political and/or religious controversies. Recollect all the tired discussions about whether a scientific ism or statement *X* is a fact or just a hypothesis.

I hope that my analysis of Quine's conception of science (ipso facto naturalism) indicates that such a distinction is not philosophically interesting nor substantive from within naturalistically conceived empirical inquiry. I do not believe that there is an epistemologically/pragmatically useful way of sharply distinguishing accepted facts from proven facts, or alternatively, accepted hypotheses from proven hypotheses. The difference is again a matter of degree here.

Be that as it may, the usual controversies linked to the sociological dimension of scientific perception apply to evolutionary theorizing with equal force. For instance, the slogan "I stand with science" would indeed not have its tribalistic bite in a form of "I stand with historically enlightened guesswork." But I think that a part of the scientific attitude should be a radical anti-tribalism. And a part of this scientific anti-tribalism should be, at least as I see it, playing with open cards. In this particular regard, evolutionists have a great example of such an attitude in the founding figure of their whole discipline. Consider this quote from Darwin's *The Descent of Man*:

[2] (Danto 1986, 224).

> Many of the views which have been advanced are highly speculative, and some no doubt will prove erroneous; but I have in every case given the reasons which have led me to one view rather than to another. It seemed worth while to try how far the principle of evolution would throw light on some of the more complex problems in the natural history of man. False facts are highly injurious to the progress of science, for they often endure long; but false views, if supported by some evidence, do little harm, for every one takes a salutary pleasure in proving their falseness: and when this is done, one path towards error is closed and the road to truth is often at the same time opened. (Darwin 1889, 606)

One need not cling to the conceptual distinctions Darwin makes here, for example, between "fact" and "view," to understand what the take-home message is. Scientific accounts should ultimately be appropriately justified (whatever that means), not just put forward as possible and/or plausible conclusions without further ado. To put it crudely, it is fair to speculate in science, but pure speculations are an inferential no-go zone. And although this conclusion sounds almost trivial, it is sometimes easier said than done.

The discussion and disagreements surrounding the concept of the just-so story also instructively reveal that different scholars have different opinions on what is an acceptable degree of speculation. My analysis of the disagreements linked to just-so story-telling thus also yields specific criteria of evaluating ultimate evolutionary explanations from within the (Neo-)Darwinian paradigm. I show that one has sufficient epistemic resources to evaluate evolutionary explanations, qua historical explanations, in view of the internal factors forming every explananda and explanans of these explanations.

Put another way, following the irrealist tenets introduced in Chap. 4, I uncover a way of evaluating evolutionary explanations qua narrative explanations. I argue that Gould and Lewontin gave us a way of filtering better or worse evolutionary origin stories. Even if one grants that evolutionary explanations are ultimately historical, it is not an obstacle for their rational evaluation. So, in this respect, I present one practical way of evaluating evolutionary explanations as better or worse, following on Roth's stipulation that historical accounts must be evaluated only comparatively.

6.2 The Double Challenge of History

Do narrative/story-like structures have a legitimate place in evolutionary science qua historical science? At least since the 1970s, with the now-canonical works of Stephen Jay Gould and Richard Lewontin, evolutionary thinkers and philosophers of biology have tackled the problem of narrative explanation qua evolutionary explanation by debating the epistemic utility of so-called just-so stories. Although the usage of the term "just-so story" is not homogenous, it is quite a popular term among scientists and scholars dealing with evolutionary theories, especially evolutionary explanations of the historical origin of phenotypic traits and the behaviors of given organisms.

This is to say that just-so storytelling is an epistemic concept that significantly shapes evolutionary-biological discourse. A crucial aspect of the discussion concerning just-so stories is that the concept emerged as an integral part of a substantial and influential criticism of adaptationism, i.e., of a specific mode of evolutionary research. It is, inter alia, for this reason that this methodological discussion often grows into heated debate about the general principles and the operation of biological evolution as well as the very structure of the life sciences. Alex Rosenberg and Daniel W. McShea sensibly comment on this fact as follows: "The best-known challenge to Darwinism from within biology was, ironically enough, framed by two eminent evolutionary biologists who between them made some of the most influential contributions to the philosophy of biology [...]," (Rosenberg and McShea 2008, 65).

Naturally, after more than 40 years of their elaboration, some of the issues related to just-so storytelling may seem old-hat. That being said, I believe that at least two issues are still vital. First, with a growing number of fields of knowledge using the attribute evolutionary—inclusive of the so-called evolutionary humanities or evolutionary social sciences—many scientists and scholars are, implicitly or explicitly, asking anew whether there could be something as sociobiology.[3] Put another way, they anew inquire into the relationship between biological evolution and cultural history. All these efforts must handle adaptationism qua metatheoretical evolutionary framework. Second, and complementarily, it is yet unsettled to what extent the biological evolution is an adaptive evolution, and whether we are epistemically able to determine such a correspondence, to a reasonable degree of scientific certainty. This proposition holds even for well-established evolutionary subfields such as evolutionary psychology (Smith 2020; cf., Gould 1991; Heyes 2018).[4]

Nowadays, discerning readers of philosophical journals and literature can sense both the re-burgeoning interest in the problem of historicity within evolutionary

[3]Consider a fascinating survey study "A Cross-Disciplinary Survey of Beliefs About Human Nature, Culture, and Science." by Carroll et al. (2017). For a historically accurate and intellectually informative treatise of the so-called Sociobiology Wars see Segerstrale (2000). I do believe that it is philosophically naïve to suppose that the controversies related to the theoretical status of sociobiology are fundamentally, let alone *only*, political in their nature. Gould himself indicates that *the point of departure shall not be the "deconstruction" of political programs but rather the demonstration of inadequate theoretical commitments*. From his historically anchored position Gould writes: "Human sociobiology is fundamentally wrong because [...] ultra-Darwinism, especially in its commitment to adaptationism, is fallacious (or at least vastly overextended)—not because many of us do not like its political implications. If any overt politics motivated the origin of human sociobiology, blame the academic jockeying of disciplinary extension, prestige, and imperialism (e.g., reduction of the social sciences to biology), not the larger social constructs of conservatism and laissez-faire. (And if you, as social scientists, wish to take a compassionate view of our efforts in biological imperialism—in a sermon-on-the-mount manner of loving one's adversaries—then do understand that we have often been subject to the same denigration by physicists and chemists. It is a sad fact that the oppressed often take up the tools of oppression.)," (Gould 1991, 51, emphasis mine).

[4]Exactly for these reasons, I am convinced that although there is a full panoply of evolutionary fields, they are more or less united through the metatheoretical framework of adaptationism.

science (e.g., Wong 2019; Beatty and Carrera 2011) as well as an interest in examining the epistemic status of historical explanation and/or narrative explanation (e.g., Morgan and Norton Wise 2017; see especially Roth 2017; Currie and Sterelny 2017; further relevant works are, e.g., Roth 2020; Bonnin 2019; Smith 2016a, b). The common denominator of all of these efforts is the endeavor to overcome epistemological problems and pitfalls related to historical research, with the main obstacle being the essential underdetermination of historical accounts (see Currie and Turner 2016). Evolutionary processes are considered to be both deterministic and stochastic: evolutionary phenomena are, thus, contingent, unpredictable and path dependent/dependent on initial conditions. Individual biological entities are, in this light, entities with long and unique natural histories. (Gould 1989; Machery 2008; Beatty and Carrera 2011; Desjardins 2011; Losos 2017). There is, thus, a wide consensus that evolutionary science is ultimately historical science (e.g., Mayr 2000; Rosenberg 2002, 2005; cf., Hull 1975).

However, from the vantage point of the philosophy of science, there is a prima facie irreconcilable schism between the camp of scholars who accept that narrative explanation is a suitable form of historical and/or evolutionary explanation,[5] and the camp of scholars who insist that narratives are not explanatory, and hence do not have a place in science, or even exemplify bad science or pseudoscience. In concrete terms narrative explanation qua historical explanation, as contrasted with natural-science-oriented models of explanation, is often conceived as "speculative," "relativist," "constructivist," or "ideological" epistemic device (for philosophically informed elaborations see primarily Roth 2020; cf., also Kuukkanen 2015; Rosenberg 2018).

I cannot here do justice to all of the philosophically important and profound issues revolving around the triad *evolutionary history, science and explanation of history, naturalism in epistemology*. I am content to show that to unravel the conceptual issue of just-so storytelling can abet theoretical research across evolutionary and scientific fields, and that it can navigate scholars to a new path of philosophical analysis which lies at the intersection of the philosophy of biology and the philosophy of history.

In this case study, I analyze the concept of just-so stories in order to demonstrate that (1) the meaning and methodological use of the term "just-so story" oscillates between a negative and a positive use and that within these two poles there exist further individual differences and nuances that rest on different ontological and methodological underpinnings; (2) with respect to the "original" meaning of the term "just-so story" developed by Gould and Lewontin at the end of the 1970s, the dichotomy of negative and positive use is not very informative and instructive, (3) Gould's and Lewontin's treatment of the term "just-so story" is more specific and sophisticated and directly reacts not only to the problems of evolutionary ontology, but also challenges some of the methodological manners and epistemic maxims of

[5] Although it must be taken into account that particular theoretical assumptions and further conceptual details may differ.

evolutionists; (4) it is an open question whether scientists and scholars should stick to the "original" meaning of the term "just-so story" or to accept that the term "just-so story" is to be perceived as an equivalent term for "implausible narrative explanation."

6.3 Evolutionary Explanations and Just-So Stories

At the end of the 1970s, geneticist Richard Lewontin (1976a) and paleontologist Stephen Jay Gould (1978) likened evolutionary theorizing to Rudyard Kipling's book of fairy tales *Just So Stories for Little Children* (1902) when they championed the notion that evolutionary scientists and scholars, instead of crafting evolutionary explanations, often provide us with "just-so stories." Springing from now-canonical criticism of sociobiology, the comparison aims to debunk an explanatory strategy that is generally referred to as adaptationism or the adaptationist programme (see the so-called "Spandrels paper," Gould and Lewontin 1979; and Lewontin's "Sociobiology as an adaptationist programe" published in 1979). A concrete example making explicit the logic of adaptive just-so storytelling can be sketched as follows:

(Q): *Why* is the color of blood red?
(A): Blood has a red color *because* red color frightens predators if needed.[6]

Gould's and Lewontin's (hereafter G&L) viewpoints, that will later be discussed in this chapter, sparked the anti-adaptationist movement within evolutionary science and the philosophy of biology. This prompted fruitful discussion regarding the nature of adaptive and/or evolutionary explanations.

"Just-so story" as a specific type of epithet can be split into two parts: the just-so part refers to the speculatively adaptive component, and the story part reflects the narrative structure of a given evolutionary account.[7] However, from our present perspective, the expression "just-so story" is used, and the problem of just-so storytelling is approached, through two distinct methodological strategies.[8]

[6] In fact, this is for philosophers of evolutionary biology a typical (albeit simplistic) example of searching for an adaptive explanation, in spite of the fact that another *non-adaptive* explanation is at hand. See Wilson (2015) and his conversation with Lewontin over the "Spandrels paper."

[7] I paraphrase here John Beatty who, while commenting on an earlier version of this text, made me aware of this twofold structure of the very expression "just-so story," which I had referred to only implicitly until then.

[8] Smith (2016a, b) presents a more nuanced and historically comprehensive examination dealing with the phrase "just-so story" which I am partly drawing on. I believe that the idealized dichotomy I offer does not contradict his analysis. Quite the opposite, I propose it as a dichotomy exposing the two main motivations underlying a variety of approaches to the problem of just-so storytelling within evolutionary science since G&L's critique.

6.3.1 *The Negative Usage of "Just-So Story"*

The term "just-so story" is used as a derogatory/pejorative term labelling weak or unwarranted "stories" postulated as fully-fledged explanations of the evolutionary origin and function(s) of individual traits and behaviours. Put another way, this approach expresses the need to contrast plausible evolutionary and/or adaptive explanations with mere just-so stories, which is an intention that squares with the original anti-adaptationist approach of G&L (see, e.g., Goldstein and Barkley 1998; Sesardic 2003; Cleland 2011; Oman 2012). I call this approach to just-so stories the "negative approach". In the following paragraphs, I cite paradigmatic[9] examples of the negative approach in chronological order beginning from G&L that, moreover, aptly reveals the successive shift in the content of the negative criticism of just-so storytelling practices:

> We are entirely in the realm of past human evolutionary history. All that can be offered are imaginative speculations about how a trait might have conferred greater fitness on its carriers. This kind of speculation is easy although it may occasionally demand some ingenuity. Sociobiologists have been aided in their construction of modern Just So Stories by expanding the concept of natural selection. (Lewontin 1976a, 145)

> When evolutionists study individual adaptations, when they explain form and behavior by reconstructing history and assessing current utility [...], [they] tell just-so stories [...]. Virtuosity in invention replaces testability as the criterion for acceptance. (Gould 1978, 530)

> Dreaming up scenarios [...] axiomatically applying the concepts "adaptation" and "natural selection" in such a fashion and to such situations that the very testing of the validity of the explanations becomes utterly hopeless. (Eldredge 1983, 361)

As it is obvious, for Lewontin, Gould, and even for Eldredge, the concept of just-so story is connected with other non-trivial concepts such as "natural selection," "adaptation," "evolutionary history," "explanation," and "speculation." These usages are thus concerned not only with the structure of (evolutionary) explanations, but with evolutionary ontology as well: they are concerned first and foremost with the speculatively adaptive just-so part. The following quotes represent the newer and nowadays more common content and meaning of the negative strategy:

> A just-so story explains nothing: it simply states authoritatively a sequence of unmotivated events by putting it in a narrational context. A just-so story is thus justificative only because of its illocutionary force, as a performative act ("if something is narrated it means that it is true") and not as an explicative one ("x happened because of general rule y"). (Valeri 2000, 254)

> A speculative story or explanation of doubtful or unprovable validity that is put forward to account for the origin of something (such as a biological trait) when no verifiable explanation is known. (Just-so story. 2024. In Merriam-Webster.com)

[9] This is to say that I consider them to be typical for the purposes of my arguments. The same holds for the examples of the *positive strategy* in the next section of this chapter. Of course, there could be slightly different and more detailed delineations.

These two usages are at first glance narrower. They are both concerned only with the concepts of "explanation," "speculation," "narrative," etc. The emphasis is no longer on evolutionary concepts ("natural selection," "adaptation"), nor is the problem of "evolutionary history" acknowledged. The emphasis is solely on the nature of an evolutionary explanation: these usages emphasize mostly the story part that is conceived as speculative and non-explicative hence implicitly unscientific. In general terms, the aim of this usage is to contrast a just-so story with, supposedly, real and robust explanation, i.e., with scientific explanation in the most robust sense of the term. Such usages, thus, must presuppose a specific notion of hard science that requires further examination and argumentation, and in fact some demonstration that it is a suitable and beneficial notion for historical inquiry.

6.3.2 The Positive Usage of "Just-So Story"

The concept of a just-so story is used as a description of explanatory strategies producing evolutionary accounts of the origin and function(s) of individual traits and behaviours. In most cases, this view of just-so stories responds to G&L's critique of the adaptationist programme, or alternatively, it corresponds to the authors taking methodological cues from G&L while evaluating evolutionary and/or adaptive accounts.[10] Hence, several scientists and scholars have suggested that the phrase "just-so story" is to be conceived as an equivalent term for a (provisional) speculative answer, evolutionary hypothesis, or heuristic, i.e., a starting point of evolutionary inquiry that might, but of course might not, promote fertile (scientific) investigation of evolutionary history (see, e.g., Woese 1980; Griffiths 1997; Gottschall 2008; King 2008; Barash and Lipton 2010; Levy 2013). For these reasons, I label this approach as the "positive approach."

On the whole, the negative approach downplays the epistemic significance of just-so storytelling and uses the term "just-so story" negatively as an antithetic term to a proper evolutionary account. In opposition, the positive approach represents a more relaxed view, insofar as it presupposes that story-like structures, or even just-so (adaptive) speculations have, for various reasons, their epistemic role in evolutionary inquiry. I argue that the efforts to neutralize negative, or explicitly ironic, connotations of the phrase "just-so story" can be seen as symptomatic: they can be seen as an unavoidable consequence of the fact that biological evolution is a historical process, and, ipso facto, that a special kind of explanation or explanatory strategies is needed.

From this perspective, it comes as no surprise that what I call the positive approach to just-so stories appears simultaneously with the widely acknowledged negative G&L's approach. Anthropologist Ward H. Goodenough, who slightly

---

[10] Naturally, then, this proposition applies vice versa. For instance, Buss et al. (1998) indicate that some of Gould's evolutionary hypotheses could be seen as just-so stories.

before G&L published their influential "Spandrel paper," had intentionally labelled his explanation of the historical origin of matrilineal clans as a just-so story. He aptly points out the peculiarities connected with the explanation of such a phenomenon:

> We have to invent "just so" stories and then accumulate for many societies the circumstantial evidence that will make them more or less plausible as time goes on. (Goodenough 1976, 34)

Daniel Dennett reacts directly to G&L's critique of the adaptationist programme and boldly infers that due to the lack of relevant historical knowledge, adaptive just-so stories are an integral part of doing evolutionary science:

> There is no alternative; the fossil record is simply not going to provide enough information to provide a rigorous, bottom-up, scientific account of all the important moments in the evolution of species. The provable history has to be embellished and extrapolated with a good deal of adaptationists' thinking. What we need is a just-so story. (Dennett 1986, 4)

These usages acknowledge the epistemological problems relating to historical research as well as suggest that to produce historical knowledge one must recruit specific strategies that are not necessarily as scientific in the most robust sense of the term: they primarily suggest that the accounts may be in a sense just so but, simultaneously, supplemented either by "circumstantial evidence," or "adaptationist thinking." Put another way, and without making any a priori/ahistorical commitments with respect to the demarcation criterion of science, these usages admit that speculations/guesses are necessary and unavoidable in historical research. Goodenough, thus, appeals to specific methodological procedures while Dennett bases his assertion on a specific ontological interpretation of evolution—on a version of adaptationism (cf., Dennett 1996).

Against this background, it is crucial to note, and subsequently demonstrate, that for G&L just-so storytelling is a practice intrinsically linked to the theoretical foundations of adaptationism and its attempts to explain the historical origin of evolutionary phenomena. I shall argue that although the negative approach to just-so stories is inspired by G&L's critique of the adaptationist programme, G&L's own approach is, in fact, more specific as well as sophisticated, and is comparable to the abovementioned methodologically positive explanatory strategy, too.

Also, the dichotomy of positive and negative approaches to just-so storytelling, that I distinguish, cannot satisfactorily capture the argumentative core of G&L's critique because individual usages, rejections included, of the term "just-so story" presuppose specific notions of science and scientific explanation, or even specific theories of evolutionary principles as will become evident in the case of G&L's critique. For these reasons it is necessary to re-visit and re-frame the foundations of the anti-adaptationist position that G&L advanced.

6.4 The Adaptationist Programme and Idle Darwinizing

G&L were not the first nor the last who criticized the adaptive reasoning within evolutionary science. At present, it is a usual practice to recognize several modes of adaptationism.[11] G&L's critique of—to use again their own words—the adaptationist programme builds upon assumptions that have their own specific location in the history of evolutionary science. The epithet "just-so stories" is firstly employed by Lewontin (see "Sociobiology – A Caricature of Selection Theory" 1976a), in a lecture where he makes a single allusion to Kipling's famous title. Two years later, Gould seconds this association with an article he entitles "Sociobiology: The Art of Storytelling" (Gould 1978) in which he develops the comparison between just-so stories and evolutionary explanations in greater detail. It is undeniable that, at least initially, G&L attack a very specific target: sociobiological interpretations of evolutionary principles, and their predominantly adaptive explanations.

In 1979, G&L developed the well-known and intentionally methodical critique of adaptationism in their joint "Spandrels paper," where they with great vigor state the following:

> We call it the adaptationist programme, or the Panglossian paradigm. It is rooted in a notion popularized [...] toward the end of the nineteenth century: the near omnipotence of natural selection in forging organic design and fashioning the best among possible worlds. This programme regards natural selection as so powerful and the constraints upon it so few that direct production of adaptation through its operation becomes the primary cause of nearly all organic form, function, and behavior. (Gould and Lewontin 1979, 3–4)

The substantial quarrel G&L have with the adaptationist programme is thus of an ontological nature. In their view, the ontology of the adaptationist programme has two legs. The first leg is the commitment to a certain kind of atomism. The second leg is the belief in the supreme creative power of natural selection: although constraints upon the operation of natural selection as well as other non-adaptive mechanisms are recognized, their evolutionary importance is typically marginalized.

Therefore, according to G&L, the adaptationist programme does not analyze the organisms as "integrated wholes" but atomizes organismal forms, often arbitrarily, into individual phenotypic traits. Furthermore, the adaptationist programme presupposes that these individual traits are the optimal solutions to specific design problems, i.e., the origin of a trait is interpreted as the direct result of natural selection. Each phenotypic trait is supposed to have its unique adaptive function. Each phenotypic trait is, in fact, an adaptation. If a trait or its parts exhibit suboptimal solution(s) to a given selective problem(s), the adaptationist programme explains their evolution and existence as an unavoidable compromise between competing optimal solution(s) and design problem(s). "Nonoptimality is thereby rendered as a result of adaptation as well," (ibid., 1).

This is exactly the reason why G&L label the adaptationist programme as the Panglossian paradigm or sometimes as panselectionism—the labels are derived

[11] See Godfrey-Smith (2001); and Lewens (2009) for post-Spandrels typologies of adaptationism.

from Voltaire's fictional character of dr. Pangloss (who is supposed to be a philosophical caricature of Leibniz), the fictional character who praises designed fashion and function of all that there is, and is genuinely convinced that the world is the best of all possible worlds.

The ontology of the adaptationist programme is the bedrock for constructing specific types of evolutionary arguments and conclusions. This ontological starting point can be translated into the null hypothesis:

(0) Every single phenotypic trait of an organism was selected for, i.e., is an adaptation.

Gould and Lewontin systematize the reasoning of the adaptationist programme into four procedures (ibid., 5–6). I quote now:

(1) If one adaptive argument fails, try another.
(2) If one adaptive argument fails, assume that another must exist.
(3) In the absence of a good adaptive argument in the first place, attribute failure to imperfect understanding of where an organism lives and what it does.
(4) Emphasize immediate utility and exclude other attributes of form.

This is the anatomy of adaptive reasoning that is, according to G&L, inherent to the adaptationist programme. Of course, G&L do not want to suggest that evolutionists consciously embrace or defend such an ontological and methodological position. As John Beatty and Eric Desjardins perceptively note, their point was that "too many evolutionary biologists practice evolutionary biology as if it were true [...]," (Beatty and Desjardins 2009, 232). G&L reconstruct the adaptationist programme by extracting and exposing a few "common styles of arguments" (Gould and Lewontin 1979, 5). These styles of arguments—this kind of adaptive reasoning—leads straight to the telling of adaptive just-so stories.

The points (0), (1), (2) and (3) all mirror the underlying ontology of the adaptationist programme and indicate that its often non-explicit proponents propose the concept of adaptation as a default explanation for the historical origin of a phenotypic trait in question.[12] At the same time, these points prevent evolutionary thinkers from considering some alternative evolutionary mechanisms to natural selection and, in the most extreme cases, evolutionary theory is perceived as a synonym for the actions of natural selection (ibid., 8).[13] These tendencies are an essential part of the process within evolutionary science that Gould a little later starts labeling as "the hardening of the modern synthesis" (see especially Gould 1983, 2002). Such a narrow notion of biological evolution goes against the pluralistic view of evolutionary principles G&L advocate in the "Spandrels paper," and at other places (see Beatty 1984a, 1984b for comprehensive analysis).

[12] See Lloyd (2015) who rigorously analyses the post-Spandrels adaptationist "programme" (i.e., a different version of *adaptationism*) with regards to its ontology, null hypotheses etc.

[13] From historical perspective, this tendency overshadowed the alternative, yet still Darwinian, evolutionary theories that tried to modify our understanding of the ontological role of natural selection. See, e.g., Portmann (1964a, b).

G&L do not suggest that the concept of natural selection, ipso facto, adaptation is to be discarded, or even that alternative evolutionary principles are equally relevant in creating and perpetuating biological design in nature. In this sense G&L are, of course, Darwinians. But, at the same time, they want us to "cherish his consistent attitude of pluralism in attempting to explain Nature's complexity," (Gould and Lewontin 1979, 10).

The procedure (4) determines another extremely important theoretical and methodological dimension of the adaptationist programme. It presupposes a particular definition of "adaptation," which sets aside the problem of (evolutionary) history. For example, Walter Bock states: "An adaptation is, thus, a feature of the organism, which interacts operationally with some factor of its environment so that the individual survives and reproduces," (Bock 1979, 39).[14] This timeless definition of adaptation/adaptive trait is one of the main causes responsible for producing adaptive just-so stories.

The adaptationist programme axiomatically postulates a correspondence between the current function of a phenotypic trait and its original past function. Moreover, in G&L's eyes, the adaptationist programme explains the historical origin of phenotypic traits in the form of a loose hypothesis imagining and invoking the principles of natural selection. As a consequence, an adaptive story that a given trait evolved just in that way is created. The important point here is that, conceptually, the current function and utility is not distinguished from the original past function and utility, if any even existed. The term "adaptation" is then used interchangeably with the term "adaptedness," and vice versa.

Therefore, according to G&L, the adaptationist programme does not employ a suitably historical method. In other words, G&L point out that the adaptationist programme exploits an evolutionarily unwarranted inferential method. The adaptationist programme relies on an inferential method that is in contemporary philosophy of biology described as "reverse engineering." Robert C. Richardson defines the substance of reverse engineering as follows:

> Having assumed the "fact" of design, they "explain" the complex structures and behaviors we see as the consequence of natural selection, often without independent evidence. They do not argue for design but from form to function and then again from function to form. (Richardson 2007, 86)

Richardson's aim is ultimately not to argue against employing natural-selective schemes for explaining evolutionary phenomena, including, for example, the exclusive features of human beings and their psyche: "Human language and human reasoning are certainly complex capacities, and they are evolved capacities, which means they were not present, or not present to the same degree, in our ancestors. Given their complexity, we might also expect them to be, directly or indirectly, the consequence of selection," (Richardson 2007, 89). Nevertheless, without additional

[14] I choose this particular definition as it is used in Gould's and Vrba's famous paper "Exaptation—a missing term in the science of form" (1982) where they persuasively show how conceptual issues affect our understanding of evolutionary ontology.

evidential reasoning, natural-selective conclusions that in effect merely assume the optimal design are not robust explanations but at best possible adaptive hypotheses—they are adaptive just-so stories in G&L's vocabulary. To put it differently, they are not merely stories but speculatively adaptive just-so accounts that a priori assume the existence of a direct link between the historical origin of a given trait and its current function. Richardson is thus rather pointing out that the fact that particular natural-selective explanations need to be supported by other "independent" evidence.[15]

To summarize, G&L and consequently many others (e.g., Gould and Vrba 1982; Mayr 1988; Griffiths 1996, 1997; cf., Williams 1966) later pointed out that the insistence on such a broad and ahistorical delineation of the term "adaptation" brings, inter alia, two important and interrelated epistemological problems. First, it is necessary to contrast the adapted with the non-adapted, i.e., with traits that may be the products of alternative evolutionary mechanisms neutral in adaptive significance, or may even be the by-products (G&L's "spandrels") of an adaptation. Second, a reconstruction of the causal history of adaptation is always needed, as one cannot simply infer the original function of a trait from its current function—the current function can be a modification or transformation (Gould's and Vrba's "exaptation") of the original, past function.[16] Hence, as for instance Lynn H. Nelson (2003) elegantly shows, there is a considerable consensus among evolutionary theorists that evolutionary research, based on the premise that evolution is to a certain degree an adaptive evolution, requires historical knowledge, or, to be precise, some usage of

[15] Evolutionary psychology is very often perceived as the exemplification of unscientific, evolutionary enterprise precisely due to its excessive adaptationist commitments. One could say that very often evolutionary psychology exemplifies the inferential structure of adaptationist programme as defined above (cf., again Richardson 2007). However, consider Buller (2006) who wants us to distinguish between "evolutionary psychology" and "Evolutionary Psychology." Evolutionary Psychology is a specific mode of evolutionary enterprise that rests on the so-called massive modularity hypothesis. The massive modularity hypothesis goes hand in hand with a specific type of adaptationist reasoning, insomuch as the proponents of massive modularity, in fact, assume that the brain/mind consists of adaptations (Heyes 2018 alternatively speaks of "high church evolutionary psychology"). On the other hand, evolutionary psychology is, for Buller, a global interdisciplinary, theoretically, and methodologically pluralistic enterprise of explaining human behavior and psychology in evolutionary terms.

[16] Gould (1997) suggests that even the by-products of adaptive evolution, i.e., spandrels, can be subsequently coopted for biological utility. Spandrels, thus, can be *exapted* as well. In other words, by-products of primary adaptations can become secondarily adaptive. Gould believes that this is a significant fact for, inter alia, the general project of evolutionary psychology (see Gould 1991). Nevertheless, even from a global perspective, Gould warns us that we shall not impose *telos* into nature via relying on optimal, and in fact direct designs and solutions to selection problems. If the biological evolution is a historical process, "you cannot demonstrate evolution with perfection because perfection need not have a history. *After all, perfection of organic design had long been the favorite argument of creationists, who saw in consummate engineering the direct hand of a divine architect. A bird's wing, as an aerodynamic marvel, might have been created exactly as we find it today*," (Gould 1980, 28, emphasis mine). In fact, there is a wide consensus that feathers originally evolved to sustain thermoregulation and became exaptations for flight later.

individual historical reconstructions. "Adaptation is a historical concept," writes Eliot Sober (1993, 84).

To put it simply, one must make a plausible case that a trait in question is an adaptation for certain function(s) by exploiting evidentiary standards that justify the very evolutionary conclusion (a trait is/is not an adaptation),[17] which, simultaneously, implies exploiting justified, i.e., suitable methodological procedures, and inferences (is reverse engineering enough?; is there other independent/circumstantial evidence?, etc.). Therefore, in the best-case scenario, the evolutionary account is a *justified* account not a *just-so* account.

It is unquestionable that G&L recognized that the explanatory challenge for evolutionists is the historical nature of their subject matter, and they devoted much energy and many pages to debating this epistemological obstacle (see, e.g., again the brilliant paper by Beatty and Desjardins 2009). However, even in the much-debated "Spandrels paper," in which both thinkers in a sense generalize their previous individual views and attacks on adaptationism, one can find a crucial and rich paragraph:

> Often, evolutionists use consistency with natural selection as the sole criterion and consider their work done when they concoct a plausible story. But plausible stories can always be told. The key to historical research lies in devising criteria to identify proper explanations among the substantial set of plausible pathways to any modern result. (Gould and Lewontin 1979, 7)

As I read them, G&L point out three interrelated issues here. First, the natural-selective and speculative, although consistent and in the end possibly credible, hypotheses are often postulated not as rough-and-ready evolutionary accounts but as robust explanations. However, this robustness is guaranteed only via consistency with the principles of natural selection. Second, G&L claim that to generate a plausible adaptive story is a rather easy enterprise. "Since the range of adaptive stories is as wide as our minds are fertile, new stones can always be postulated," (ibid.) Third, they acknowledge that evolutionists must (1) put on historians' shoes, and thus produce specifically historical explanations; and must (2) face the fact that historical research is characterized by generating multiple possible explanations/stories.

Overall, the unwillingness of the adaptationist programme to pay attention to the specificity of historical inquiry and historical explanations is responsible for G&L's accusation that evolutionists provide us with mere stories instead of any kind of explanation. The usage of the term "story" is thus of a rhetorical rather than technical nature. In any case, G&L focus predominantly on the just-so aspect: they parse the speculatively adaptive reasoning of evolutionists of that time.

It is, then, necessary to acknowledge that G&L's critique of the adaptationist programme is not, at its core, a Popperian or positivistic critique of utilizing

[17] See Andrews et al. (2002) for biologically informed post-Spandrel discussion relating to testing and justification of adaptive accounts (cf., Griffiths 1996).

story-like structures to explain evolutionary phenomena (see Richardson 2007 for a similar reading).[18] It is instead a critique of a certain interpretation of biological evolution, in the broadest possible sense of the term and, secondarily, of specific methodological manners and epistemic maxims related to evolutionary inquiry qua historical inquiry.

Broadly speaking, G&L neither deny the possibility that storytelling per se can play a role in scientific and/or evolutionary research, nor do they primarily argue that adaptive just-so stories are unscientific merely because "stories" exemplify an unverifiable and/or unfalsifiable form of explanation. For instance, Lewontin at one point honestly admits that evolutionary thinkers can be "forced to the extreme adaptationist programe because the alternatives, although they are undoubtedly operative in many cases, are untestable in particular cases," (Lewontin 1978, 230).

The term "just-so story," it turns out, is to be conceived as a terminus technicus for the mis- and overuse of the concept of adaptation; or equivalently, as a terminus technicus for unwarranted reliance on natural-selective hypothesizing. Indeed, G&L use "just-so story" as a negative and derogatory term. Nevertheless, as I have indicated, it is certainly not true that they did not recognize that evolutionary science is historical science and that evolutionists need a special kind of explanation and/or explanatory strategies to intelligibly grasp historical events and entities. According to G&L, the adaptationist programme is not merely an unscientific enterprise. It is, in fact, much more: it is an exemplification of what Lewontin denotes as "idle Darwinizing" (Lewontin 1976b, 24) that oversimplifies the functioning of natural selection (paradoxically sometimes by overestimating its power) and, ipso facto, downplays any other evolutionary mechanisms, inclusive of the very important factor(s) of historical randomness.[19] That is the meaning of the just-so allegation that G&L raised against their evolutionary colleagues.

I argue that this interpretation of adaptive just-so storytelling can navigate us to new paths of the epistemological analysis of evolutionary theorizing qua an essentially historical enterprise. G&L tried to draw attention to these paths when they recognized that the key task for evolutionists is to "devise" explanatory criteria for historical research. As time went by, Gould gradually struggled to demonstrate why a special kind of evolutionary explanation—narrative explanation—is for evolutionary thinkers, qua natural historians, indispensable, and fully appreciated the "plurality of scientific modes" (Gould 2002, 1332ff). In *Wonderful Life* Gould claims:

---

[18] Lewontin, at least at the end of 1970s, explicitly admits that his critique of sociobiology is not based on a particular notion of science, Popperian included; he appeals to more general criteria of evaluation. Lewontin says: "I am not a Popperian and I do not believe that strict falsifiability is necessary for a valid scientific theory. There is, however, a more elementary demand. Even if a theory has no potential falsifier it must at least have the potential of failure of confirmation. It must be possible to construct an experiment or observation which would fail to confirm the theory in that instance," (1976b, 26–27).

[19] Gould, along the same line, very often speaks of "naive or simplistically fundamentalist ultra-Darwinism[…]." (1997, 10754).

> But historical science is not worse, more restricted, or less capable of achieving firm conclusions because experiment, prediction, and subsumption under invariant laws of nature do not represent its usual working methods. The sciences of history use a different mode of explanation, rooted in the comparative and observational richness of our data. We cannot see a past event directly, but science is usually based on inference, not unvarnished observation (you don't see electrons, gravity, or black holes either). (Gould 1989, 279)

Indeed, Gould's pluralistic view of science and his optimism in regard to scientific explanation of the past are in contrast to the more modest and rather skeptical views of Lewontin who, as Beatty and Desjardins put it, "had sarcastically, but pointedly suggested that while evolution by natural selection may be an historical process, nonetheless the relevant history may be unrecorded and unrecoverable [...]," (Beatty and Desjardins 2009; cf., Lewontin 1974, 1978, 1990, 1991). This tension between Gould and Lewontin is grounded in the fact that the possibility and the willingness to speak of scientific knowledge about the past depends on the very notion of science and scientific explanation, and additionally, in the fact that evolutionist like historiographers do not have direct epistemic access to the past.

6.5 Some (Just-So?) Stories Are Better Than Others

It is an open question whether contemporary evolutionary scientists and scholars should adhere to G&L's "original" meaning of adaptive just-so storytelling, or whether the term "just-so story" is to be perceived more broadly, or even as a term equivalent to "narrative explanation" (see Smith 2016a for this argument). As for the first option, I contend that G&L's analysis of adaptive reasoning is still a valid and valuable theoretical tool for evaluating evolutionary accounts, and can, thus, prevent evolutionists from the idle Darwinizing that is an inherent interpretative danger of any Darwinian theory of evolution. Put another way, G&L's analysis of a particular mode of adaptive reasoning can be used to evaluate and identify evolutionary just-so accounts that rely on speculatively adaptive hypothesizing.[20]

As for the second option, the tendency to use the term "just-so story" in a broad and, in fact, neutral manner is questionable due to its rather derogatory nature—one could even argue that exactly for this reason the term should be discarded as epistemically worthless (see, e.g., Love 2003). Moreover, historically-minded scientists and scholars who conceive narratives as indispensable for evolutionary inquiry and explanation would defend the narrative, story part of the expression "just-so story,"

[20] See, for instance, Dutilh Novaes (2018) and her meticulous evaluation of the adaptive explanation concerning human rationality submitted by Mercier and Sperber in their *The Enigma of Reason* (2017). Dutilh Novaes in my view rightly assesses that *The Enigma of Reason* manifests the methodological flaws that stem from unwarranted adaptationist commitments and that substantially mirror G&L's diagnosis of idle Darwinizing. Furthermore, compare it again with the recently published, but already widely discussed, paper "Is Evolutionary Psychology Possible?" written by Subrena E. Smith (2020), which also challenges the hyper-adaptationist mode of a specific version of evolutionary psychology.

while intentionally trying to avoid the speculative (and sometimes speculatively adaptive) just-so part.

For these reasons, I suggest that a just-so story is to be perceived as an implausible narrative explanation as it relies on just-so adaptive reasoning. This is to say that just-so stories are a subset of a set of implausible narrative explanations. Yet another option is to conceive them as implausible how-possibly explanations in Brandon's sense (see Brandon 1990; cf., Pearson 2018).

From such a theoretical shift both sides could eventually benefit. G&L's analysis of adaptationism can still be employed to identify poorly construed evolutionary (just-so) narratives, and, at the same time, scientists and scholars can gain a theoretical tool for the endeavor to devise general and more rigorous criteria suitable for evolutionary inquiry qua historical/narrative inquiry. As I also already pointed out in Chap. 4, Roth (2020, 75) claims: "There will be factors both internal to a narrative explanation (assessing the sequencing) and external to it (comparison with competing narratives, if any), I suggest that evaluation in the end can only be on a case-by-case basis. The fact that narratives cannot be expected to aggregate will be a limiting factor; the extent that the events explained have been standardized, at least to some extent, will abet comparative evaluation."

I believe that G&L's articulation of the adaptationist programme clearly and convincingly shows that there will always be a possibility to identify poorly construed adaptive, "just-so" accounts that would exemplify what Lewontin denotes as idle Darwinizing. That is to say, G&L show a crucial way of evaluation concerning the internal factors of evolutionary narratives qua historical narratives.

Besides that, I argue that in this particular sense, G&L were wrong when they stated that "plausible stories can always be told." Their own critique of adaptationism illustrates that some stories are better than others; they indirectly showed that it is not easy work to generate a plausible adaptive evolutionary story. Nonetheless, further interdisciplinary evolutionary research and theoretical analysis at the intersection of philosophy of biology and history is needed in order to examine the connection between just-so speculation and the narrative/story-like structure of historical explanation.

References

Andrews, Paul W., Steven W. Gangestad, and Daniel Matthews. 2002. Adaptationism—How to carry out an exaptationist program. *Behavioral and Brain Sciences* 25 (4): 489–504.

Barash, David P., and Judith Eve Lipton. 2010. *How women got their curves and other just-so stories: Evolutionary enigmas*. New York: Columbia University Press.

Beatty, John. 1984a. Chance and natural selection. *Philosophy of Science* 51 (2): 183–211.

Beatty, John. 1984b. Pluralism and panselectionism. In *PSA: Proceedings of the Biennial Meeting of the Philosophy of Science Association*, ed. Paul D. Asquith and Philip Kitcher, vol. 2, 113–128. East Lansing: Philosophy of Science Association.

Beatty, John, and Isabel Carrera. 2011. When what had to happen was not bound to happen: History, chance, narrative, evolution. *Journal of the Philosophy of History* 5 (3): 471–495.

Beatty, John, and Eric C. Desjardins. 2009. Natural selection and history. *Biology & Philosophy* 24 (2): 231–246.

Bock, Walter J. 1979. The synthetic explanation of macroevolutionary change—A reductionist approach. *Bulletin of Carnegie Museum of Natural History* 13:20–69.

Bonnin, Thomas. 2019. Evidential reasoning in historical sciences: Applying Toulmin schemes to the case of Archezoa. *Biology & Philosophy* 34 (2): 30.

Brandon, Robert N. 1990. *Adaptation and environment*. Princeton: Princeton University Press.

Buller, David J. 2006. *Adapting minds: Evolutionary psychology and the persistent quest for human nature*. Cambridge: MIT Press.

Buss, David M., Martie G. Haselton, Todd K. Shackelford, April L. Bleske, and Jerome C. Wakefield. 1998. Adaptations, exaptations, and spandrels. *American Psychologist* 53 (5): 533–548.

Carroll, Joseph, et al. 2017. A cross-disciplinary survey of beliefs about human nature, culture, and science. *Evolutionary Studies in Imaginative Culture* 1 (1): 1–32.

Cleland, Carol E. 2011. Prediction and explanation in historical natural science. *The British Journal for the Philosophy of Science* 62 (3): 551–582.

Currie, Adrian, and Kim Sterelny. 2017. In defence of story-telling. *Studies in History and Philosophy of Science* 62:14–21.

Currie, Adrian, and Derek Turner. 2016. Introduction: Scientific knowledge of the deep past. *Studies in History and Philosophy of Science A* 55:43–46.

Danto, Arthur C. 1986. Art, evolution, and the consciousness of history. *The Journal of Aesthetics and Art Criticism* 44 (3): 223–233.

Darwin, Charles. 1889. *The descent of man and selection in relation to sex*. New York: D. Appleton. (Originally published 1971).

Dennett, Daniel C. 1986. Julian Jaynes's software archeology. *Canadian Psychology / Psychologie canadienne* 27 (2): 149–154.

Dennett, Daniel C. 1996. *Darwin's dangerous idea: Evolution and the meanings of life*. New York: Simon & Schuster.

Desjardins, Eric. 2011. Historicity and experimental evolution. *Biology & Philosophy* 26 (3): 339–364.

Eldredge, Niles. 1983. À la recherche du docteur Pangloss. *Behavioral and Brain Sciences* 6 (3): 361–362.

Godfrey-Smith, Peter. 2001. Three kinds of adaptationism. In *Adaptationism and optimality*, ed. Steven H. Orzack and Elliott Sober, 335–357. Cambridge: Cambridge University Press.

Goldstein, Sam, and Russell A. Barkley. 1998. ADHD, hunting, and evolution: 'Just so' stories. *ADHD Report* 6 (5): 1–4.

Goodenough, Ward H. 1976. On the origin of matrilineal clans: A 'just so' story. *Proceedings of the American Philosophical Society* 120 (1): 21–36.

Gottschall, Jonathan. 2008. *Literature, science, and a new humanities*. New York: Palgrave Macmillan.

Gould, Stephen Jay. 1978. Sociobiology: The art of storytelling. *New Scientist* 80 (1129): 530–533.

Gould, Stephen Jay. 1980. *The panda's thumb*. New York: W.W. Norton.

Gould, Stephen Jay. 1983. The hardening of the modern synthesis. In *Dimensions of Darwinism: Themes and counterthemes in twentieth-century evolutionary theory*, ed. Marjorie Grene, 71–93. Cambridge: Cambridge University Press.

Gould, Stephen Jay. 1989. *Wonderful life*. New York: W.W. Norton.

Gould, Stephen Jay. 1991. Exaptation: A crucial tool for an evolutionary psychology. *Journal of Social Issues* 47 (3): 43–65.

Gould, Stephen Jay. 1997. The exaptive excellence of spandrels as a term and prototype. *Proceedings of the National Academy of Sciences* 94 (20): 10750–10755.

Gould, Stephen Jay. 2002. *The structure of evolutionary theory*. Cambridge: Harvard University Press.

Gould, Stephen Jay, and Richard C. Lewontin. 1979. The spandrels of San Marco and the Panglossian paradigm: A critique of the adaptationist programme. *Proceedings of the Royal Society of London B: Biological Sciences* 205 (1161): 581–598.

Gould, Stephen Jay, and Elisabeth S. Vrba. 1982. Exaptation—A missing term in the science of form. *Paleobiology* 8 (1): 4–15.

Griffiths, Paul E. 1996. The historical turn in the study of adaptation. *The British Journal for the Philosophy of Science* 47 (4): 511–532.

Griffiths, Paul E. 1997. *What emotions really are: The problem of psychological categories*. Chicago: University of Chicago Press.

Heyes, Cecilia. 2018. *Cognitive gadgets: The cultural evolution of thinking*. Cambridge: Harvard University Press.

Hull, David. 1975. Central subjects and historical narratives. *History and Theory* 14 (3): 253–274.

Kipling, Rudyard. 1902. *Just so stories for little children*. London: Macmillan.

Kuukkanen, Jouni-Matti. 2015. *Postnarrativist philosophy of historiography*. New York: Palgrave Macmillan.

Levy, Gabriel. 2013. The prospects and pitfalls of 'just-so' storytelling in evolutionary accounts of religion. *Method & Theory in the Study of Religion* 25 (4–5): 451–459.

Lewens, Tim. 2009. Seven types of adaptationism. *Biology & Philosophy* 24 (2): 161–182.

Lewontin, Richard C. 1974. *The genetic basis of evolutionary change*. New York: Columbia University Press.

Lewontin, Richard C. 1976a. Sociobiology—A caricature of Darwinism. In *PSA: Proceedings of the Biennial Meeting of the Philosophy of Science Association*, ed. Frederick Suppe and Paul D. Asquith, vol. 2, 22–31. East Lansing: Philosophy of Science Association.

Lewontin, Richard C. 1976b. Sociobiology—A caricature of selection theory. In *PSA: Proceedings of the Biennial Meeting of the Philosophy of Science Association*. East Lansing: Philosophy of Science Association. https://www.matierevolution.fr/spip.php?article286.

Lewontin, Richard C. 1978. Adaptation. *Scientific American* 239 (3): 212–230.

Lewontin, Richard C. 1979. Sociobiology as an adaptationist program. *Behavioral Science* 24 (1): 5–14.

Lewontin, Richard C. 1990. The evolution of cognition. In *Thinking: An invitation to cognitive science*, ed. Daniel N. Osherson and Edward E. Smith, 229–246. Cambridge: MIT Press.

Lewontin, Richard C. 1991. Facts and the factitious in natural sciences. *Critical Inquiry* 18 (1): 140–153.

Lloyd, Elisabeth A. 2015. Adaptationism and the logic of research questions: How to think clearly about evolutionary causes. *Biological Theory* 10 (4): 343–362.

Losos, Jonathan B. 2017. *Improbable destinies: Fate, chance, and the future of evolution*. New York: Penguin.

Love, Glen A. 2003. *Practical ecocriticism: Literature, biology, and the environment*. Charlottesville: University of Virginia Press.

Machery, Edouard. 2008. A plea for human nature. *Philosophical Psychology* 21 (3): 321–329.

Mayr, Ernst. 1988. *Toward a new philosophy of biology: Observations of an evolutionist*. Cambridge: Harvard University Press.

Mayr, Ernst. 2000. Darwin's influence on modern thought. *Scientific American* 283:78–83.

Mercier, Hugo, and Dan Sperber. 2017. *The enigma of reason*. Cambridge: Harvard University Press.

Morgan, Mary S., and M. Norton Wise. 2017. Narrative science and narrative knowing. Introduction to special issue on narrative science. *Studies in History and Philosophy of Science Part A* 62:1–5.

Nelson, Lynn Hankinson. 2003. The descent of evolutionary explanations: Darwinian vestiges in the social sciences. In *The Blackwell guide to the philosophy of social science*, ed. Paul A. Roth and Stephen Turner, 258–290. Oxford: Blackwell Press.

Novaes, Catarina Dutilh. 2018. The enduring enigma of reason. *Mind & Language* 33 (5): 513–524.

Oman, Charles M. 2012. Are evolutionary hypotheses for motion sickness 'just-so' stories? *Journal of Vestibular Research* 22 (2): 117–127.

Pearson, Cassandra. 2018. How-possibly explanation in biology: Lessons from Wilhelm His's 'simple experiments' models. *Philosophy, Theory, and Practice in Biology* 10:1–15.
Portmann, Adolf. 1964a. *Animals as social beings*. Trans. Olive W. Coburn. New York: Harper & Row.
Portmann, Adolf. 1964b. *New paths in biology*. Trans. Ann J. Pomerans. New York: Harper & Row.
Richardson, Robert C. 2007. *Evolutionary psychology as maladapted psychology*. Cambridge: MIT Press.
Rosenberg, Alexander. 2002. Reductionism in an historical science. In *Promises and limits of reductionism in the biomedical sciences*, ed. M. H. Van Regenmortel and David L. Hull. Hoboken: John Wiley.
Rosenberg, Alexander. 2005. Lessons from biology for philosophy of the human sciences. *Philosophy of the Social Sciences* 35 (1): 3–19.
Rosenberg, Alexander. 2018. *How history gets things wrong: The neuroscience of our addiction to stories*. Cambridge: MIT Press.
Rosenberg, Alexander, and Daniel W. McShea. 2008. *Philosophy of biology: A contemporary introduction*. New York: Routledge.
Roth, Paul A. 2017. Essentially narrative explanations. *Studies in History and Philosophy of Science Part A* 62:42–50.
Roth, Paul A. 2020. *The philosophical structure of historical explanation*. Evanston: Northwestern University Press.
Segerstrale, Ullica. 2000. *Defenders of the truth: The battle for science in the sociobiology debate and beyond*. Oxford: Oxford University Press.
Sesardic, Neven. 2003. Evolution of human jealousy: A just-so story or a just-so criticism? *The Annals of the American Academy of Political and Social Science* 33 (4): 427–443.
Smith, Robert J. 2016a. Explanations for adaptations, just-so stories, and limitations on evidence in evolutionary biology. *Evolutionary Anthropology: Issues, News, and Reviews* 25 (6): 276–287.
Smith, Robert J. 2016b. Freud and evolutionary anthropology's first just-so story. *Evolutionary Anthropology: Issues, News, and Reviews* 25 (2): 50–53.
Smith, Sarah E. 2020. Is evolutionary psychology possible? *Biological Theory* 15 (1): 39–49.
Sober, Elliott. 1993. *Philosophy of biology*. Boulder: Westview Press.
Williams, George C. 1966. *Adaptation and selection*. Princeton: Princeton University Press.
Wilson, David Sloan. 2015. The spandrels of San Marco revisited: An interview with Richard C. Lewontin. *The Evolution Institute*. https://evolution-institute.org/the-spandrels-of-san-marco-revisited-an-interview-with-richard-c-lewontin/
Wittgenstein, Ludwig. 1975. *Philosophical remarks*, ed. Brian Blackwell and Rush Rhees. Trans. Raymond Hargreaves and Roger White. Oxford: Blackwell.
Woese, Carl R. 1980. Just so stories and rube Goldberg machines: Speculations on the origin of the protein synthetic machinery. In *Ribosomes: Structure, function and genetics*, ed. G. Chambliss et al., 357–373. Baltimore: University Park Press.
Wong, Timothy W. 2019. The evolutionary contingency thesis and evolutionary idiosyncrasies. *Biology & Philosophy* 34 (2): 22.

Chapter 7
A Pragmatist Gambit

Abstract This concluding chapter not only summarizes the content of the main chapters, but also makes explicit how naturalism defended here relates to other key isms (especially historicism, pragmatism, irrealism, and pluralism), as well as how it transforms our understanding of the human epistemic condition. I specifically argue that if naturalism is conceived as a methodologically and metaphysically opportunistic epistemological project, it is not set in opposition to what is often called "historicism," meaning the approaches that operationalize histories and/or genealogies when explaining socio-cultural practices or processes. Also, towards the end, I reflect on whether such a naturalism has any practical force (for doing philosophy). This ultimately leads me to embrace the notion of "conversational pragmatism," originally introduced by Rorty.

Keywords Naturalism · Historicism · Pragmatism · Pluralism · Human condition

> History cannot escape its own process. It will, therefore, always be rewritten. As the new present arises, the past is the past of a different present. Judgment in which emphasis falls upon the historic or temporal phase of redetermination of unsettled situations is thus a culminating evidence that judgment is not a bare enunciation of what already exists but is itself an existential requalification. That the requalifications that are made from time to time are subject to the conditions that all authentic inquiry has to meet goes without saying.
>
> —John Dewey[1]

> "So you are saying that human agreement decides what is true and what is false?" – It is what human beings say that is true and false; and they agree in the *language* they use. That is not agreement in opinions but in form of life.
>
> —Ludwig Wittgenstein[2]

In this book, I have attempted to build a case for specific understanding of naturalism. Let me now quickly go over the key points I put forward in the first three

[1] (Dewey 1938, 239).

[2] (Wittgenstein 1986, 88).

M. Hubálek, *Dissolving Naturalism and Historicism Into Each Other*, Studies in History and Philosophy of Science 67,
https://doi.org/10.1007/978-3-032-07535-2_7

chapters (i.e., Chaps. 2, 3, and 4). In Chap. 2, I re-visited Quine's philosophical system, which I consider to be both a viable source of inspiration for contemporary naturalistic philosophers, and a source of confusion for contemporary understanding of naturalism qua ism.

My contribution to a clearer and more substantial understanding of Quine's system lies in a focused analysis of his conception of science. My main claim in Chap. 2 is that, for Quine, naturalism is a deliberately chosen trademark for re-configuring the relation between what is commonly called "science" to "epistemology" (traditionally viewed as a branch of philosophy), and also to "rationality" as such. As I have already stated, according to Quine, our common sense, science, and rationality itself comprise one cognitive and epistemic continuum. They are all part of our social and cultural evolution, part of the development of our web of belief—our total theory of the world. I argue that, in this particular regard, naturalism is a re-assessment of the *condition humaine*.

The pivotal motive of this re-assessment is obviously a broadly construed Darwinism, i.e., a special scientific explanation of us as beings with a natural history (cf. again with Chap. 3, and Sect. 4.3). However, I intentionally put emphasis on the phrase "broadly construed." To avoid potential misunderstandings, consider also this helpful minimalistic outline of Darwinism, advanced by Rorty:

> By "Darwinism" I mean a story about humans as animals with special organs and abilities (e.g., certain features of the human throat, hand, and brain that let humans coordinate their actions by batting marks and noises back and forth). According to this story, these organs and abilities have a lot to do with who we are and what we want [...]. (Rorty 1998, 47–48)
> Darwinism requires that we think of what we do and are as continuous with what amoebas, spiders, and squirrels do and are. (ibid., 295)

It comes unsurprisingly then that the traditionally recognized intellectual disciplines (from historiography to natural science, and even inclusive of philosophy) are considered by Quine to be grounded historically. It cannot be otherwise, when one takes seriously the assumption that individual intellectual disciplines are institutionalized social/community/linguistic practices. The conception of science Quine works with is thus fundamentally holistic and dialectical; it is empirical inside out. Mariana Szapuová summarizes this point suitably in the following way:

> As Quine proposes, we should look at the relationship between science and empirical data from a naturalist perspective as "an input-output relation within flesh-and-blood denizens of an antecedently acknowledged external world" [Quine 1992, 19] and inquiries into knowledge should focus on the research of the people who are doing the inquiring. Here empirical inquiry becomes relevant as it shows that "seeking the truth" cannot be separated from human needs, interests, emotions, or even prejudice and bias, which is good reason to reject the ideals of objectivity and neutrality. (Szapuová 2021, 40)[3]

[3] By "objectivity" and "neutrality" Szapuová here obviously means *something more* than "intersubjectivity," i.e., a social account of objectivity, if you will. Helen E. Longino offers a handy minimalistic characterization of objectivity defined in social terms: "Social interactions determine what values remain encoded in inquiry and which are eliminated, and thus which values remain encoded in the theories and propositions taken as expressing scientific knowledge at any given time. Values are not incompatible with objectivity, but objectivity is analyzed as a function of community prac-

In this view, naturalism is a methodologically and metaphysically opportunistic epistemological project. Naturalism so understood has strong pragmatist flavors, and I tried to spotlight this fact several times in this book (cf., also Quine 1981). At the very least, Quine himself understood very well that the standards of various scientific practices of organizing knowledge/beliefs are always tied to related, specific purposes. This indicates clearly that naturalized epistemology can have (in fact must have) its normative dimension; the acknowledgment of specific purposes is what gives the sciences their practical force. And Quine's talk of empirical checkpoints (or, alternatively, of empirical/observable/behavioral consequences) confining ultimately our theories (or alternatively beliefs/hypotheses) taken as a whole unit could be elegantly squared with what is sometimes called the "pragmatic theory of truth."

However, to properly adjudicate how Quine's philosophical system as such relates to the tradition of American pragmatism, as well as to its contemporary form, often dubbed "neo-pragmatism," is an assignment to be completed on a different occasion.[4]

At any rate, I contend that to perceive naturalism as methodologically and metaphysically opportunistic makes explicit the tension between the underlying cognitive/epistemic continuum of all human inquiry and the ordinary disciplinary/methodological pluralism parasitizing on this continuum. In addition, such a perception does not set naturalism in opposition to what is often called "historicism" (meaning the approaches operationalizing *histories/genealogies* when explaining the sociocultural practices or processes). All of our intellectual outputs (e.g., beliefs, hypotheses, theories) are likewise free to be scrutinized from *inside out*, inter alia, since the context of discovery cannot be unambiguously detached from the context of justification.

I have also revealed (in Sect. 2.3) that Quine himself tried to address the epistemic issues related to historical research itself. However, his solutions and/or suggestions will, I reckon, leave the vast majority of practicing historians unmoved. It is quite understandable. Quine's solutions and suggestions related to answering historical what- and why-questions are still too broad, as well as clothed in unusual language. For some historians, it might be problematic to think of the historical inquiry as consisting of devising "theories," relying indirectly on "predictions," or forming "hypotheses." Now, my aim was not to argue that one should take on board and cling to everything Quine says about writing history. It is, I believe, rather the

tices rather than as an attitude of individual researchers towards their material or a relation between representation and represented," (Longino 1990, 216).

[4] Dewey, for instance, offers the following characterization of the pragmatic theory of truth: "Naturally, the pragmatist claims his theory to be true in the pragmatic sense of truth; it works, it clears up difficulties, removes obscurities, puts individuals into more experimental, less dogmatic, and less arbitrarily sceptical relations to life; aligns philosophy with scientific method; does away with self-made problems of epistemology; clarifies and reorganizes logical theory, etc.," (Dewey 1985, 9). I am quite confident that Quine would subscribe to such a characterization.

whole naturalistic framework Quine gradually establishes that should be taken more subtly and seriously.

The pivotal lesson is that there is no need for an a priori (ahistorical) demarcation criterion of science and scientific explanation. Thus, naturalism qua ism does not give us any specific (let alone timeless) solutions to our practical/technological and intellectual goals. It is the therapy one no longer needs when appreciating that there is no extra-scientific perspective, and that specific inquiries, inclusive of their normative implications, must be conducted and confronted from an intra-scientific perspective.

In other words, *epistemology naturalized* denies that there could be some theory-free (i.e., assumption-free) vantage point. As I read Quine, he continuously struggles to argue that there are no epistemic certainties nor unconceptualized givens. There is just us—epistemic agents. And we are doing anything *humanly* possible to produce and justify knowledge claims (and any other claims for that matter) as a going concern.

And this is, by the way, again an argument supported by our actual (scientific) practices. In 1991, Quine published "Two Dogmas in Retrospect," in which he once more returns to the problem of analytic truths. He, of course, still holds that, although the notion of analyticity has some "obvious" and "useful" functions, they are all "epistemologically insignificant." The very notion of semantic definition is dynamic in Quine's account. He gives this example: "When in relativity theory momentum is found to be not quite proportional to velocity, despite its original definition as mass times velocity, there is no flurry over redefinition or contradiction in terms, and I don't think there should be. The definition served its purpose in introducing a word for subsequent use, and the word was thereafter ours to use in the evolving theory, with no lingering commitments. Definition is episodic. Mostly in natural science we are not even favored with definitions, much less bound by them. New terms are just introduced by partial descriptions: electrons, neutrinos, quarks," (Quine 1991, 271, emphasis mine). Nevertheless, I still think that Quine's own attempts (however awkward sometimes) to include historiography into our science (understood as our total world theory) are in numerous ways instructive. I have thus paid a lot of attention to Quine's procedures on this matter (in Sect. 2.3).

To cut a long story short here, Quine tries in a systematic (and quite stubborn) way to illustrate how one can accommodate theories/hypotheses with *indirect* empirical consequences into our scientific world-picture. Put another way, how one can rationalize beliefs whose connection to direct observation is minimal or seemingly non-existent.[5]

It is at this point that Quine efficiently highlights that even our wildest guesses (speculations, hypotheses), for instance, concerning particular historical events, can

[5] Consider in this context also this passage: "What now of those parts of mathematics that share no empirical meaning, because of never getting applied in natural science? What of the higher reaches of set theory? We see them as meaningful because they are couched in the same grammar and vocabulary that generate the applied parts of mathematics. We are just sparing ourselves the unnatural gerrymandering of grammar that would be needed to exclude them," (Quine 1992, 94).

in principle always be rationally guided, ipso facto, bounded. Beyond our innate capabilities and Occam's razor-like principles, our web of belief (our total world-picture) as such always bounds us, whether as laymen or scientists. We are always a part of a certain epistemic and linguistic culture/community, relying on all kinds of standards. In other words, our beliefs are always entrenched in many other (background) beliefs.[6]

Again, one of the many important consequences, and I claim advantages as well, of this epistemic outlook is that our evaluative practices are also naturalized. So they are by the same token somewhat universalized, as our evaluative practices manifest our human epistemic condition—the rhizome from which our special sciences and their specific goals grow. Quine and Ullian (1978, 133), for instance, at one point pose the following question: "What course could one hope to find that would guarantee being always right? What makes the best strategy reasonable is that it does cause us to be right – our beliefs to be true – more generally than alternative strategies do. When the evidence available points the wrong way, away from the truth, the reasonable person is wrong along with it." And then they conclude:

> To learn to distinguish the plausible from the implausible is to develop one part of wisdom; it leads as well as anything can toward true belief. But wisdom's better part bids us to remain aware that we have less than the whole truth about even those matters we understand best. Such awareness can never be misplaced, since "the whole truth" about anything is but a fanciful ideal. (ibid.)[7]

However, I must note over again here that Quine himself never purposely elaborates any of the above-mentioned tenets with respect to actual historiographical practice. As a philosopher, he had general and theoretical goals, after all. That being said, Quine presents us with a naturalistic view of science (total theory of the world) that works with historiography as being unproblematically united with all human inquiry.

This is one of the reasons why the work of Roth on the concept of historical explanation (Chap. 4) fills the lacuna of our naturalistic understanding of *historical significance*. Alternatively, describing the situation from the other side, one could also say that his work fills the lacuna of our contemporary understanding of the scientific status of historiography, by paying closer attention to actual historiographical texts and practices. Put another way, even though Roth primarily addresses scholars interested in the theory and philosophy of history/historiography, his work

[6] Compare with this remark: "To maintain our beliefs properly even for home consumption we must attend closely to how they are supported. *A healthy garden of beliefs requires well-nourished roots and tireless pruning*. When we want to get a belief of ours to flourish in someone else's garden, the question of support is doubled: we have to consider first what support sufficed for it at home and then how much of the same is ready for it in the new setting," (Quine and Ullian 1978, 126–127, emphasis mine).

[7] Quine and Ullian also insist that: "The line between evaluation and belief is not sharp, and neither is that between means and ends. *Yet both distinctions can be worth making*. The more an evaluation is seen as an evaluation of some end, the greater is its contrast with belief. Think of aesthetic matters. There is no disputing about tastes: no disputing in the sense in which one can dispute beliefs. There is only a training of tastes," (ibid., 135, emphasis mine).

offers practicing historians a way of construing historiography as a proper scientific inquiry. But again in two complementary senses.

Since Roth takes on board all the naturalistic tenets I find in Quine, Roth also thinks that: "at the most general epistemological and metaphysical levels, no principled distinction emerges between empirical knowledge generally and knowledge of the past. The forms of inference required to have empirical knowledge at all – inductive, abductive, and deductive – arise for all such cases of knowledge." Historical inquiry is a regular form of human inquiry understood naturalistically; historiography is a science (see again Roth 2020, 61, 145ff). That is the first sense. To explain it with reference to one of my examples and distinctions, according to Roth, no principled distinction (i.e., no distinction that would categorically matter in practice) exists between answering a what-question (inquiring in the matters of fact) and answering a why-question (providing an explanation of given fact[s]) with respect to present happenings and past happenings.

It is also worth remembering that to have some means of answering a question in the form of "what happened?" (or when/where it happened) is sometimes enough, and it certainly sometimes has a historical significance in itself, without asking the (often additional) why-question (why something happened). To ask whether or not Vincent van Gogh committed suicide is different from asking why he committed suicide (see again Chap. 5).

Then, one might wonder why Roth makes a case for historical irrealism and not just irrealism. That is because Roth wants to point out that there is something peculiar about answering historical why-questions; about historical explanations. That is the second sense in which historiography (and any other historical science) is a special science in the traditional disciplinary sense. Roth recognizes historical inquiry (or at least one of its indispensable parts) as having a distinct "explanatory form" (Roth 2020, 145). Why do historical scientists need a special explanatory form? Following Danto and Mink, Roth argues that historians often deal with temporalities by examining/explaining "narrative sentences."

For example, "The Thirty Years War began in 1618." or "The Holocaust began in 1933."[8] Put another way, historians deal with retrospective descriptions (hindsights) having a narrative structure (beginning-middle-end structure). And, to put it crudely, historians need a temporal language/perspective to account for these kinds of truths. In this respect, *hindsight trumps foresight*. According to Roth, historical explanation is in effect narrative explanation which "concerns itself with a developmental process that emerges only in retrospect, (ibid., 20)." Of course, not every developmental process (or simply temporal change) is historical. Historical processes are not predetermined. Thus, one is entitled to speak of the historical relationship between two particular events when the relationship is not causal in the strict sense (meaning here that the relationship in question is not a direct result of laws or law-like

[8] However, Roth is not arguing, e.g., that historians generally formulate their explananda on purpose as narrative sentences. I read him as saying that, *very often*, one can reformulate historical explananda (historical why-questions) as narrative sentences.

determinants).[9] Historical relation is viewed in this way rather as *genealogical* relation. Consider this potent passage Roth finds in Hayden White, in which White writes à la Danto that:

> A given historical event can be viewed as the fulfillment of an earlier and apparently utterly unconnected event when the agents responsible for the occurrence of the later event link it "genealogically" to the earlier one. The linkage between historical events of this kind is neither causal nor genetic. […] The relationships between earlier and the later phenomena are purely retrospective, consisting of the decisions on the parts of a number of historical agents. […] The linkage is established from the point in time experienced as a present to a past, not, as in genetic relationships, from the past to the present. (White 1999, reconstructed by Roth 2020, 146)

This passage is also telling, as White perspicuously shows how our conceptual and other commitments co-create our retrospective insights; our narrative explanations. Roth shares with White this "constructivist" assumption, as they both, although each in his own way, emphasize the non-trivial freedom we have in our choice of vocabularies and modes of description by which we interconnect historical events. On the other hand, logically speaking, narrative explanation structurally still resembles "ordinary" abductive inference (alternatively, inference to the best explanation); one is now faced with an unexplained end-state: an event *E*. If postulating a past event *B* would render the present event *E* as a matter of course, then one has a reason for postulating the event *B*.

Furthermore, this kind of metaphysical and methodological freedom is not only consistent with the naturalism I defend in this book, but even one of its *terminus a quo*.

In Sect. 4.2.2, plus Sect. 4.3 and Sect. 4.4, I additionally attempt to demonstrate how any excessively relativistic interpretations of Roth's historical irrealism can be easily dismissed as non-naturalistic. That is, as ignoring the actual garden variety of practices, paradigms, and commonsensical presuppositions of historiography and other relevant special sciences, when debating particular subject-matters and/or explanations. Put in Quine's language, as ignoring our web of belief—our total theory of the world—that always serves us as a ground for all reasoning as well as evaluation.

Roth himself advocates openly for strictly comparative means of evaluating historical explanations. The underlying reason for this is, of course, a form of non-representationalism he subscribes to (see again Sect. 4.3). And it is this maneuver that makes Roth also a pragmatist theorist (at least in my book). As he articulates: "evaluating explanations that have narrative form essentially […] will primarily be a function of assessing competing explanations, and so draw on evaluative criteria more akin to theory appraisal than to hypothesis confirmation," (Roth 2020, 66).

[9] So, unless we find a way of foreseeing/predicting "historical events," e.g., predicting in 1618 that The Thirty Years War just started, there will always be a need for historical/narrative explanations. I put aside the question whether in that case we would actually still speak of history, insomuch as this kind of foresight would also presuppose the possibility to theoretically standardize events such as "war" in some stable way.

Again crudely, facing all the underdeterminations we, the epistemic and cultural agents, decide what will count as a better or worse explanation against our particular goals, values, and interests, as well as with respect to particular explananda.

I thus also think that Roth's treatment of historical significance suitably displays that naturalism is, in effect, possible to see as a pluralist approach to all forms of explication and explanation. That is why I examine the practical force of the naturalism defended here in this book also "indirectly" via two independent case studies (Chaps. 5 and 6). So, although these case studies have several aims on their own, they serve me also as practical illustrations and applications of all the announced naturalistic attitudes and viewpoints.

Last but not least, I am of the view that Roth offers a compelling case for perceiving the well-known attempt to historicize science (ipso facto rationality itself) by Kuhn as exemplifying the utilization of historical/narrative explanation (Sect. 4.2.3). If one accepts this, historical explanation is indispensable for accounting what counts and what should count as the rational standards within a certain epistemic community. As I already wrote, one needs to narratively describe the transition, within a given mature science, from normal science *S1* to normal science *S2*. In other words, to put forward a developmental sequence.

This is not just another argument for acknowledging that "reason/science has a history." It is, in addition, a reminder that one must employ essentially historical means (explanatory form) to understand this history. So, as I say together with Roth: Roth naturalizes simultaneously both the concept of science and the concept of historical explanation. Without retrospective histories of scientific evolution, one cannot make sense of a (normal) science. "Understanding Kuhn's work as a narrative naturalizes narrative explanation through a form of mutual containment—since narrative helps constitute what counts as normal science, narrative cannot be separated from an understanding of what science is," (Roth 2020, 162).

A particular science (e.g., astronomy or chemistry) is ultimately distinguished from a non-science (or a pseudo-science, e.g., astrology or alchemy) by its particular history, i.e., by a particular synchronic understanding of diachronically established human practices and theoretical paradigms. Only there, within this dialectical process, can we find epistemic and other related standards of and for our reasoning (cf., Roth 2020, 19). And, additionally, this is how naturalism in epistemology accounts for itself. Eschewing apriorism leads to the recognition that our epistemic concepts are historically dynamic; they are historically coopted norms of scientific inquiry that were deemed useful and valuable by a certain community at a certain time. Naturalized epistemology is, as Quine writes, simply science self-applied. We can do no better than this and, to use again one of Quine's quips, it is both liberalization as well as limitation of our epistemic condition.

On the other hand, various sciences dispose of various degrees of theoretical standardization. For some sciences—typically for the human/social ones—developing a paradigm is a relatively cheap venture (cf., ibid., 138). Thus, although Roth claims that historiography is a science, it is hardly imaginable that historians could spontaneously converge to constitute a normal science in Kuhn's sense (cf., again ibid., 9–14, 63–67, 74).

But since historians create their own explananda, it is possible, then, to see as paradigms the particular historical explanations/accounts. Hilberg's explanation of the Holocaust is paradigmatic, at minimum, for any further attempts to analyze the policies of Nazi Germany, just like "the Van Gogh madness narrative" is paradigmatic for explaining, among other things, his death. The low degree of standardization is not an obstacle for evaluating (competing) historical explanations (and for any other explanations, for that matter), quite the opposite. Only after our interest-driven theorizing/explaining of given facts can we even have discussions and disagreements about historical events.

With respect to historiography, Roth illuminatively states that "any charge that narrative form [...] imposes a fictionalized structure on history misses that fundamental epistemic insight narrative sentences reveal, viz., that human histories exist only as a product of a very special sort of retrospective description. This is why to say that events such as the beginning of the Holocaust or the onset of the Black Death emerge only from that sort of perspective in no way implies some lack of 'reality.' Rather, it acknowledges the unavoidable fact that such events exist only by virtue of humans who shape experience in certain ways for certain purposes," (ibid., 69).

This perspectival and conceptual freedom we have is indeed only relative; the language games we play are still constrained by the facts (or by the world, or by our experiences, if you will). Our ever-rewiring web of belief still grounds our declarative sentences as well as inferences, even in the absence of any other formal method. "Say what you choose, so long as it does not prevent you from seeing the facts. (And when you see them there is a good deal that you will not say.)," Wittgenstein urges us (1986, 37).

In view of this, I contend that naturalism qua ism does not need any further specification; any specifying adjective.

There is no need to distinguish between, e.g., "metaphysical" and "methodological" naturalism, since metaphysical commitments are the results of our conceptual choices and the pushbacks of the causal contexts (the world). Also, there is no need to argue, e.g., for "relaxed" naturalism (à la Daniel D. Hutto); or to be scared of, e.g., "bald" naturalism (à la John McDowell) or of "radical" naturalism (à la Philip Kitcher), since naturalism is theoretically and methodologically opportunistic position. Naturalism is an ism guiding us across all the traditionally recognized philosophical domains; assuming, of course, that one takes on board the *science/epistemology-first approach* to all forms of explication and explanation.

In 1967, Danto opens the entry "Naturalism" in *The Encyclopedia of Philosophy* (Danto 1967, 448–450) by stating this:

> **Naturalism**, in recent usage, is a species of philosophical monism according to which whatever exists or happens is natural in the sense of being susceptible to explanation through methods which, although paradigmatically exemplified in the natural sciences, are continuous from domain to domain of objects and events. Hence, naturalism is polemically defined as repudiating the view that there exists or could exist any entities or events which lie, in principle, beyond the scope of scientific explanation. *In all other respects naturalism is ontologically neutral in that it does not prescribe what specific kinds of entities there must*

> *be in the universe or how many distinct kinds of events we must suppose to take place. Accordingly, naturalism is merely compatible with the various forms of materialism it has been confused with; materialism is logically distinct from naturalism and requires independent support unless (as is not the case) materialism is the sole ontology compatible with the ubiquitous employment of scientific method.* There is thus room within the naturalistic movement for any variety of otherwise rival ontologies, which explains the philosophical heterogeneity of the group of philosophers who identify themselves as naturalists: *it is a methodological rather than an ontological monism to which they indifferently subscribe, a monism leaving them free to be dualists, idealists, materialists, atheists, or nonatheists, as the case may be*. (ibid., 448, long emphasis mine)

In particular for all of us reading this delineation from the post-positivistic vantage point of today with its liberalized view of (natural) science, as well as of the scientific method(s), it seems to be quite a fitting delineation of naturalism in philosophy as elsewhere. And, for what it is worth, my hope is that the arguments I have proposed and gathered in this book also make this delineation of naturalism not only reasonable but also the most favorable.

I would like to end with a note on the issue of naturalistic vocabulary *in philosophy*. Is there a vocabulary philosophical naturalists should favor? One could also go one step further and pose the question this way: is there a particular vocabulary that is the most suitable for exploiting the metaphysical and methodological plasticity of naturalism for the aims of philosophy? (i.e., for its synoptic and synthetizing aims, cf., Sect. 1.1). And is not naturalism on this account doomed to become a kind of unproductive "discursive eclecticism?"

I think that all these questions are to be answered negatively. I believe that the necessary intimate relationship of naturalism with the actual practices and paradigms of current special sciences will always sufficiently ground as well as bound philosophical discourse (and any other discourse for that matter). So, at least in my opinion, the best description of the philosopher's job in the historically self-aware culture, such as ours, was given by Rorty in his seminal book *Philosophy and the Mirror of Nature* (originally published in 1979). Rorty says there that the role of a philosopher "is that of the informed dilettante, the polypragmatic, Socratic intermediary between various discourses. In his salon, so to speak, hermetic thinkers are charmed out of their self-enclosed practices. Disagreements between disciplines and discourses are compromised or transcended in the course of the conversation," (Rorty 1980, 317).

Adhering to this kind of *conversational pragmatism* also seems to be the only imaginable rebuttal of the potential charge that the naturalism defended here in this book is too limiting; that, absent any a priori metaphysical and methodological commitments, it does not have an actual practical force (for doing philosophy). But espousing naturalism is only the opening move.

It means espousing a platform where we can freely have conversations about various isms and theories within and against their contexts of origin and practical consequences.

It means doing what is humanly possible to achieve the goals we have set for ourselves within and against our history.

It means recognizing that there is no transcendental vantage point beyond our private and public settings. The only transcendence we are left with is our relationship to our past and posterity.

References

Danto, Arthur C. 1967. *Analytical philosophy of history*. Cambridge: Cambridge University Press.

Dewey, John. 1938. *Logic: The theory of inquiry*. New York: Henry Holt and Company.

Dewey, John. 1985. A short catechism concerning truth. In *The middle works of John Dewey: 1899–1924*, ed. Jo Ann Boydston. Carbondale: Southern Illinois University Press. (Originally published 1910).

Longino, Helen E. 1990. *Science as social knowledge: Values and objectivity in scientific inquiry*. Princeton: Princeton University Press.

Quine, Willard Van Orman. 1981. *From a logical point of view*. 2nd ed. Cambridge, MA: Harvard University Press.

Quine, Willard Van Orman. 1991. Two dogmas of empiricism. In *From a logical point of view*, 2nd ed., 20–46. Cambridge, MA: Harvard University Press.

Quine, Willard Van Orman. 1992. *The pursuit of truth*. Cambridge, MA: Harvard University Press.

Quine, Willard Van Orman, and Joseph S. Ullian. 1978. *The web of belief*. New York: Random House.

Rorty, Richard. 1980. *Philosophy and the mirror of nature*. Princeton: Princeton University Press.

Rorty, Richard. 1998. *Philosophy and social hope*. London: Penguin.

Roth, Paul A. 2020. *The philosophical structure of historical explanation*. New York: Routledge.

Szapuová, Marina. 2021. Quinean inspiration in feminist epistemology: On the potential alliance between naturalism and feminism. *Filosofický časopis (Philosophical Journal)* 69 (Special issue 3): 28–43.

White, Hayden. 1999. *Figural realism: Studies in the mimesis effect*. Baltimore: Johns Hopkins University Press.

Wittgenstein, Ludwig. 1986. *Philosophical investigations*. Trans. G. E. M. Anscombe. Oxford: Basil Blackwell.

References

Correction to: Dissolving Naturalism and Historicism Into Each Other

Correction to:
M. Hubálek, *Dissolving Naturalism and Historicism Into Each Other*, Studies in History and Philosophy of Science 67, https://doi.org/10.1007/978-3-032-07535-2

The original version of the book was inadvertently published with an incorrect city for the volume author in the front matter (listed as Prague). This has been corrected on page IV and in the Preface to the accurate location: Hradec Králové.

The updated version of this book can be found at
https://doi.org/10.1007/978-3-032-07535-2

M. Hubálek, *Dissolving Naturalism and Historicism Into Each Other*, Studies in History and Philosophy of Science 67,
https://doi.org/10.1007/978-3-032-07535-2_8

Index

M. Hubálek, *Dissolving Naturalism and Historicism Into Each Other*, Studies in History and Philosophy of Science 67,
https://doi.org/10.1007/978-3-032-07535-2

MIX
Papier aus verantwortungsvollen Quellen
Paper from responsible sources
FSC® C105338

If you have any concerns about our products,
you can contact us on
ProductSafety@springernature.com

In case Publisher is established outside the EU,
the EU authorized representative is:
Springer Nature Customer Service Center GmbH
Europaplatz 3, 69115 Heidelberg, Germany

Printed by Libri Plureos GmbH
in Hamburg, Germany

Meet the editors

Dr. Juber Akhtar has more than 16 years of research and teaching experience. He obtained his BPharm and MPharm from Jamia Hamdard, New Delhi, India, and Manipal University, Karnataka, India, respectively. He obtained his Ph.D. from Integral University, Lucknow, India. Dr. Akhtar is nominated as the head of the Faculty of Pharmacy, Integral University. He also served as chairman cum biological scientist for the Institutional Animal Ethics Committee (IAEC). He works at the Causality Assessment Committee (CAC) and is in charge of the Animal House facility at the Integral University. He has experience teaching abroad and was previously a professor at Buraydah College of Pharmacy and Dentistry, Kingdom of Saudi Arabia. Dr. Akhtar has more than 100 publications in reputed journals to his credit. He served as an academic book editor for international publishers and is an editorial board member for many esteemed journals. He has supervised a dozen Ph.D. and MPharm students in research projects. Dr. Akhtar's area of research interest is the development of nanoparticulate drug delivery systems.

Dr. Badruddeen obtained both a BPharm and MPharm from Hamdard University, Delhi, India. He obtained a Ph.D. in Pharmacology from Integral University, Lucknow, India, in 2014. He has expertise in cardiovascular, neurodegenerative, and endocrine diseases. He has experience in researching the safety, toxicity, and clinical efficacy of herbal and allopathic drugs. He has many original articles in reputed journals to his credit. Dr. Badruddeen is an academic book editor and an editorial board member for many esteemed journals. He has supervised a dozen research scholars and is a member of the Indian Science Congress Association, member secretary of the Institutional Animal Ethics Committee (IAEC), coordinator of the Adverse Drug Monitoring Center (AMC), chairman of the Causality Assessment Committee (CAC), and a member of the Animal House facility, Integral University.

Dr. Mohammad Irfan Khan obtained an MPharm from Hamdard University, Delhi, India. He also holds a Ph.D. in Phytochemistry. He has expertise in Ayurvedic medicine and nutraceuticals and has worked with some of the world's major pharmaceutical companies. He has more than 13 years' experience in research and formulation development of Ayurvedic, herbal, and nutraceutical products and successfully developed more than 100 formulations. He has more than fifty original papers in international journals to his credit. Dr. Khan has supervised many research scholars at post-graduate and Ph.D. levels.

Dr. Mohammad Ahmad holds a specialization in Pharmacology from Jamia Hamdard, New Delhi, India. He received his Ph.D. from Integral University, Lucknow, India. Presently, he is an Assistant Professor of Pharmacology, Faculty of Pharmacy, Integral University. He has been teaching PharmD, BPharm, and MPharm students and conducting research in the field of diabetes and cancer. From 2011 to 2014, he worked as a research assistant on a project sponsored by the Council for Science and Technology, Uttar Pradesh, India. He has published more than twenty original articles in reputed journals. He is a member of the British Society of Nanomedicine, Indian Science Congress Association, and International NanoScience Community, Budapest, Hungry. He is also a social nominee of the Committee for the Purpose of Control and Supervision of Experiments on Animals (CPCSEA). Dr. Ahmad is actively involved in research and his areas of interest include pharmacology and drug and pharmaceutical regulatory affairs. He is also engaged in research pertaining to diabetes mellitus, oral delivery of anticancer agents, and chemoprevention using natural bioactive compounds.

Contents

Preface

This book, *Vitamin B and Vitamin E – Pleiotropic and Nutritional Benefits*, is an insightful exploration of a group of essential vitamins that hold the key to optimal human health and vitality. It is a comprehensive overview of the multifaceted world of vitamin B, a critical water-soluble vitamin, and vitamin E, an oil-soluble vitamin, both of which play integral roles in bodily functions and overall well-being. Understanding the importance of these vitamins is paramount in a world where lifestyle choices and dietary habits often fall short of meeting our nutritional needs. The vitamin B complex encompasses a spectrum of vitamins, each with unique properties and contributions to health ranging from energy metabolism to neurological health and from cell division to the synthesis of red blood cells.

This book is organized into three sections. Chapters discuss cellular metabolic reactions in which B vitamins act as cofactors, how B vitamins are supplied primarily through dietary intake, gut health and gut microbiota, vitamin B and its association with the disease pellagra, and the medicinal significance and complications associated with vitamin E deficiency.

The book is designed for a diverse readership, including scholars and researchers in the field of nutrition and health care as well as students and individuals interested in optimizing their wellbeing. Our hope is that the content herein will serve as a valuable resource, empowering readers with knowledge that can positively influence their dietary choices and lifestyle decisions.

We extend our sincere gratitude to the authors and contributors who have shared their expertise and insights, making this book possible. We also express our heartfelt appreciation to our readers for embarking on this educational journey with us.

May this book inspire curiosity, spark intellectual discourse, and pave the way for a healthier, well-nourished world.

Sincerely,

Dr. Juber Akhtar
Professor,
Faculty of Pharmacy,
Integral University,
Lucknow, India

Badruddeen, Mohammad Irfan Khan and Mohammad Ahmad
Integral University,
Lucknow, India

Section 1

Role of Vitamin B in Body Health and Gut Flora

Chapter 1

Vitamin B Complex and Body Weakness

Hayder Lateef Al-msaid, Hydar Muhsin Khalfa and Hasan Hadi Ali

Abstract

B vitamins are crucial for metabolism. They are chemically unique vitamins with a variety of uses that are often present in the same meals. The vitamin B often operates in concert to provide the body with a multitude of health advantages. The metabolism has been demonstrated to be supported and speeded up by vitamin B. Maintain toned muscles and healthy skin. Boost immune and nervous system performance. Improved red blood cell development and division help to avoid anemia. Together, these factors also assist in battling the signs and causes of stress, depression, and cardiovascular disease. Water-soluble and found throughout the body, all vitamin B. Any excess that is expelled in the urine daily replenishes them, and a vitamin B shortage may result in a wide range of health issues.

Keywords: vitamin B, increasing vitamin B, decreasing vitamin B, vitamin B and body activities, body weakness

1. Introduction

Eight water-soluble vitamins known as B vitamins are crucial for maintaining healthy cell metabolism. In the past, the B vitamins were considered to be a single vitamin known simply as the B vitamin (most people thought it was somewhat similar to vitamin C or vitamin D) [1]. They are chemically separate vitamins that are often present in the same foods and serve a variety of purposes, according to research. Vitamin B complexes are dietary supplements that include all eight B vitamins [2].

The exact name of each vitamin indicates which B vitamin supplements are being taken individually (e.g., B1, B2, and B3), The B vitamins often operate in concert to provide the body with a multitude of health advantages [3]. The metabolism has been demonstrated to be supported and speeded up by B vitamins. Maintain toned muscles and healthy skin, and boost immune and nervous system performance. Improved red blood cell production helps to avoid anemia. Together, these factors also serve to battle the signs and causes of stress, depression, and cardiovascular disease [4].

Water soluble and found throughout the body, all B vitamins. Any extra excreted in the urine is daily replaced with them. A large number of different health issues may be brought on by vitamin B deficiency [5]. Important new research in this field is presented in this book.

Contrary to popular belief, oral vitamin B12, 1 mg daily, is used to treat the majority of vitamin B12 deficient patients worldwide.

There are at least six distinct ways that vitamin B3 (niacin) insufficiency may manifest, although they all overlap. The intake may be inadequate even when the food provides an adequate amount of niacin and there are no issues with absorption or storage [6]. This is because certain people may need unusually high levels of vitamin B-3, which a conventional diet cannot provide, due to hereditary factors. Up to one-third of genetic mutations cause the matching enzyme to have a decreased affinity for its own enzyme, which lowers the pace of the reaction [7].

Therefore, extremely large dosages of the matching enzyme may be employed to cure the roughly 50 hereditary illnesses of humans caused by these faulty enzymes. Vitamin B3 serves as a cofactor in many of the enzymes that are implicated in various hereditary diseases. These include increased risks for cancer and alcoholism brought on by poor aldehyde dehydrogenase binding, phenylketonuria II, and hyperphagia brought on by inadequate dihydropteridine reductase binding [8].

The HM74A and HM74B subtypes of niacin-responsive receptors were only recently identified. Niacin stimulates prostaglandin production *via* the high-affinity receptor HM74A. The HM74A protein is dramatically reduced in areas of the brain of schizophrenia patients, indicating a niacin-related deficiency that results in significantly increased vitamin B3 needs [9].

Niacin deficiency is most often brought on by diets with little to no vitamin B3 in them. For instance, pellagra is often detected in people who have consumed a lot of maize, a grain for which niacin is difficult to come by patients who have issues with absorption and storage are also at risk for vitamin B3 insufficiency. The body's supply of this vitamin and certain antibiotics will be depleted by excessive eating of foods like sweets and carbs [10].

Niacin deficiency is often caused by addiction and may be remedied by taking large amounts of this vitamin. Niacin, for example, is needed as a cofactor for aldehyde dehydrogenase, one of the key enzymes involved, making vitamin B3 essential for the breakdown of alcohol. Niacin and nicotine have chemical similarities; hence, nicotine may occupy receptor sites. High dosages of vitamin B3 have undoubtedly assisted some individuals in kicking their nicotine addiction [11].

The excessive oxidative stress that results in unusually high metabolic demand for this vitamin may also lead to a niacin deficit. It seems that diseases like Parkinson's, amyotrophic lateral sclerosis, and multiple sclerosis are caused by an excessive amount of dopamine being broken down, which produces neurotoxins like dopachrome [12].

This process may be speed up by vitamin B3, although the body often has low levels of the vitamin. Similar to this, individuals with schizophrenia create excessive amounts of adrenaline, as well as its hazardous metabolites, adrenochrome, and other chromium indoles. They, therefore, have niacin depletion, which is now recognized as the disease's diagnostic sign [13]. With aging, the body's capacity to absorb nutrients often decreases. As a consequence, niacin deficiency, among other vitamin inadequacies, is most prevalent among the elderly. These deficiency-related conditions react to high-dose niacin, including lipid imbalances, cardiovascular diseases, stroke, and arthritis. The recommended daily therapeutic intervention varies from 10 mg in newly diagnosed instances of pellagra from 6 to 10 grams for cholesterol normalization, cardiovascular disease, and stroke, according to the literature and Dr. Abram Hoover's experience with more than 5000 patients [14].

2. Health and folic acid

There is no evidence that folic acid is also important for the development of the central nervous system. Folic acid has long been recognized to cause a kind of anemia known as megaloblastica, but there is no evidence that it may cause problems early in pregnancy and at the time of conception [15].

The neural tube develops with birth abnormalities as a result. Inadequate amounts of folic acid have recently been linked to increased blood levels of the amino acid homocysteine (Hcy). Hcy is a well-known risk factor for conditions affecting the neurological and cardiovascular systems, as well as for dementia, Alzheimer's disease, osteoporotic fractures, and problems during pregnancy [16].

Folic acid has also been linked to reducing the risk of a variety of cancers. Recent epidemiological studies, for instance, show a negative correlation between folate status and the frequency of colorectal adenomas and carcinomas, indicating that lowering this risk may require maintaining appropriate folate levels [17].

On the other side, a number of studies indicate that excessive intakes, often linked to folic acid supplements, may raise the risk of breast cancer in postmenopausal women, particularly in those who consume moderate amounts of alcohol. Additionally, there is a chance that extensive folic acid fortification may conceal a vitamin B12 shortage, which might result in neurological impairment. Similar to folic acid insufficiency, vitamin B12 deficiency results in anemia and irreparable harm to the central and peripheral neurological systems [18].

However, over many generations, folate fortification may also have an impact on genetic selection for potentially deleterious genotypes and epileptic seizure management. There is a lot of interest in learning if dietary supplements and food products include folic acid, which is increasingly being considered on a global scale as an essential functional food element [19].

Folic acid fortification may be beneficial or damaging to health. Crohn's disease (CD), which may affect any portion of the gastrointestinal system, and ulcerative colitis (UC), which can affect the colon, are the two primary manifestations of inflammatory bowel disease (IBD), which is a chronic relapsing–relapsing inflammatory disorder of uncertain cause. The cornerstone of therapy for the majority of IBD patients is medical care with aminosalicylates (5-ASA), steroids, and immunosuppressive or immunosuppressive drugs. Surgery is only performed on people who have serious illnesses that are resistant to medical treatment or who have problems [20].

In the development, management, and therapy of IBD, nutrition is crucial. Inflammatory bowel disease patients often experience malnutrition, particularly those with Crohn's disease (CD). Patients with inflammatory bowel illness have been reported to have a variety of vitamin and mineral deficiencies. Despite their pathogenic significance in clinical symptoms and the many consequences associated with IBD, nutritional disorders are often disregarded in the care of patients. There are several factors that contribute to malnutrition in IBD, including inadequate oral nutritional intake, malabsorption, and nutrient loss [21], excessive nutritional needs, iatrogenic justification for surgery or treatment. Members of the "B vitamin complex" include thiamine (vitamin B1), riboflavin (vitamin B2), niacin, pyridoxine (vitamin B6), pantothenic acid, biotin, folic acid (vitamin B9), and vitamin B12. These are substances that dissolve in water and are crucial for the metabolic functions of living cells [22].

They function as coenzymes or as prosthetic groups attached to enzymes. When one of these vitamins is improperly ingested, the utilization of the other vitamins may be compromised. Folic acid and vitamin B12 deficiencies are often reported in IBD patients, and they are also linked to the anemia, thrombosis, and carcinogenesis that are associated with the disease [23].

Patients with IBD have also been shown to have low blood concentrations of other "vitamin B complex" members because of their deficiencies. By changing branched-chain carbonic acids into straight-chain branched carbonic acids, adenosylcobalamin-dependent CoA-carbonyl mutations accelerate the 1,2-rearrangement of carbonyl groups. Only two mutants of this enzyme family are now known, isobutyryl-CoA and methylmalonyl-CoA (MCM, EC 5.4.99.2). Both of these mutants have undergone substantial research [24].

All of the substances are significant water and soil contaminants; therefore, the novel cobalamin and CoA-carbonyl mutations play a hitherto unrecognized function in both natural and induced bioremediation processes. Consequently, the pathways that have not yet been connected to CoA-carbonyl mutation action. Additionally, it is probable that the enzyme structure and the fold that were used to predict the development of substrate specificity are related. Finally, the potential biological and kinetic effects of using a cobalamin-dependent enzyme to bend the routes are explored [25].

The water-soluble, catalytically active form of vitamin B6 is called pyridoxal 5′-phosphate (PLP), and it functions as a cofactor for several crucial human enzymes. Unique to a range of reactions to amino acids are PLP-dependent enzymes. They have the capacity to catalyze (transport, decarboxylation, or substitution/deletion). Various reactions happen without an enzyme. However, the protein portion simultaneously directs the enzyme's catalytic power toward a particular process. This specificity is not absolute, however. Physiologically significant side reactions that most PLP enzymes catalyze may also provide fascinating stereochemical and mechanical details about the structure of the enzyme's active site [1].

Dental spirochetes include cytolysin, a PLP-dependent C-S lyase whose most significant interaction is the removal of, from L-cysteine to create pyruvate, ammonia, and H2S. The latter is most likely in charge of the catalytic enzyme's hemolytic and hemeoxide-related action. One of the best examples of PLP-dependent enzymes' very versatile catalysts is cystalysin. In reality, it has only recently been shown that cetalicin is present [26].

Additionally, it may catalyze the transition between L- and D-alanine with turn numbers estimated in minutes, as well as the cracking of both isoforms of alanine, desulfurization of sulfuric acid L-cysteine, and decarboxylation of L-aspartate and oxalacetate. The cofactor binding mode, substrate specificity, formation of intermediate reactions typical of most PLP enzymes, and involvement of some active site residues in primary and secondary catalytic reactions are just a few of the intriguing characteristics of cystalysin that have been revealed through extensive biochemical investigations [27].

3. High homocysteine levels

Plasma from patients with renal failure, hypothyroidism, and methyltetrahydrofolate reductase polymorphism as well as those with homocystinuria, a hereditary disorder with a recessive pattern, was examined. The most crucial cerebral impairment, osteoporosis, lens shift, and arterial and venous thrombosis are among clinical symptoms of elevated plasma homocysteine levels. About 50% of deaths in patients with chronic renal failure are caused by cardiovascular illnesses, which are the main source of morbidity and mortality in the general population. Hyperhomocysteinemia is decreased by vitamin B6, vitamin B12, and folic acid. Transformation of sulfur and the remethylation process [28].

Although vitamin B medications often fail to correct plasma homocysteine levels, their long-term benefits are helpful in lowering the life-threatening vascular dangers that patients with homocystinuria face.

Patients with chronic renal failure, particularly those with stage V chronic kidney disease, are found to have hyperhomocysteinemia. The outcomes of observational clinical investigations on the consequences of increased plasma homocysteine levels on cardiovascular disease in hemodialysis patients have varied. In fact, several studies have shown a relationship between hypohomocysteinemia and cardiovascular disease fatalities in addition to hyperhomocysteinemia. The strong correlation between homocysteine and inflammatory indicators of malnutrition may be the cause of these contradicting observations [29].

Malnutrition-atherosclerotic syndrome is a serious clinical disease that often affects dialysis patients, interfering with homocysteine levels. My colleagues and I recently noticed in a clinical study that dialysis patients who were receiving vitamin B treatment and had high protein catabolism and low homocysteine levels outlived the other three groupings by a substantial margin. Recent prospective clinical trials that looked at the effects of moderate hyperhomocysteinemia patients receiving vitamin B treatment to decrease homocysteine on cardiovascular events revealed no therapeutic benefits. Because some of the patients had normal homocysteine levels, the follow-up period may have been too little, and confounding variables were not taken into consideration, these findings may be deceptive [30].

The study discusses the interesting mystique surrounding homocysteine and highlights the most relevant information about the impact of vitamin B therapy on cardiovascular events.

4. Vitamin B12

Two important enzyme pathways the conversion of homocysteine to methionine and the conversion of methylmalonyl coenzyme A to succinyl coenzyme are affected physiologically by it. Elevated blood homocysteine and methylmalonic acid result from disruption of any of these processes brought on by vitamin B12 deficiency. When folic acid levels are low, homocysteine levels also increase. It has been proposed that serum homocysteine rather than serum vitamin B12 analysis is more sensitive to intracellular functional vitamin B12 insufficiency. As a result, in the so-called one-carbon cycle, homocysteine, vitamin B12, and folic acid are strongly tied to one another [31]. The suggested mechanism is connected to methylation processes involving the nervous system's metabolism of homocysteine. The coenzyme that is required for its effective functioning is vitamin B12 methionine can only be produced when 5-methyletrahydrofolate donates its methyl group to tetrahydrofolate. On the other side, folate stimulates the remethylation of homocysteine, a cytotoxic molecule that contains sulfur, making it a cofactor in one-carbon metabolism [32].

Amino acids have the ability to cause DNA strand breaks, oxidative damage, and cell death. What or what appears to be consistent with the widespread belief that vitamin B12 and folic acid, either directly through the maintenance of two functions, DNA synthesis and methylation reactions, or indirectly, as a result of their deficiency, which results in methylation reactions mediated by SAM that are inhibited by its product SAH, and from through the related toxic effects of homocysteine causing direct damage to the vascular endothelium and inhibition of N-methyl-D [33].

This vitamin functions in a rising number of organs and bodily systems, the list of which is constantly expanding. The peripheral and central neurological systems, bone marrow, skin, mucous membranes, bones, and blood vessels are all impacted, in addition to children's typical growth. In addition to a sophisticated chemical structure, vitamin B12 (cobalamin) also includes the trace element cobalt, which is vital for human health. Vitamin B12 has significant immunological and neurotrophic effects in addition to playing a significant role in DNA synthesis. The human body's many systems must remain in balance. Even under the worst illness stressors, the individual is the ideal illustration of a system that always strives to attain optimum control. Consider that everything is standard and replaceable (as needed) if vitamin B12 is one of these things, why is it important, it is conceivable that therapy with vitamin B12 may rectify abnormalities brought on by other physiologically active chemicals even when the blood cobalamin level is normal [34]. This has been shown to be effective in treating recurrent aphthous stomatitis with vitamin B12 (regardless of blood level!) in the authors' research. This phenomenon is referred to as the "master switch" effect. Deficiency in vitamin B12 is a widespread issue that affects the entire population. Clinically significant is the early identification of vitamin B12 insufficiency, and there is evidence that deficiency occurs more often than anticipated. Patients who are unable to absorb vitamin B12 from food and people with dietary habits that exclude animal foods are both at risk for vitamin B12 insufficiency [35]. In addition, owing to the association between meat, cholesterol, and cardiovascular disease, there is a widespread inclination to steer clear of foods rich in vitamin B12, such as beef. Additionally, there is a propensity toward vegetarianism due to ideological reasons, particularly among the younger population. Two key causes of the low consumption of animal products, particularly red meat, are changes in lifestyle among population groups with high socioeconomic status and the prevalence of poverty. As a result, the amount of vitamin B12 in the general population is declining, which has an impact on the pathology caused by vitamin B12 deficiencies (e.g., neurological and hematological disorders). If more studies corroborate the link between homocysteine and vitamin B12, the authors may also report an increase in cardiovascular disease. Instead of these changes, major health issues should be avoided [36].

5. Vitamin B6

This vitamin is made up of a number of pyridoxal-containing substances, including pyridoxol, pyridoxamine, pyridoxaldehyde, and its derivatives. Structure the pyridoxine transporter at the sinusoidal pole has a determining hepatocyte absorption of pyridoxal, the catalytically active form of vitamin B6. Because pyridoxal may be transported to cells by an organ transport system when pyridoxine transporters in hepatocytes can specifically detect and bind to the structure. Thus, pyridoxine may be used as a liver-targeting group and added to big molecules and low molecular weight drugs to be used as contrast in MRI agents and anticancer comparisons [37]. The study of medication transport to the liver advances. The insertion of pyridoxine into these compounds has been shown to boost their liver absorption, and the molecules that include pyridoxine groups have liver-targeting characteristics.

DOI: http://dx.doi.org/10.5772/intechopen.109486

6. Conclusions

B vitamins are crucial for metabolism. They are chemically unique vitamins with a variety of uses that are often present in the same meals.

The B vitamins often operate in concert to provide the body with a multitude of health advantages. The metabolism has been demonstrated to be supported and speeded up by vitamin B. Maintain toned muscles and healthy skin. Boost immune and nervous system performance. Improved red blood cell development and division help to avoid anemia. Together, these factors also assist to battle the signs and causes of stress, depression, and cardiovascular disease.

Water soluble and found throughout the body, all B vitamins. Any excess that is expelled in the urine daily replenishes them, and a vitamin B shortage may result in a wide range of health issues.

Author details

Hayder Lateef Al-msaid*, Hydar Muhsin Khalfa and Hasan Hadi Ali
Department of Biology, Kufa University, Najaf, Iraq

*Address all correspondence to: haiderl.ligan@uokufa.edu.iq

References

[1] Hrubša M, Siatka T, Nejmanová I, Vopršalová M, Kujovská Krčmová L, Matoušová K, et al. Biological properties of vitamins of the B-complex, part 1: Vitamins B1, B2, B3, and B5. Nutrients. 2022a;**14**(3):484

[2] Vidović B, Đuričić B, Odalović M, Milošević Georgiev A, Tadić I. Dietary supplements use among Serbian undergraduate students of different academic fields. International Journal of Environmental Research and Public Health. 2022;**19**(17):11036

[3] Peterson CT, Perez Santiago J, Iablokov SN, Chopra D, Rodionov DA, Peterson SN. Short-chain fatty acids modulate healthy gut microbiota composition and functional potential. Current Microbiology. 2022;**79**(5):128

[4] Sarojnalini C, Hei A. Fish as an important functional food for quality life. U: Functional Foods, (Lagouri, V, Ured). 2019;**2019**:77-97

[5] Fratoni V, Brandi ML. B vitamins, homocysteine and bone health. Nutrients. 2015;**7**(4):2176-2192

[6] Hanna M, Jaqua E, Nguyen V, Clay JB. B vitamins: Functions and uses in medicine. The Permanente Journal. 2022;**26**:89-97

[7] Bellows L, Moore R, Anderson J, Young L. Water-soluble vitamins: B-complex and vitamin C. Food and nutrition Series Health. 2012;**9**:312

[8] Huang Y, Shan Y, Zhang W, Lee AM, Li F, Stranger BE, et al. Deciphering genetic causes for sex differences in human health through drug metabolism and transporter genes. Nature Communications. 2023;**14**(1):1-15

[9] Miller CL, Dulay JR. The high-affinity niacin receptor HM74A is decreased in the anterior cingulate cortex of individuals with schizophrenia. Brain Research Bulletin. 2008;**77**(1):33-41

[10] Gries CL, Scott ML. The pathology of thiamin, riboflavin, pantothenic acid and niacin deficiencies in the chick. The Journal of Nutrition. 1972;**102**(10):1269-1285

[11] Johnson S. The multifaceted and widespread pathology of magnesium deficiency. Medical Hypotheses. 2001;**56**(2):163-170

[12] Sauve AA. NAD+ and vitamin B3: From metabolism to therapies. Journal of Pharmacology and Experimental Therapeutics. 2008;**324**(3):883-893

[13] Foster HD. What Really Causes Schizophrenia. Canada: Trafford Publishing; 2003

[14] Nordell AD, McKenna M, Borges ÁH, Duprez D, Neuhaus J, Neaton JD, et al. Severity of cardiovascular disease outcomes among patients with HIV is related to markers of inflammation and coagulation. Journal of the American Heart Association. 2014;**3**(3):e000844

[15] Reynolds E. Vitamin B12, folic acid, and the nervous system. The Lancet Neurology. 2006;**5**(11):949-960

[16] Reynolds EH. The neurology of folic acid deficiency. Handbook of Clinical Neurology. 2014;**120**:927-943

[17] Ruiz RB, Hernández PS. Diet and cancer: Risk factors and epidemiological evidence. Maturitas. 2014;**77**(3):202-208

[18] Green R, Kara N, Cocks H. Vitamin B12 deficiency: An unusual cause of vocal

fold palsy. The Journal of Laryngology & Otology. 2011;**125**(12):1309-1311

[19] Heasman M, Mellentin J. The Functional Foods Revolution: Healthy People, Healthy Profits? UK: Earthscan; 2001

[20] Bressler B, Marshall JK, Bernstein CN, Bitton A, Jones J, Leontiadis GI, et al. Clinical practice guidelines for the medical management of nonhospitalized ulcerative colitis: The Toronto consensus. Gastroenterology. 2015;**148**(5):1035-1058

[21] Balestrieri P, Ribolsi M, Guarino MPL, Emerenziani S, Altomare A, Cicala M. Nutritional aspects in inflammatory bowel diseases. Nutrients. 2020;**12**(2):372

[22] Lykstad J, Sharma S. Biochemistry, Water Soluble Vitamins. USA. 2019

[23] Gold SL, Manning L, Kohler D, Ungaro R, Sands B, Raman M. Micronutrients and their role in inflammatory bowel disease: Function, assessment, supplementation, and impact on clinical outcomes including muscle health. Inflammatory Bowel Diseases. 2022;**1**:1-257

[24] Erbach J, Bonn F, Diesner M, Arnold A, Stein J, Schröder O, et al. Relevance of biotin deficiency in patients with inflammatory bowel disease and utility of serum 3 hydroxyisovaleryl carnitine as a practical everyday marker. Journal of Clinical Medicine. 2022;**11**(4):1118

[25] Stogios PJ, Bordeleau E, Xu Z, Skarina T, Evdokimova E, Chou S, et al. Structural and molecular rationale for the diversification of resistance mediated by the Antibiotic_NAT family. Communications Biology. 2022;**5**(1):263

[26] Gupta S. Recent reports on pyridoxal derived Schiff base complexes. Reviews in Inorganic Chemistry. 2022;**42**(2):161-177

[27] Mukherjee S, Banerjee O, Singh S. The role of B vitamins in protecting mitochondrial function. In: Molecular Nutrition and Mitochondria. USA: Academic Press; 2023. pp. 167-193

[28] Ibodov B, Shamsieva U, Akramova D. Hyperhomocysteinemia is a prior Diator of congenital anomalies. Eurasian Medical Research Periodical. 2023;**16**:15-20

[29] Fujita M, Wander K, Tran T, Brindle E. Characterizing the extent human milk folate is buffered against maternal malnutrition and infection in drought-stricken northern Kenya. American Journal of Biological Anthropology. 2022;**179**(2):171-183

[30] Percival R. The Meat Paradox:'Brilliantly Provocative, Original, electrifying'Bee Wilson. Hachette UK: Financial Times; 2022

[31] Malalasekara L, Escalante-Semerena JC. A method for the isolation of α-ribazole from vitamin B12, and its enzymatic conversion to α-ribazole 5′-phosphate. In: Methods in Enzymology. Vol. 668. USA: Academic Press; 2022. pp. 125-136

[32] McCabe CF, LaBarre JL, Domino SE, Treadwell MC, Baylin A, Burant CF, et al. Maternal and neonatal one-carbon metabolites and the epigenome-wide infant response. The Journal of Nutritional Biochemistry. 2022;**101**:108938

[33] Pandya M, Shah S, Dhanalakshmi M, Juneja T, Patel A, Gadnayak A, et al. Unravelling Vitamin B12 as a potential inhibitor against SARS-CoV-2: A computational approach. Informatics in Medicine. 2022;**30**:100951

[34] Jomova K, Makova M, Alomar SY, Alwasel SH, Nepovimova E, Kuca K, et al. Essential metals in health and disease. Chemico-Biological Interactions. 2022;**2022**:110173

[35] Cicero AF, Minervino A. Combined action of SAMe, folate, and Vitamin B12 in the treatment of mood disorders: A review. European Review for Medical and Pharmacological Sciences. 2022;**26**(7):2443-2459

[36] Liu Y, Geng T, Wan Z, Lu Q, Zhang X, Qiu Z, et al. Associations of serum folate and vitamin B12 levels with cardiovascular disease mortality among patients with type 2 diabetes. JAMA Network Open. 2022;**5**(1):e2146124-e2146124

[37] Gai Y, Li Y, Wu S, Xu L, Lu Y, Lan X, et al. Preparation and In vitro evaluation of a gadolinium-containing Vitamin E TPGS micelle as a potential contrast agent for MR imaging. Pharmaceutics. 2023;**15**(2):401

Chapter 2

Vitamin B, Role of Gut Microbiota and Gut Health

Satrio Wibowo and Almira Pramadhani

Abstract

The human gastrointestinal system is constantly exposed to pathogenic microorganisms and beneficial compounds, such as food components and commensal bacteria. Vitamin B are a class of water-soluble organic compounds obtained through diet, supplementation, and gut microbiota synthesis. B vitamins are absorbed for host metabolism in the small intestine, whereas microbes produce and absorb B vitamins in the large intestine. The authors have accumulated evidence from various studies that each B vitamin plays an essential role in gastrointestinal health and has a reciprocal relationship with the gut microbiota. Previous studies have also proven that microbial imbalance in the gut lead to competition for the utilization of B vitamins between the host and microbes, affecting the gut microbial composition, gut health, and host metabolism. This review aims to explain further the types of B vitamins in human digestion, the mechanism of B vitamin synthesis, and the role of B vitamins in the composition of the gut microbiota and the health of the gastrointestinal tract. Thus, it can help practitioners to consider administering B vitamins to maintain the patient's gut health.

Keywords: B vitamins, microbiota, intestine, biosynthesis, gut health

1. Introduction

The human gastrointestinal system is continuously exposed to toxic compounds, such as pathogenic microorganisms, and beneficial compounds, such as food components and commensal bacteria. Therefore, the immunity of the gastrointestinal system must be balanced between an active and suppressive immune response. Vitamins are micronutrients essential for normal human metabolism because they have various physiological effects, one of which is immunity. So, vitamin deficiency causes an increased risk of infectious diseases, allergies, and inflammation that can damage the gastrointestinal system. Over the past decade, a large number of studies have investigated the role of vitamins in various gastrointestinal diseases, including their potential in the prevention or treatment of various malignancies, inflammatory diseases, and hepatobiliary disorders [1, 2].

One of the vitamins that have a role in the gastrointestinal system is vitamin B, a group of water-soluble organic compounds with various functional roles, including cofactors in many enzymatic reactions, cellular energy-producing reactions, neurotransmitter synthesis, cell signaling, nucleic acid biosynthesis, and immune function. The intestine is an important organ, especially for storing and absorbing food. The gut also has the densest microbiota in the body, consisting primarily of four major phyla, that is, Bacteroidetes, Firmicutes, Proteobacteria, and Actinobacteria. The microbiota will maintain a symbiotic relationship with the host and protect against harmful pathogens. Some gut microbiota also produces limited amounts of B vitamins, and some require B vitamins for growth. Much evidence has shown that a healthy gut is associated with a healthy microbiota ecology, and B vitamins play an essential role in modifying the gut microbiota [3].

Although B vitamins from food are mainly absorbed in the small intestine, excess B vitamins cannot be absorbed in the small intestine and are transferred to the more distal intestine. Thus, the biosynthesis of B vitamins is mostly supplied by the distal gut microbiota. B vitamins in the distal intestine are dominant in performing important physiological functions in the body, including acting as nutrients for the host and microbiota, regulating immune cell activity, mediating drug efficacy, suppressing colonization of pathogenic bacteria, and modulating colitis. Therefore, the deficiency of B vitamins will certainly damage normal cell metabolism and trigger the development of several pathogenic microorganisms in the intestines [3, 4].

2. Types of B Vitamins in human digestion

Vitamins are essential micronutrients that all living cells need to carry out biochemical reactions. Vitamins are classified into fat-soluble and water-soluble. The fat-soluble vitamins are vitamins A, D, E, and K, while the water-soluble vitamins include vitamin C, biotin (vitamin H or B7), and a series of B vitamins—thiamin (B1); riboflavin (B2); niacin (B3); pantothenic acid (B5); pyridoxine, pyridoxal, pyridoxamine (B6); folic acid (B9), and cobalamin (B12) [5]. Fat-soluble vitamins act as essential elements of cell membranes, and the excess is stored by cells. Water-soluble vitamins function as coenzymes for specific chemical biochemical reactions, and the excess is excreted in the urine [6].

2.1 Vitamin B1

Vitamin B1 (thiamin) is a cofactor for several enzymes, including pyruvate dehydrogenase and a-ketoglutarate dehydrogenase, which are involved in the tricarboxylic acid (TCA) cycle. The thiamin molecule consists of a pyrimidine ring (4-amino-2-methylpyrimidine) and a thiazolium (4-methyl-5-(2-hydroxyethyl)-thiazolium), linked by a methylene bridge between the C3 carbon atom of the pyrimidine ring and the N3 nitrogen atom of the thiazolium ring. Vitamin B1 is found in high concentrations as thiamin pyrophosphate (TPP) [2, 7]. Thiamine strengthens the immune system, degrades glucose, helps nerve communication, and maintains processes in cells and tissues [8].

The intestinal epithelium absorbs free thiamine through thiamine transporters, that is, THTR-1 and THTR-2, which are transported to the blood for distribution throughout the body. Free thiamine is converted back to TPP and used for energy metabolism in the TCA cycle. The mechanism of thiamine absorption from food and microbiota is relatively similar. However, TPP produced by the gut microbiota is not converted to free thiamin because alkaline phosphatase is not secreted in the large intestine. Thus, TPP from the microbiota is absorbed directly by the large intestine.

The TPP transporter is widely expressed in the apical membrane of the colon. Absorbed TPP enters the mitochondria via MTPP-1 and is used as a cofactor for ATP formation. This suggests that TPP microbiota is vital for energy generation in the large intestine, by a mechanism that differs between vitamin B1 from the diet and from the microbiota [2].

2.2 Vitamin B2

Vitamin B2 (riboflavin) and its active forms (flavin adenine dinucleotide FAD] and flavin mononucleotide FMN]) are cofactors for enzymatic reactions in the TCA cycle and electron transport mediators of fatty acid oxidation (β-oxidation). Riboflavin is also involved in the metabolism of folate, vitamin B12, and vitamin B6, so it helps maintain the integrity of the mucosa, skin, eyes, and nervous system. Riboflavin is essential in the early development of the brain and postnatal gastrointestinal tract, as it can modulate some metabolic activities such as DNA repair, iron absorption and distribution, inflammation, and immune responses [2, 9].

The process of riboflavin absorption in the small intestine and colon is specific and mediated by the transporters RFVT-1, RFVT-2, and RFVT-3. All three are products of the SLC52A1, SLC52A2, and SLC52A3 genes expressed in the human gut, respectively, with RFVT-3 expression being the dominant one [9]. Exogenous vitamin B2 in the form of FAD or FMN is converted into free riboflavin by FAD pyrophosphatase and FMN phosphatase in the small intestine [2].

2.3 Vitamin B3

Niacin and niacinamide, known as nicotinic acid (NA) and nicotinamide (Nam), are different forms of vitamin B3. Vitamin B3 is a biosynthetic precursor for nicotinamide adenine dinucleotide (NAD+) and nicotinamide adenine dinucleotide phosphate (NADP+), which are coenzymes in respiratory oxidation processes, the Krebs cycle, the formation and inhibition of reactive oxygen species (ROS), post-translational protein and regulatory protein modification, and the formation of a second messenger. Vitamin B3 also functions for DNA proliferation. Thus, vitamin B3 is the center of homeostasis and cellular growth [2, 8, 10].

In contrast to other B vitamins, vitamin B3 can be produced by mammals via the endogenous enzymatic pathway from tryptophan and stored in the liver. However, vitamin B3 must also be obtained from food [2]. Vitamin B3 deficiency is endemic in some areas of the world where malnutrition is common. In more developed countries, vitamin B3 deficiency is caused by poor food choices, adverse drug reactions, alcoholism, and infectious or autoimmune diseases [6, 10].

2.4 Vitamin B5

Vitamin B5 (pantothenic acid) is a coenzyme A (CoA) precursor, an important cofactor for the TCA cycle and fatty acid oxidation. CoA has a role in various human biochemical reactions, such as cell growth, intermediate metabolism, and neurotransmitter synthesis. The structure of CoA functions as an activating carbonyl group and as an acyl group carrier to help facilitate these reactions. Like vitamins B1 and B2, vitamin B5 is also involved in controlling host immunity through energy production by immune cells [2, 11].

From dietary sources, vitamin B5 is found in high concentrations as CoA or phosphopantethein. CoA and phosphopantetheine are then converted to free pantothenic acid by endogenous enzymes such as phosphatase and pantetheinase in the small intestine. Whereas, from the microbiota, Vitamin B5 is produced in the form of free pantothenic acid, which is directly absorbed in the large intestine, converted to CoA, and distributed in the same way as vitamin B5 from food [2].

2.5 Vitamin B6

Vitamin B6 has the basic structure of 2-methylpyridine, 3-hydroxypyridine, and 5-hydroxymethyl pyridine. Vitamin B6 consists of various forms, that is, pyridoxal (aldehyde, eCHO), pyridoxine (alcohol, eCH2OH), and pyridoxamine (amine, eCH2NH2). These forms of vitamin B6 are precursors of the coenzymes pyridoxal phosphate (PLP) and pyridoxamine phosphate (PMP), which are involved in various cellular metabolic processes, including amino acid, lipid, and carbohydrate metabolism. Vitamin B6 also plays a role in nucleotide synthesis, neurotransmitter metabolism, and heme synthesis. So, this vitamin affects almost all aspects of metabolic function and cellular homeostasis. Vitamin B6 deficiency causes inflammation such as allergies and rheumatoid arthritis, as well as nerve dysfunction [2, 12].

Dietary vitamin B6 is available in the form of PLP or PMP, which is then converted into free vitamin B6 by the endogenous enzyme pyridoxal phosphatase and subsequently absorbed by the small intestine. Absorption of B6 from food sources occurs primarily in the small intestine jejunum, with the absorption rate varying according to the B6 species present. Meanwhile, microbial-synthesized vitamin B6 in the form of PLP is converted into free vitamin B6 in the large intestine, then absorbed through passive transport, transported to the blood, and distributed throughout the body [2, 12, 13].

2.6 Vitamin B7

Biotin (vitamin B7 or vitamin H) is a B-complex vitamin that acts as an essential coenzyme for five carboxylases: pyruvate carboxylase, 3-methylcrotonyl-CoA carboxylase (MCC), propionyl-CoA carboxylase (PCC), and acetyl-CoA carboxylase 1 and 2. This carboxylase helps several chemical processes in cells, including gluconeogenesis, amino acid metabolism, and fatty acid synthesis. Acetyl-CoA carboxylase 1 is found in the cytoplasm and catalyzes the binding of bicarbonate to acetyl-CoA during the synthesis of fatty acids, while another carboxylase is found in mitochondria. PCC catalyzes a critical step in the metabolism of propionyl-CoA, which is derived from the catabolism of odd-chain fatty acids and several other nutrients. Meanwhile, MCC catalyzes the metabolism of the amino acid leucine. Pyruvate carboxylase catalyzes the conversion of pyruvate into oxaloacetate, which is a key step in gluconeogenesis [14, 15].

The enzyme holocarboxylase synthetase (HLCS) catalyzes the binding of biotin to all five carboxylases, thus playing an essential role in biotin-dependent metabolic pathways. In addition, HLCS also functions in gene regulation at the chromatin level. Meanwhile, the biotinidase enzyme in the small intestine catalyzes the release of biotin from the breakdown product of carboxylase, thus playing an important role in biotin recycling. Free biotin is then absorbed via the biotin transporter SMVT [2, 15].

2.7 Vitamin B9

Vitamin B9 (folate), in its active form as tetrahydrofolate, is a cofactor in several metabolic reactions, including the synthesis of DNA and amino acids. A folic acid is a synthetic form of folate found in supplements. Bacteria use folate to synthesize the nucleic acids that make up their DNA. In an animal model of endometrial carcinoma study, folate was found to be the most important B-complex vitamin for nucleic acid synthesis, amino acid conversion, and antioxidant properties for eliminating free radicals [16]. In addition to DNA synthesis, folate also functions as a cofactor in homocysteine methylation and reduces the risk of *neural tube defects* [8]. Folate supplementation studies have also demonstrated a role in preventing other diseases, such as neurological disorders and cognitive and psychiatric disorders, and protection against degeneration in ulcerative colitis [17].

Vitamin B9 in food is available in monoglutamate and polyglutamate folate. In the intestinal epithelium, the folate transporter PCFT deconjugates polyglutamate folate to monoglutamate folate, which is then absorbed in the small intestine. Before being transported to the blood, monoglutamate is converted to tetrahydrofolate (THF), an active form and cofactor. Intestinal bacteria produce vitamin B9 as THF from GTP, erythrose 4-phosphate, and phosphoenolpyruvate. Bacterial THF is absorbed directly in the large intestine via PCFT and circulated throughout the body by the blood [2]. Folic acid is converted by the body into DHFR (DiHydro-Folate Reductase), which commensal and pathogenic bacteria can use to form nucleic acids, thus being the basis of their survival and reproduction cycles [2, 17].

2.8 Vitamin B12

Vitamin B12 (cobalamin) is composed of a corrin ring, with a cobalt center, and has upper and lower ligands as coordinates. The active forms of this vitamin are methylcobalamin and adenosylcobalamin, which catalyze the synthesis of methionine. Food vitamin B12 is decomposed from protein into free vitamin B12 by pepsin in the stomach. Free vitamin B12 is then absorbed by small intestinal epithelial cells via intrinsic factor (IF), a gastric glycoprotein. In epithelial cells, the IF-vitamin B12 complex is decomposed into free vitamin B12 by lysosomes and then released into the blood, becoming an active form and distributed throughout the body. Cobalamin and cobamide contain cobalt, but cobamide has a lesser ligand consisting of 5, 6-dimethylbgensiidazole (DMB). Mechanically, the lower DMB ligand is essential for binding Vitamin B12 to intrinsic factor (IF); then, it can be recognized by cubilin and megalin, which facilitate endocytosis in intestinal epithelial cells. Bacterial Vitamin B12 is synthesized from precorrin-2 to produce adenosylcobalamin, which is absorbed directly by the large intestine and distributed throughout the body [2, 18].

Vitamin B12 functions for nucleotide synthesis, branched-chain amino acid regulation, long-chain fatty acid metabolism, and cellular development. Vitamin B12 is also a vital cofactor in cytoplasmic methionine synthase and in mitochondrial methylmalonyl-CoA mutase, leading to homocysteine methylation to methionine and the conversion of methylmalonyl-CoA to succinyl-CoA. Furthermore, Vitamin B12 is a cofactor that other gut microbiota use to regulate the breakdown of short-chain fatty acids such as butyrate, propionate, and acetic acid. Vitamin B12 has also been shown to maintain healthy nerve cells and help red blood cell synthesis. This vitamin plays a role to immune homeostasis, utilization of microbial metabolites, and cellular metabolism, making it an essential factor in immunity against pathogens [2].

3. Synthesis of Vitamin B and role of microbiota

Humans cannot synthesize vitamins except for vitamin D. So, other vitamins must be obtained exogenously from food or the gut microbiota. Commensal bacteria found in the gut, such as Lactobacillus and Bifidobacterium, can synthesize de novo vitamins in the human body. Members of the gut microbiota can synthesize vitamin K and most of the water-soluble B vitamins, such as vitamins B12, B9, B6, B2, and B1 [6]. Each of the B vitamins has a different synthesis mechanism, and the microbiota synthesize most.

3.1 Vitamin B1

Various gut microbiota, mainly in the large intestine, produce vitamin B1 as free thiamin and TPP. *Bacteroides fragilis* and *Prevotella copri* (phylum Bacteroidetes); *Clostridium difficile*, some *Lactobacillus spp*., *Ruminococcus lactaris* (Firmicutes); *Bifidobacterium sp.* (Actinobacteria); and *Fusobacterium varium* are microbiota that can produce vitamin B1 through thiazole and pyrimidine synthesis pathways [2]. It is estimated that synthesizing thiamin by gut microbes supplies about 2.3% of the daily requirement of human vitamin B1. The enzymes involved in the thiamin biosynthetic pathway are dominantly found in enterotype 2, which is one of the gut microbiota clusters rich in *Prevotella sp.* [4]. In the intestine, there is *Faecalibacterium spp.* (Firmicutes), which lacks the vitamin B1 synthesis pathway and requires vitamin B1 for growth. Therefore, these bacteria must obtain vitamin B1 from other bacteria or from the host's diet via thiamin transporters in the mucosa. This indicates that there is competition for vitamin B1 requirements between the host and specific microbiota [5].

3.2 Vitamin B2

Not much different from vitamin B1, the synthesis of vitamin B2 is also widely played by the microbiota. Riboflavin from food, plus riboflavin produced by commensal bacteria, causes excess riboflavin in the distal intestine. In addition to the common lactic acid bacteria producing riboflavin in the gut, genomic analysis of 256 species of human gut microbiota has found 56% of the microbiota possess a cluster of genes for de novo riboflavin biosynthesis [4]. *B. fragilis, P. copri, C. difficile, Lactobacillus plantarum, Lactobacillus fermentum,* and *R. lactaris* have important factors for the synthesis of vitamin B2, so these bacteria are important sources of vitamin B2. *Bacillus subtilis* and *Escherichia coli* can also carry out riboflavin biosynthesis. Riboflavin biosynthesis requires guanosine 5′-triphosphate (GTP) and ribulose 5-phosphate as precursors. The first step of the branch of the GTP-dependent biosynthetic pathway is encoded by ribA, which catalyzes the 3,4-dihydroxy-2-butanone 4-phosphate configuration of ribulose 5-phosphate [5]. In clinical trials, it was found that daily consumption of 200 g of yogurt for 2 weeks can contribute to total vitamin B2 in the body, which is reflected in an increase in plasma-free riboflavin levels. This clinical trial also showed that most of the Lactobacilli strains consume riboflavin, thereby reducing its bioavailability in fermented products. Tempe, a traditional Indonesian fermented soybean meal, has been shown to increase the concentration of B vitamins such as riboflavin due to its microbial biosynthesis. The last article also reported that bacterial isolation from tempeh, which was shown to belong to Streptococcus and Enterococcus, significantly increased the concentration of riboflavin in this fermented product [19].

3.3 Vitamin B3

Unlike other B vitamins, vitamin B3 can be synthesized by the body from the amino acid tryptophan through endogenous enzymatic pathways and then stored in the liver. Niacin is a group of nicotinamide and nicotinic acid. These two metabolites are precursors for nicotinamide adenine dinucleotide (NAD), so nicotinamide and nicotinic acid can also be produced by recycling NAD in cells. An organism is considered a niacin producer when it contains the de novo synthesis pathway of NAD. These organisms include *B. fragilis, P. copri, R. lactaris, C. difficile, Bifidobacterium infantis, Helicobacter pylori,* and *F. varium* [2, 20]. Colonocytes have mechanisms that mediate niacin uptake transporters. Supplementation of microcapsules containing niacin will release its contents in the ileocolonic area, which increases the serum niacin concentration according to the dose consumed. Intake of 900 to 3000 mg niacin microcapsules will significantly increase the Bacteroidetes population [4].

3.4 Vitamin B5

For the synthesis of vitamin B5, bacteria synthesize from 2-dihydropantoate and beta alanine via the de novo synthesis pathway. Bacteria that have a biosynthetic pathway for vitamin B5 include *B. fragilis*, *P. copri*, some *Ruminococcus spp.*, *Salmonella enterica*, and *H. pylori*. Some bacteria synthesize free pantothenic acid, which is directly absorbed in the large intestine to be converted into Co, and distributed in the same way as vitamin B5 from food. Various bacteria, including *E. coli*, *Salmonella typhimurium*, and *Corynebacterium glutamicum*, can synthesize vitamin B5. Some of these bacteria use aspartate and intermediate metabolites of valine biosynthesis to produce vitamin B5. In contrast, most of the *Fusobacterium spp.*, *Bifidobacterium spp.*, some strains of *C. difficile*, *Faecalibacterium spp.*, and *Lactobacillus spp.* do not have this pathway, and some express pantothenic acid transporters to utilize vitamin B5 as an energy source. This indicates that these bacteria compete with the host for vitamin B5 [2, 3, 6, 21].

3.5 Vitamin B6

There are various forms of vitamin B6 in the body, that is, pyridoxine, pyridoxal, and pyridoxamine. Vitamin B6 in food is converted into free vitamin B6 by endogenous enzymes, such as pyridoxal phosphatase, before being absorbed by the small intestine. Microbiota with a biosynthetic pathway for vitamin B6 includes *B. fragilis*, *P. copri*, *Bifidobacterium longum*, *Collinsella aerofaciens*, and *H. pylori*. In contrast, most of the Firmicutes genera (*Veillonella, Ruminococcus, Faecalibacterium,* and *Lactobacillus spp.*) lack the vitamin B6 biosynthetic pathway [2].

3.6 Vitamin B7

As much as 40% of the human gut microbiota can synthesize de novo vitamin B7. The microbiota produce this vitamin as free biotin, which is synthesized from malonyl CoA or pimellate via pimeloyl CoA. Microbiota with a biosynthetic pathway for vitamin B7 include *B. fragilis, P. copri, F. varium,* and *Campylobacter coli.* The production of vitamin B7 affects others microbiota; for example, *B. longum*, which produces pimelate, a precursor of vitamin B7 that certainly increases the production of vitamin B7 by other gut microbiota. Biotin absorption in the small and large intestines occurs via a transporter-mediated process encoded by the SLC5A6 gene. Lipopolysaccharides inhibit colonic biotin uptake by impairing the expression of these membrane transporters. Biotin utilization also occurs between the host and bacteria, like *Lactobacillus murinus*, which consumes and reduces the biotin available in the intestine [2, 4].

3.7 Vitamin B9

The gut microbiota synthesizes vitamin B9 as tetrahydrofolate (THF) from GTP, erythrose 4-phosphate, and phosphoenolpyruvate. Dihydropteroate synthase catalyzes Folate biosynthesis, which reduces dihydrofolate to tetrahydrofolate [5]. THF is then directly absorbed in the large intestine and distributed throughout the body. Microbiota such as *B. fragilis*, *P. copri*, *C. difficile*, *L. plantarum*, *Lactobacillus reuteri, Lactobacillus delbrueckii ssp. bulgaricus,* and *Streptococcus thermophilus* and several species of *Bifidobacterium spp.*, *F. varium*, and *S. enterica* have folate biosynthetic pathways. It shows that almost all species in all phyla of microbiota produce folate. However, the ability of some of these bacteria to produce and utilize folate varies widely due to strain-dependent properties. Some Bifidobacteria do not produce folate when this vitamin is present, whereas others produce it regardless of vitamin concentration. Microorganisms are also able to increase folate content in a wide variety of foods [2, 19].

3.8 Vitamin B12

Vitamin B12 is a vitamin that contains cobalt, with the active forms being methylcobalamin and adenosylcobalamin. Dietary vitamin B12 forms a complex with dietary protein, further decomposed into free vitamin B12 by pepsin in the stomach. Free vitamin B12 is then absorbed by small intestinal epithelial cells via a gastric glycoprotein, intrinsic factor (IF). In epithelial cells, the IF-vitamin B12 complex is decomposed into free vitamin B12 by lysosomes and then released into the blood, which is further converted into the active form and distributed throughout the body [2].

Vitamin B12 is produced by aerobic and anaerobic pathways, with 30 different enzymes required for its biosynthesis. Several genes express enzymes essential for de novo vitamin B12 biosynthesis [18]. Bacterial vitamin B12 is synthesized from precorrin2 to produce adenosylcobalamin, which is then absorbed directly by the large intestine and distributed throughout the body. Microbiota with a biosynthetic pathway for vitamin B12 include *B. fragilis*, *P. copri*, *C. difficile, Faecalibacterium prausnitzii*, *R. lactaris*, *Bifidobacterium animalis, B. infantis*, *B. longum*, and *F. varium*. In addition to the microbiota, *L. plantarum* and *Lactobacillus coryniformis* also produce vitamin B12 from fermented foods, and *B. animalis* synthesizes vitamin B12 during milk fermentation [2, 21].

4. Sources of B Vitamins from the diet

Sources of B vitamins can be obtained apart from the biosynthetic mechanism, that is, exogenously from food. Vitamin B1 is found in high concentrations as thiamin pyrophosphate (TPP) in meat, especially pork and chicken; egg; cereals and rice; as well as nuts. The *World Health Organization* (WHO) / *Food and Agriculture Organization* (FAO) recommends a daily intake of 1.1–1.2 mg of vitamin B1 for adults [2].

Vitamin B2 is found in many animal sources, such as poultry, fish, liver, and eggs. Dairy products and cheese can also be sources of riboflavin, which makes a significant contribution to children and adults. Vegetable sources such as cereals, grain products, and bread can be a source of riboflavin-rich foods in some developing countries. Leafy greens, such as broccoli, mustard, and turnips, are also good sources of riboflavin. Natural grain products tend to be relatively low in riboflavin, but when intake is increased, these foods increase the bioavailability of riboflavin. WHO/FAO recommends a daily intake of 1.0–1.3 mg of vitamin B2 for adults [2, 15].

Animal foods such as fish and meat contain vitamin B3 as nicotinamide, and plant foods such as nuts and mushrooms contain vitamin B3 as nicotinic acid. WHO/FAO recommends a daily intake of 11–12 mg of vitamin B3 for adults. Vitamin B5 is high in concentrations of CoA or phosphopantetheine in the liver, eggs, milk, chicken, beef, salmon, cereals, grains, and fermented soybeans. WHO/FAO recommends a daily intake of 5 mg of vitamin B5 for adults [1, 2, 11].

Vitamin B6 is abundant in salmon, chicken, tofu, sweet potatoes, potatoes, bananas, and avocados. Vitamin B6 found in food is in the form of PLP or PMP. WHO/FAO recommends a daily intake of 1.3–1.7 mg of vitamin B6 for adults. Vitamin B7 is abundant in foods such as egg yolks, heart, cereals such as oats, nuts, vegetables such as spinach and mushrooms, rice, and vegetable oil. Dairy and breast milk products also contain biotin. Raw egg whites contain large amounts of avidin, which binds tightly to vitamin B7 and prevents its absorption in the intestines. WHO/FAO recommends a daily intake of vitamin B7 of 30 g for adults [1, 2, 14].

Foods such as beef liver, spinach, and asparagus are high in vitamin B9. Legumes, red meat, and liver also contain high levels of folate. However, up to 70% of folate is lost during cooking due to thermal degradation or dissolution in the cooking water. WHO/FAO recommends a daily intake of 400 g of vitamin B9 for adults. Most humans get vitamin B12 from food, especially from animal protein and fermented foods. Vitamin B12 is found in beef liver, bivalves, salmon, chicken, eggs, beans, and spinach. WHO/FAO recommends a daily intake of 2.4 g of vitamin B12 for adults [2, 18, 22].

From the explanation of the acquisition of vitamins from these foods, malnutrition will significantly inhibit normal metabolism in infants and the elderly. In the case of dietary supplements, vitamin-containing probiotic products are generally produced as freeze-dried powders and formulated as capsules, powders, and tablets [23]. Parameters influencing the manufacture of this dietary supplement include viable cell count and water activity. For the application of probiotics as food and drink, the product is prepared in the form of vegetative cells and added to food products [24]. Probiotic products require special care for the profile of a product, mitigation of health risks due to pathogens, and maintenance of aseptic conditions [25].

5. Role of B Vitamins in gut microbiota

B vitamins are biosynthetic precursors of essential cofactors in various metabolic pathways and play an important role in immunity. In addition to being needed by the host, B vitamins are also needed by some intestinal microbiota for metabolism. Thus, in addition to producing B vitamins, the gut microbiota also consumes B vitamins for primary enzymatic reactions. The gut microbiota is one of the densest microbial communities in the human body. This microbial community consists primarily of four main phyla, that is, Bacteroidetes, Firmicutes, Proteobacteria, and Actinobacteria. In the large intestine, the microbiota are grouped into 2, namely, B-vitamin-producing bacteria and auxotrophic bacteria. The survival of bacteria of auxotrophic species is highly dependent on B vitamins. Although most B vitamins are absorbed in the small intestine, B vitamins are produced and absorbed by bacteria mainly through the large intestine. Many B-vitamin transporters are expressed in the large intestine. As a result, there is competition between the host and bacteria. So that the host experiencing dysbiosis—the gut microbiota is unbalanced, and significantly affects the metabolism of B vitamins in the gut [13].

5.1 Vitamin B1

Thiamine produced in the gut microbiota has a specific role in the composition or function of the gut microbiome. Thiamin is required by certain gut bacteria, *Bacteroides thetaiotaomicron* and *Faecalibacterium spp.*, which have potential consequences on the host thiamine. Thiamine biosynthesis and its transport system are essential for the growth of B. thetaiotaomicron. Although *Faecalibacterium spp.* has a vitamin B1 synthesis pathway, this species requires more vitamin B1 for growth than production. Therefore, these bacteria must obtain vitamin B1 from other bacteria or host food via transporters, such as *B. thetaoimicron* [2, 4]. Intestinal dysbiosis may lead to a predominance of thiamin-only-consuming bacteria, which may contribute to decreased thiamin availability to the host. *E. coli* in the human fecal microbiota was found to be negatively correlated with fecal thiamin [26].

5.2 Vitamin B2

Riboflavin is vital in the growth of bacteria that are very sensitive to oxygen as an electron transfer agent. Examples of bacteria that are sensitive to oxygen are *F. prausnitzii* and Roseburia. *F. prausnitzii* is a major butyrate producer in the human microbiota and has anti-inflammatory properties and gut-protective functions. Vitamin B2 is also an essential precursor to flavin mononucleotide and flavin adenine dinucleotide (FAD), coenzymes of glutathione reductase that protect cells from reactive oxygen species (ROS). Thus, vitamin B2 can act as an indirect antioxidant and modifies the gut microbiota condition through ROS reduction. This condition is also proven to reduce the population of *E. coli*. Previous studies showed that supplementation of dietary riboflavin for 14 days increased *F. prausnitzii* and Roseburia and concomitantly reduced *E. coli*. Phylum Actinobacteria and Firmicutes express riboflavin transporters and enzymes required to form FAD and FMN from free riboflavin [2, 4, 27]. Vitamin B2 also directly affects the fecal microbiome, namely, the genera Alistipes and Clostridium. However, this increase occurred only with high doses of riboflavin supplied directly to the large intestine and may not apply to intakes from foods or dietary supplements that are absorbed mostly in the small intestine [27].

5.3 Vitamin B3

Vitamin B3 is the only B vitamin that humans can synthesize. It is a precursor to nicotinamide adenine dinucleotide (NAD), a coenzyme in cellular oxidation–reduction reactions with a central role in aerobic respiration. Niacin acts as an agonist for the cell surface receptor niacin receptor 1, which pairs with G-proteins. Niacin also has strong antioxidant and anti-inflammatory properties and can function as a modulator of gut protection and prevent bacterial endotoxin production. Thus, niacin deficiency causes intestinal inflammation and diarrhea, which has a direct impact on the gut microbiota population. Vitamin B3 deficiency impacts the diversity and low number of Bacteroidetes, especially in obese individuals. Supplementation with tryptophan and niacin has been shown to restore the composition of the gut microbiota through the angiotensin I (peptidyl-dipeptidase A)-converting enzyme. Furthermore, intake of niacin microcapsules (900 to 3000 mg) significantly increased the Bacteroidetes population. These results suggest that niacin has a beneficial effect on gut microbial composition in humans [2–4].

5.4 Vitamin B5

Lactobacillus spp., *Streptococcus spp.*, and *Enterococcus spp*. are members of the phylum Firmicutes that do not produce pantothenate but require pantothenic acid for their growth. This indicates that there is a symbiosis in the distal intestine between pantothenic acid-eating bacteria and pantothenic acid-producing bacteria. A study has demonstrated the uptake of pantothenic acid and biotin by the sodium-dependent multivitamin transporter (SMVT, SLC5A6) across the intestinal loop. Mostly Fusobacterium and *Bifidobacterium spp*., some strains of *C. difficile*, and *Faecalibacterium spp*., lack the vitamin B5 synthesis pathway but express the pantothenic acid transporter to utilize vitamin B5 as an energy producer [2, 4].

5.5 Vitamin B6

Vitamin B6 can play an essential role in shaping microbiota composition and metabolic capacity. In bacteria such as *E. coli*, vitamin B6 is synthesized in the PLP form from various precursors, including, deoxyxylulose 5-phosphate, 4-phosphohydroxy-L-threonine, glyceraldehyde-3-phosphate, and D-ribulose 5-phosphate. PLP produced by commensal bacteria works with ribonucleotide metabolism to facilitate the effects of 5-fluorouracil, a drug used to treat colorectal cancer. Vitamin B6 deficiency results in a marked decrease in intestinal arginine biosynthesis, and disruption of this metabolic pathway leads to the selective growth of certain gut bacteria, namely, the Bacteroidaceae family, and an increase in Lachnospiraceae. An increase in vitamin B6-producing bacteria such as *Bacteroides acidifaciens* has been shown to weaken the colonization of S. typhimurium and promote recovery from intestinal inflammation [4, 28].

5.6 Vitamin B7

Free biotin can affect the composition of the gut microbiota because it is required for the growth and survival of some microbiota. *Prevotella spp.*, *Bifidobacterium spp.*, Clostridium, Ruminococcus, Faecalibacterium, and *Lactobacillus sp.* do not have the vitamin B7 synthesis pathway because it lacks the essential biotin biosynthetic gene. However, they express a free biotin transporter, indicating a need for biotin. These results indicate that these bacteria also utilize biotin from food and bacteria to compete with the host. Therefore, it is necessary to control diseases related to some of these microbes. Another study showed that enzymes in the biotin biosynthetic pathway were overexpressed in the Bacteroides enterotype. Biotin uptake in the small and large intestines occurs via a carrier-mediated process involving the SMVT system encoded by the SLC5A6 gene [2, 4]. SMVT dysfunction reduces biotin in the intestine, causing dysbiosis, and induces Nox and ROS, which cause damage to enterocyte apoptosis. This mechanism causes the intestinal villi to shrink and increases intestinal permeability, inflammation, and dysplasia, all of which induce dysbiosis [29].

5.7 Vitamin B9

Folate is essential for several metabolic processes, including carbon transfer, thymidylate synthesis, purine synthesis, and the synthesis of several amino acids. Once absorbed, folate also participates in nucleotide synthesis, DNA repair, and methylation [30]. This function applies to both the host and the microbiota that require folate. The biosynthesis and expression of folate transporters in the gut microbiota are influenced by gut microbes, such as Bifidobacteria. In commensal bacteria, a vitamin B9 metabolite, 6-formylpterin (6-FP), is produced by the photodegradation of folic acid. This metabolite cannot activate MAIT cells (mucosal-associated invariant T), suppressing excessive MAIT cell responses and preventing excessive allergic and inflammatory responses [4].

5.8 Vitamin B12

Microorganisms use various forms of cyanocobalamin in many reactions, including methionine synthesis, carbon skeleton mutation, elimination reactions, amino mutations, and acetate and methane synthesis [31]. As many as 83% of the microbiota (260/313 species) encode cobalamin-dependent enzymes. Most of these species also lack the genes needed to synthesize cobalamin. In another report, 75.9% of bacteria utilized cobalamin, and only half possessed the cobalamin biosynthetic pathway. An example is *B. thetaiotaomicron*, which does not encode the cobalamin biosynthetic pathway gene but has three homologous cobalamin transporters. This statement indicates that this microbiota depend on the cobalamin absorption mechanism to maintain survival. Supplementation of 3.94 g/ml cyanocobalamin was shown to increase fecal cobalamin, with a lower Bacteroides population condition. Cobalamin and its derivatives also determine pathogenicity in the host gut. The bacterial transcription factor EutR requires ethanol amine, cobalamin precursors, and cobalamin-derived adenosylcobalamin for transcription of virulence factors required for host infection and spread of *enterohemorrhagic E. coli* (EHEC) serotypes O157:H7 and Salmonella [4].

6. The role of B Vitamins on human digestive health

The B vitamins in our body have several crucial physiological functions. The functional roles of these micronutrients are diverse, ranging from cellular energy-producing reactions such as the mitochondrial citric acid cycle to respiratory oxidation, immunity, neurotransmitter synthesis, cell signaling, and nucleic acid biosynthesis. Thus, a deficiency of B vitamins will impair normal cell metabolism and trigger the development of several chronic diseases in humans [4, 32]. Apart from being essential for the human body, B vitamins are also crucial in shaping the diversity and richness of the gut microbiota. A wealth of evidence has shown that a healthy gut lies in a healthy microbial ecology [3, 4].

6.1 Vitamin B1

Thiamin is a precursor of thiamin pyrophosphate, which is essential for carbohydrate metabolism and nerve function. Energy metabolism, particularly the balance between glycolysis and the citric acid cycle, is related to the functional control of immune cells, which is referred to as immunometabolism. Immunometabolism by vitamin B1 is vital for glycolysis-dependent digestive cells, especially Peyer's patch. In the gut, nave immunoglobulin (Ig)M+ B cells differentiate into IgA+ B cells in the Peyer patch, and then, IgA+ B cells differentiate into IgA-producing plasma cells in the lamina propria. The naive B cells in Peyer's patch prefer to use the vitamin B1-dependent citric acid cycle to generate ATP. However, once B cells differentiate into IgA-producing plasma cells, they switch to using glycolysis to generate ATP [2, 3].

Consistent with the importance of vitamin B1 in generating energy in the gut, mice fed a vitamin B1-deficient diet showed impaired maintenance of naive B cells in Peyer's patches. In addition to reducing the number of naive B cells in Peyer's patch, thiamin deficiency also reduces the size of B cell follicles, evidenced by the reduction of naive B cells in female Balb/c experimental animals. The researchers also showed that feeding the mice a vitamin B1-deficient diet caused the vitamin B1 deficiency to last for only a week. Vitamin B1 deficiency that affects the host immune response through the regulation of differentiation and proliferation of these immune cells ultimately affects the gut microbiota [2, 3].

6.2 Vitamin B2

Riboflavin is required for the development of the gastrointestinal tract after birth and is linked to crypt hypertrophy, crypt bifurcation dysfunction, and a loss of proliferative potential in intestinal cells. These changes are visible during the postnatal and post-weaning stages. These changes were irreversible, even after the experimental administration of riboflavin in vivo and in vitro. Riboflavin deficiency has been shown to reduce the number of villi but, on the other hand, can increase the length of the villi. Reduction of riboflavin in humans has also been associated with shortened duodenal crypts and reduced cell division. In vitro studies using Caco-2, HCT116, and HT29 cells demonstrated a potential mechanism for the riboflavin deficiency phenotype, which led to the result that riboflavin inhibited cell growth by reducing cellular ATP generation and increasing oxidative stress. This impairs mitosis and accumulates aneuploidy cells. These changes in gut morphology may also be associated with an adaptive response to stress-induced deficiency [4].

Riboflavin is also essential for methylation reactions, nucleotide synthesis, and DNA stability and repair. A cohort study in the Netherlands on a diet in cancer suggested that riboflavin was likely to be associated with a reduced risk of proximal colon cancer among women (RR = 0.61; P-trend = 0.07). This finding is reinforced by results from the Women's Health Initiative cohort observational study, which showed that higher total riboflavin intake was associated with a reduced risk of colorectal cancer (HR = 0.81; 95% CI: 0.66–0.99) [1].

6.3 Vitamin B3

Human and mouse colonic epithelial cells possess efficient and specific mechanisms for vitamin B3 absorption. This vitamin plays an essential role in reducing inflammation, so a deficiency will lead to inflammatory bowel diseases such as ulcerative colitis. Vitamin B3 controls inflammation by inhibiting vascular permeability in intestinal tissue by activating PGD2/DP1 signaling in endothelial cells. This vitamin also modulates the inflammatory response by increasing the rate of ATP generation in Caco-2 cells. In addition, vitamin B3 is involved in various cellular oxidation–reduction metabolic reactions and rapamycin signaling pathways, thereby suppressing colon inflammation [3, 32].

Vitamin B3 synthesized by the gut microbiota contributes to local colonocyte nutrition and maintains intestinal stem cell morphology. Vitamin B3 is also known to protect colonic epithelial cells against dextran-sulfate-sodium (DSS)-induced apoptosis and promote cell proliferation in experimental animals. The mechanism of protection of the intestinal epithelium by vitamin B3 is by activating the prostanoid D 1 (DP1) receptor on macrophages and endothelial and colonic epithelial cells. One study found that retention of vitamin B3-containing enemas effectively promoted mucosal healing in patients with ulcerative colitis, with possible mechanisms of downregulation of colonic inflammatory cytokines and suppression of pro-inflammatory gene expression [3, 4].

6.4 Vitamin B5/pantothenic acid

Vitamin B5, or pantothenic acid, is an essential coenzyme A (CoA) precursor and acts as a carrier protein. This vitamin is involved in various metabolic pathways, such as the citric acid cycle, cell growth, neurotransmitter synthesis, and fatty acid oxidation [4]. Dietary pantothenic acid supplementation also affects the gut microbial profile. Increased intake of pantothenic acid increased the relative numbers of Prevotella and Actinobacteria [3].

6.5 Vitamin B6/pyridoxine

Vitamin B6 functions primarily as a cofactor for the biosynthesis and catabolism of amino acids. In addition, this vitamin is also involved in fatty acid biosynthesis and neurotransmitter biosynthesis and as an antioxidant [3]. Relative pyridoxine deficiency is found in 10–25% of Inflammatory Bowel Disease (IBD) cases. Plasma levels of B6 are considered a risk factor for thrombosis in patients with IBD because they have an inverse relationship with homocysteine [1].

A cross-sectional study has shown an association between the severity of intestinal irritation and low dietary vitamin B6 intake. The mechanism of this phenomenon is that the lack of vitamin B6 affects the balance of anti-inflammatory and pro-inflammatory cytokines. Vitamin B6 deficiency also reduces microbial diversity and significantly alters gut metabolites such as short-chain fatty acids, which also play an essential role in triggering this irritation. In addition, studies of vitamin B6 deficiency in animals have shown a significant reduction in the number of mucus-secreting cells, an important factor in maintaining gut health. Vitamin B6 also decreases cell calcium transport but does not affect the basic morphology of enterocytes, such as cell viability, cell volume, membrane permeability, and protein content [4].

Vitamin B6 can influence colorectal carcinogenesis through its role in DNA synthesis and methylation. Animal studies have shown that this vitamin can inhibit angiogenesis, suppress nitric oxide, and reduce oxidative stress [1].

6.6 Vitamin B7/Biotin

Vitamin B7 acts as a coenzyme for several biochemical reactions, such as glycolysis, cell signaling, and epigenetic regulation. This vitamin also controls the expression of genes, including nuclear factor kappa B (NF-kB), through a histone-binding mechanism known as biotinylation. Therefore, this vitamin may also have an anti-inflammatory effect on the gastrointestinal mucosa [32].

6.7 Vitamin B9/Folate

Vitamin B9 is essential for the replication and regeneration of nucleic acids, affecting cells' survival rate. Folate is involved in synthesizing S-adenosylmethionine (SAM), which is required for cellular biosynthesis and DNA methylation. This vitamin is essential for replicating and recovering nucleic acids, influencing survival rates through cell proliferation and regeneration. In addition, folate regulates gene activity, regenerates the intestinal lining, reduces lymphocyte growth, and reduces NK cell cytotoxicity [3]. Thus, every living cell, including gastrointestinal cells, requires folate to carry out these various biochemical and biosynthetic processes [4].

Because of its essential role in producing methyl donors, folate deficiency significantly impairs DNA replication. Folate deficiency causes an increase in the crypt depth of the intestinal mucosa in the duodenum and jejunum, resulting in a decreased villi-to-crypt ratio. In experimental animals, induced methyl donor deficiency by feeding a folate-deficient diet accompanied by the antibiotic succinylsulfathiazole 1% has also increased crypt depth and altered gut cell differentiation. In this study, folate deficiency also caused megaloblastic changes in the epithelial cell nuclei and reduced crypt mitosis. These changes are seen more prominently in the ileum with elongation of the crypts, an increase in goblet cells, and a decrease in Paneth cells. Deficiencies of folate, riboflavin, vitamin B6, and vitamin B12 concomitantly alter Wnt signaling in the experimental colon and decrease apoptosis in epithelial cells. Unexpectedly, these changes are irreversible, even with an excess of folate [4].

Folate deficiency significantly alters intestinal cell morphology and is associated with increased intestinal carcinogenesis [3]. Two case–control studies demonstrated that folate supplementation and high red blood cell folate levels significantly reduced the risk of dysplasia and neoplasia in patients with ulcerative colitis. Folate supplementation, in combination with sulfasalazine administration, has a protective effect on the development of colorectal cancer in patients with chronic ulcerative colitis. The mechanism of folate protection against colorectal cancer is to prevent aberrations in DNA synthesis and aberrations in DNA methylation. However, it should be noted that folate deficiency can also occur due to IBD therapy, such as methotrexate and sulfasalazine. A recent meta-analysis of folate supplementation has found that folate plays a role in preventing pancreatic cancer. Individuals with a high dietary folate intake were 34% less likely to develop pancreatic cancer than individuals with a low folate intake [1].

6.8 Vitamin B12

Like folate, cobalamin is involved in the synthesis of methyl donors. These donors are essential for the nucleic acid synthesis and protein and lipid metabolism. Vitamin B12 also acts as a cofactor for methionine synthase in sulfur amino acid metabolism to recycle homocysteine into methionine. The effect of vitamin B12 deficiency on colonic morphology is similar to that of folate deficiency because of its association with several cellular metabolic reactions. Vitamin B12 deficiency protects against DSS-induced inflammation in C57BL/6 mice. Other studies have shown reduced cell differentiation and gut protective factors in vitamin B12-deficient mice. In addition, in patients with vitamin B12 deficiency, the villi become shorter with a reduced villi/crypt ratio. Deficiency or excess of vitamin B12 affects the growth of gut microbiota.

However, vitamin B12 deficiency did not relatively change the gut microbiota composition in healthy mice but did change it in DSS-induced colitis mice [4, 6, 32].

Vitamin B12, together with vitamins B9 and B6, influence the occurrence of colorectal carcinoma through its role in DNA synthesis and methylation. In addition, these three vitamins have been shown to inhibit angiogenesis, suppress nitric oxide, and reduce oxidative stress in animal models. However, an experimental study that administered a combination of folic acid (2.5 mg), vitamin B6 (50 mg), and vitamin B12 (1 mg) in 1470 subjects did not reduce the risk of colorectal carcinoma after a follow-up period of 9.3 years. Patients with celiac disease also have higher total plasma homocysteine levels than the general population, indicating lower serum levels of vitamins B6, B9, and B12 [1].

The liver is the physiological reservoir of cyanocobalamin in humans. Vitamin B12 deficiency is observed in several liver diseases such as hepatitis, cirrhosis, and hepatocellular carcinoma. Vitamin B12 inhibits HCV through the inhibition of ribosome entry sites. Vitamin B12 is also associated with aphthous stomatitis (**Figure 1**) [1].

7. Conclusion

B vitamins act as cofactors for several cellular metabolic reactions. In addition to vitamin B3, other B vitamins must be obtained through dietary intake, supplementation, and synthesis of the gut microbiota. The biosynthesis of B vitamins in the gut

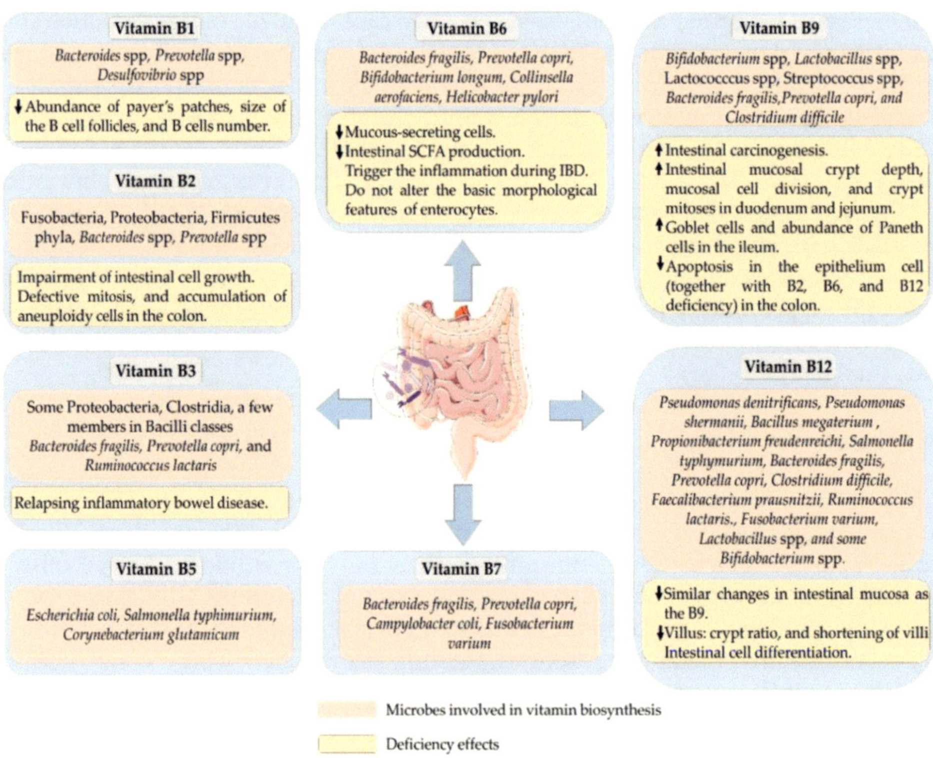

Figure 1.
Summary of the main gut bacteria synthesizing B vitamins and the effects of their deficiency on gut health [3].

is influenced by several factors, including exposure to antibiotics and free radicals, genetic makeup, dietary habits, and lifestyle. Competition between gut microbiota, pathogenic microbes, and the host leads to deficiency conditions, especially if exogenous intake is not optimal. A deficiency of B vitamins ultimately affects the gut microbial composition, gut health, and overall host metabolism.

Author details

Satrio Wibowo[1*] and Almira Pramadhani[2]

1 Department of Pediatrics, Faculty of Medicine, University of Brawijaya, Malang, Indonesia

2 University of Brawijaya, Malang, Indonesia

*Address all correspondence to: satrio_wibowo@ub.ac.id

References

[1] Masri OA, Chalhoub JM, Sharara AI. Role of vitamins in gastrointestinal diseases. World Journal of Gastroenterology. 2015;**21**(17):5191-5209. DOI: 10.3748/wjg.v21.i17.5191

[2] Yoshii K et al. Metabolism of dietary and microbial vitamin B family in the regulation of host immunity. Frontiers in Nutrition. 2019;**1**:48. DOI: 10.3389/fnut.2019.00048

[3] Hossain KS, Amarasena S, Mayengbam S. B vitamins and their roles in gut health. Microorganisms. 2022:**10**(6):1168. DOI: 10.3390/microorganisms10061168

[4] Uebanso T et al. Functional roles of B-vitamins in the gut and gut microbiome. Molecular Nutrition and Food Research. 2020;**64**(18):2000426. DOI: 10.1002/MNFR.202000426

[5] Gu Q, Li P. Biosynthesis of vitamins by probiotic bacteria. 2016. 10.5772/63117

[6] Jiang Q et al. Metagenomic insights into the microbe-mediated B and K2 vitamin biosynthesis in the gastrointestinal microbiome of ruminants. Microbiome. 2022;**10**(1):1-16. DOI: 10.1186/S40168-022-01298-9/FIGURES/5

[7] Tylicki A et al. Thiamine and selected thiamine antivitamins-biological activity and methods of synthesis. Bioscience Reports. 2018;**38**:20171148. DOI: 10.1042/BSR20171148

[8] Karunaratne DN et al. Nutrient delivery through nanoencapsulation. In: Nutrient Delivery. New York: Elsevier; 2017. pp. 653-680. DOI: 10.1016/b978-0-12-804304-2.00017-2

[9] Thakur K et al. Riboflavin and health: A review of recent human research. Critical Reviews in Food Science and Nutrition. 2017;**57**(17):3650-3660. DOI: 10.1080/10408398.2016.1145104

[10] Makarov M, Trammell SA, Migaud ME. The chemistry of the vitamin B3 metabolome. 2019. 10.1042/BST20180420

[11] Sanvictores T, Chauhan S. Vitamin B5 (pantothenic acid). In: Clinical Guide to Nutrition & Dietary Supplements in Disease Management. Florida: StatPearls Publishing; 2022. pp. 699-701. DOI: 10.1016/B978-0-443-07193-5.50106-0

[12] da Silva VR, Gregory JF. 'Vitamin B6. In: Present Knowledge in Nutrition. New York: Elsevier; 2020. pp. 225-237. DOI: 10.1016/B978-0-323-66162-1.00013-5

[13] Putnam EE, Goodman AL. B vitamin acquisition by gut commensal bacteria. 2020. 10.1371/journal.ppat.1008208

[14] Saleem F, Soos MP. Biotin deficiency. Journal of Nutrition Education. 2019;**13**(3):96. DOI: 10.1016/S0022-3182(81)80040-4

[15] Zempleni J, Kuroishi T. Biotin. Advances in Nutrition. 2012;**3**(2):213-214. DOI: 10.3945/AN.111.001305

[16] Dalto DB, Matte JJ. Pyridoxine (Vitamin B6) and the glutathione peroxidase system; a link between one-carbon metabolism and antioxidation. Nutrients. MDPI AG. 2017. DOI: 10.3390/nu9030189

[17] Scaglione F, Panzavolta G. Folate, folic acid and 5-methyltetrahydrofolate are not the same thing. 2013. 10.3109/00498254.2013.845705

[18] Roth W, Mohamadzadeh M. Vitamin B12 and gut-brain homeostasis in the pathophysiology of ischemic stroke. eBioMedicine. 2021;**73**:103676. DOI: 10.1016/J.EBIOM.2021.103676

[19] Leblanc JG et al. B-Group vitamin production by lactic acid bacteria-current knowledge and potential applications. 2011. DOI: 10.1111/j.1365-2672.2011.05157.x

[20] Dimitrov D et al. Systematic genome assessment of B-vitamin biosynthesis suggests co-operation among gut microbes. Frontiers in Genetics. 2015;**1**:148. DOI: 10.3389/fgene.2015.00148

[21] Indira M et al. Bioactive molecules of probiotic bacteria and their mechanism of action: A review. Biotech. 2019;**9**(8):1-11. DOI: 10.1007/s13205-019-1841-2

[22] Ami N et al. Folate and neural tube defects: The role of supplements and food fortification. Pediatrics & Child Health. 2016;**21**(3):145-149. DOI: 10.1093/PCH/21.3.145

[23] Fenster, K. et al. (2019) 'microorganisms The Production and Delivery of Probiotics: A Review of a Practical Approach'. 10.3390/microorganisms7030083.

[24] Mishra, S. et al. (2019) 'microorganisms Probiotics and Prebiotics for the Amelioration of Type 1 Diabetes: Present and Future Perspectives'. 10.3390/microorganisms7030067

[25] Khaneghah AM, Fakhri Y. Probiotics and Prebiotics as Functional Foods: State of Art. Agri Res & Tech: Open Access J. 2018;**13**(3):10.19080/ARTOAJ.2018.13.555884

[26] Barra NG et al. Micronutrients impact the gut microbiota and blood glucose. Journal of Endocrinology. BioScientifica Ltd. 2021:R1-R21. DOI: 10.1530/JOE-21-0081

[27] Pham VT et al. Vitamins, the gut microbiome and gastrointestinal health in humans. Nutrition Research. 2021;**95**:35-53. DOI: 10.1016/J.NUTRES.2021.09.001

[28] Mayengbam S, Chleilat F, Reimer RA. Vitamin B6 deficiency impairs gut microbiota and its metabolites in rats. 2020

[29] Neophytou C, Pitsouli C. Biotin controls intestinal stem cell mitosis and host-microbiome interactions. Cell Reports. 2022;**38**(10):110505. DOI: 10.1016/J.CELREP.2022.110505

[30] Mukhopadhyay B. et al. Microbial metabolic capacity for intestinal folate production and modulation of host folate receptors. 2019. 10.3389/fmicb.2019.02305

[31] Stabler SP. Vitamin B12. In: Present Knowledge in Nutrition. New York: Elsevier; 2020. pp. 257-271. DOI: 10.1016/B978-0-323-66162-1.00015-9

[32] Xu Y et al. Cobalamin (Vitamin B12) induced a shift in microbial composition and metabolic activity in an in vitro colon simulation. Frontiers in Microbiology. 2018;**9**(Nov):1-9. DOI: 10.3389/fmicb.2018.02780

Section 2

Vitamin B and Pellagra

Chapter 3

Pellagra: Down Not Out If Down and Out – Part 1

Adrian C. Williams, Christina Wood and Lisa J. Hill

Abstract

Pellagra is caused by a dietary deficiency of milk and meat leading to insufficient nicotinamide (vitamin B3), the precursor to nicotinamide adenine dinucleotide (NAD). "Pellagra sine pellagra" was well recognised and may be common as supplementation was never globally implemented and a screening test never developed. Meat and milk intake varies 30-fold globally so there are perhaps 2 billion at risk of deficiency. Such patients will have physical and cognitive stunting, poor conduct and be prone to acute and chronic infections, including TB, and premature ageing, including dementia. Resilience may be poor to NAD-consuming insults whether chemical, microbial or traumatic that conspires to cause brain injury but comes with the opportunity for pre-conception dietary corrections breaking cycles of deprivation and poor educational outcomes. Such individuals may otherwise be subject to discrimination as was the pellagra ridden "Butterfly" caste causing racial and ethnic tensions. Poor countries with many having to spend 50-80% of income on food and very little on animal produce cannot develop properly unlike wealthier meat rich empires, past and present. The many benefits of experiments with food programmes and basic income support are because, as Engels curves predict, more is spent on milk and meat enabling demographic, epidemiological, and economic transitions and modernity.

Keywords: NAD, pre-conception nutrition, addiction, opiates, cycles of deprivation, empires, meat transitions, demographic transitions, epidemiological transitions, TB, poverty traps, metabolic syndrome, class ceilings

1. Introduction

Cycles of life, right from lifes origin to humankind, depend upon nuclear fusion reactions and low entropy photon- energy from the sun. Photosynthesis splits water in to the oxygen that we breathe and hydrogen that combined with CO2 forms the carbohydrates that we eat and is carried by NAD to mitochondria forming proton-motive forces during oxidative phosphorylation producing usable energy as ATP and reduced nicotinamide adenine dinucleotide phosphate (NADPH) for growth [1–3]. Increasing use of oxygen and NAD supplies allowed animal evolution from the Cambrian (when animals started eating animals) to mammals (increasing milk supplies) and man [4].

RESPIRATION / COMBUSTION

	NATURAL VARIANCE	UNNATURAL VARIANCE	COST / COMMONS	POLLUTION / DANGERS
SUNLIGHT	POLAR	← LAND OWNERSHIP →		FEW
O2	Altitude	NONE	Free commons	Air pollution in poor areas
H20	High in rivers or deserts helped by irrigation	High dam use / desertification	Variable from common good to meters	Pollution in poor areas ↘ Water borne infections
CHO	LOW	Medium Monocultures	CHEAP	RARE
NAD	Medium for hunting / High by natural animal domesticates	High enclosures / appropriations	EXPENSIVE	Bush meat Wet Markets ↓ Food poisoning Pandemics
EXTERNAL ENERGY	Significant in availability of animals, wood, water, coal, gas, oil	Exacerbated by commodification from slave manpower to electricity	OFTEN EXPENSIVE	C02 pollution high without sustainable solar / tidal nuclear / wind usage

Figure 1.
Basic elements of the energy cycle had a natural variance based on geography. Human history often exagerated these variances first by making hunting meat easier in some geographies and quite dangerous for the hunters even as they shared in an equitable way as social insurance - and then with the advent of agriculture natural domesticates varied by continent and meat became expensive everywhere often shutting out the poor. Meat was more available in the middle-east and Europe and less available in the Americas and Africa - that also had to contend with many carnivorous competitors and more parasitic disease in the tsetse belt.

High meat intake and sharing by hunters was a major feature of our evolution, early economics and relationships with animals and each other in our "social leap [5–11]. Combined with selected plants, some psychedelic, and use of fire (cooking releases nicotinamide from meat and from (maize) plants if cooked carefully and learnt culturally) - meat was an important factor in how our brains expanded and our minds changed [12–14]. Oxygen is freely available, water less so and NAD supplies needed for energy and as metabolic master molecules are much harder to obtain. This difficulty has worsened while we "niche constructed" agriculture as sources became expensive creating self-reinforcing transgenerational "poverty traps" that are hard to escape from and contrast with rich others as "meat elites" that "hit the jackpot": more recently we moved away from mixed farming and agroecology to industrial farming with damaging effects on carbon, nitrogen and phosphorus cycles and made inequalities of meat consumption worse in places (**Figures 1–3**) [15–22].

2. Engel's curves and social metabolisms

As Cicero said (43 CE) in his essay De Natura Deorum *"we create, as it were, a second world within the world of nature"* but our conscious keystone species currently has this world divided geographically effecting a NAD and proton-related lottery of our life chances and (brain) health – as well as contributing to climate change endangering biodiversity from many mammalian and other extinctions [23–25]. Life is energetically expensive and constantly needs to overcome the second law of

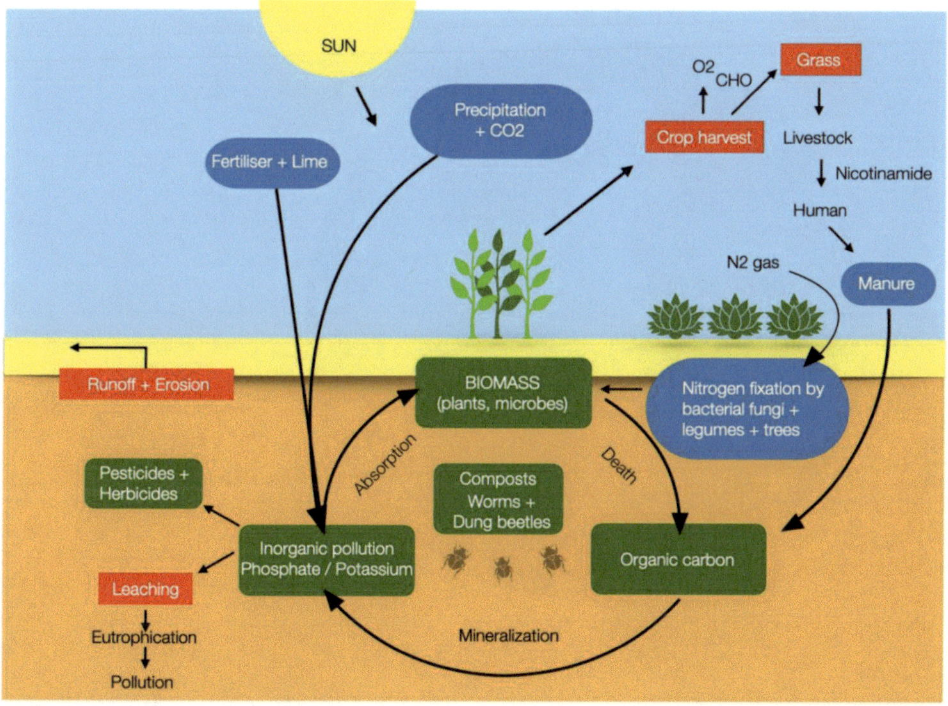

Figure 2.
Many natural cycles such as the nitrogen and carbon cycle were enhanced by good farming techniques but later these got disrupted in the short-term quest for high yields. For instance animal (and often human) manure is usually no longer returned directly to the soil – The classic metabolic rift. Use of natural gas to make ammonia-based fertiliser damages the carbon and nitrogen cycle with methane leaks and CO2 and N20 emissions compounded by eutrophic effects on biodiversity compared with natural nitrogen fixation by clover and legumes. This creates a second metabolic rift as artificial fertilisers mean that animals and plants need not be grown together. Ruminants contribute to CH4 emissions and new pastureland usually requires deforestation. On the other hand perennial grasses are often on land that cannot be used for crops and if well-managed, often with silviculture, can act as a carbon sink, as can the oceans at the same time as supplying sources of nicotinamide.

thermodynamics (as Schrodinger pointed out in "*What is Life*" in 1943) and combat increasing entropy but preferably not, as is happening, at other people's expense [26–31]. This ties in with concepts of a social metabolism and free energy that interacts with internal homeostatic mechanisms (Claude Bernard's 1850 *"milieu interieur"* that can easily be overwhelmed both for the individual, as our discussion of pellagra will demonstrate, and for the anthroposphere [26, 32–34].

Much has been said on the damage done to health and happiness from relative inequality and the spectre of "status anxiety" but absolute disposable income in relation to the cost of a balanced diet may be the more important foundation: Engel's curves predict that once calorie intake is satisfied from starches and sugars families spend more on fresh vegetables and meat (**Figure 4**) [35]. This rise up the food chain that improves both health - and wealth is not a zero-sum game as it is the basis for expansion of markets for actual luxuries and modernisation [36–42]. Low incomes makes this impossible and is exacerbated by relative prices of healthy versus unhealthy foods often being worse with added "poverty premiums" in poor countries and neighbourhoods and the distraction of technological luxuries in a post-industrial world - let alone increasing costs for energy or rent leading to "slow micronutrient starvation" and metabolic risk [43–47]. The true cost (hours of work needed)

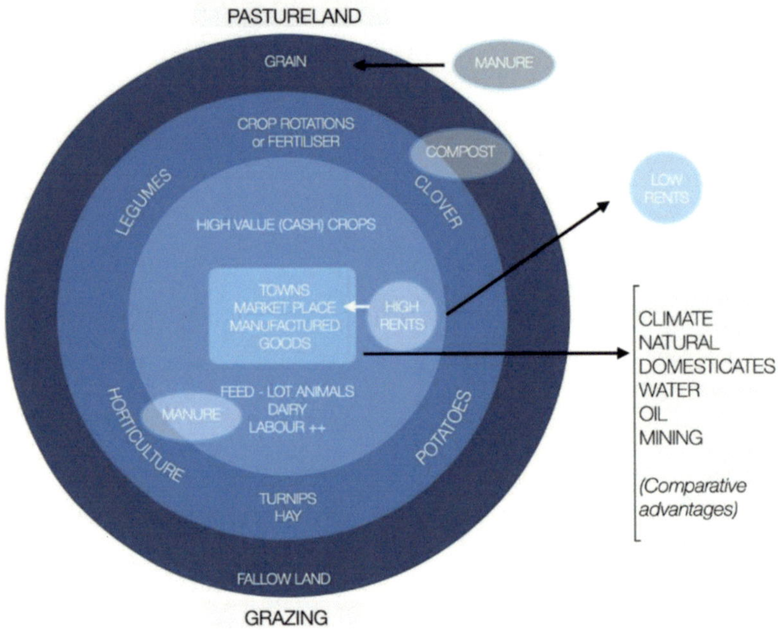

Figure 3.
Many cultures closely integrated a balance between cereals vegetables and animal products. Mixed farming is shown but trading relationships between pastoralists and agriculturalists were common. Where a balance was difficult behaviours included warfare to obtain land or cattle or slaves for farm labour or captives to cannibalise. Hunting was maintained for longer if domesticates were inadequate or sometimes if climate change meant that it had to be re-introduced. Class fights over pastureland with attempts to enclose it and severe punishment for poaching was common. Mixed farming leading to reasonable diets for all through frequent feast days was easier when transport constraints contained food products locally. Industrialisation and urban rural splits developed and even if there were some benefits to the development of water and railway networks to supplies this usually favoured the rich core, with more meat/milk or out of season or tropical foods, not the periphery or third world where the export of cash crops farmed on plantations often destroyed local mixed farming. Distance created space for complexity in supply chains with monopolies and corporate and financial middle-men and though lowering prices non-accountable costs to health and well-being and the climate with environmental degradation and labour exploitation rose. Many pellagra outbreaks were linked to market failures such as that of cash crops such as cotton.

to pay for a breakfast of bacon and eggs has fallen by over 90% over a hundred years for workers but this advantage has not been shared with all within or across countries [48, 49].

In part 1 we talk about pellagra largely in the context of economic class; in part 2 we talk about the development of high meat, and therefore nicotinamide, divergences and convergences between countries and continents. As Charles Dickens said (1859) in another crucial and informative era *"it was the best of times, it was the worst of times, it was the age of wisdom, it was the age of foolishness..."* – we argue it is time to correct a basic and foolish inequality from which many other inequalities derive but we need to review nutritional science and the importance of NAD and the history of pellagra first [50–58].

3. Nutritional science: 19th and 20thC history and public health

Following on from Hippocrates and dietary and exercise regimens and the importance of diet for human progress and "perfectibility" Condorcet, Godwin and

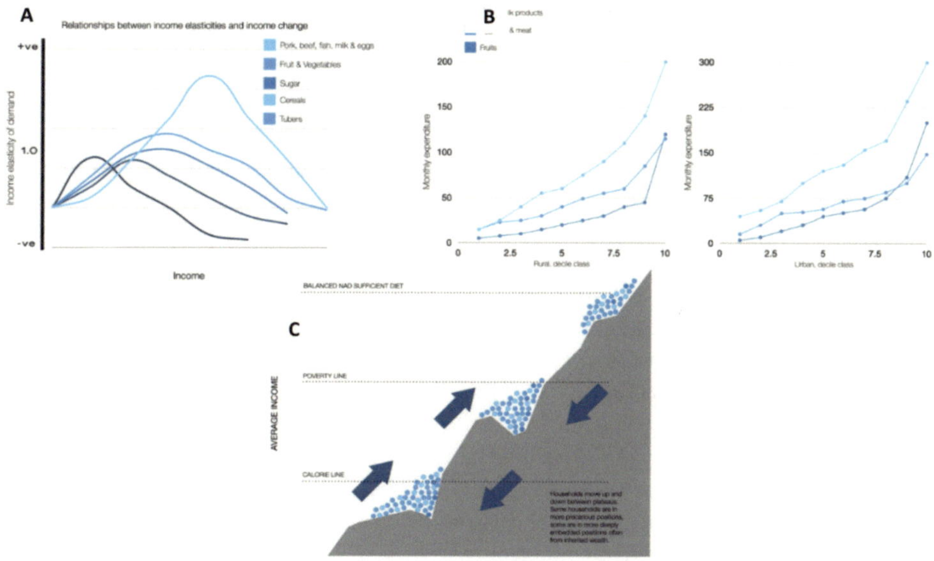

Figure 4.
A) Engel's curves illustrate how the poor will favour cheap starches and cereals for their calories and only when their income rises and a lower proportion of income is spent on food will they eat much meat or vegetables that are more expensive and involve more waste. Income has to pay for other necessities such as rent and (from wood or charcoal to electricity) heat for warmth and cooking as defined first by Maslow's hierarchy of needs. Increasing (technological) "Veblen" luxuries may compete with a healthy diet as do excessively cheap calories from sugars and oils in ultra-processed foods. B) Engel's curves also often differ between rural and urban populations and as we will see later between rich and poor countries. Rich countries are more likely to have subsidised meat and fortified with vitamins, including vitamin B3, cereals. C) an optimal nicotinamide rich, but not too rich, (around 10–15 mg/day) will be well above the poverty line but by no means unachievable if it was made a priority.

Frank (who in 1970 commented "*starvation and sickness are pictured on the face of the entire working class. You recognise it at first sight*") took a more optimistic stance than Malthus. Magendie ("pas d'azote"), Mulder, von Liebig (who developed meat extracts), von Voit, Chittenden ("Nutrition of Man" 1907) and Atwater began to understand the difference between nitrogenous protein and non-nitrogenous calorific fuels and made specific arguments for greater rationality over food choices, including meat. Physicians, such as See (1887) made dietary and sanitary recommendations that included milk to strengthen resistance and reduce exposure to (infectious) disease and infant diarrhoea as did early public health officers such as Newman (who wrote on the importance of nutrition in his book "the building of a Nations health" (1939) and Newsholme. All was who helped by the rise of statistics from Petty (1655) to Graunt and Farr's Life tables who championed the proverb "prevention is better than a cure" [59]. Villerme (1818) and de Chateauneuf showed large differences in mortality and diet between rich and poor in Paris as did a contemporaneous survey in Naples and called for protection against the high price of food: as did Paton (1901) in England and a BMJ leader in 1913 "*..or is it that the wages they command are so low that they cannot purchase the necessities of life*" [60]. Even at the height of "Germ" theory (Robert Koch (1890) and the success of earlier sanitation projects (Edwin Chadwick 1840s) and the first antibiotics some such as William Allison (1830s) and Rene Dubos (1940s) ("think globally, act locally") and Thomas McKeown (1970s) championed a role for diet and the ever- changing ecology to which man has to adapt with microbes being necessary but not sufficient to cause disease [61, 62]. Nutritional interactions with physical degeneration and infection are complex and bidirectional as was pointed out early by Price (1930s) and Scrimshaw (1960s) and others before and since and here [63–65].

Hopkins (1906) led the charge following others, such as Eijkman, on "unsuspected dietetic factors" then named "vitamins" by Funk presenting the idea that beriberi, rickets, xeropthalmia, scurvy and pellagra were nutritional deficiency diseases. Later Kwashiorkor (described in 1933 on maize diets) was initially felt to be more related to protein malnutrition by many but is cured by milk and theory has reverted to it being a form of juvenile pellagra and a general concentration on the importance of micronutrients. Inadequate income rather than lack of education was a controversy partly eliminated by Boyd-Orr (1936) who with others at the FAO(Food and Agricultural Organisation) founded in 1945 and the World bank (influenced by Berg's "the nutrition factor: its role in national development") tried hard to recognise this as a global not just a national issue such as the self-interested response to the poor height and health of army recruits at the time of South African Wars. Cod liver oil usage (vitamins A & D) became common and fortification of some foods including niacin became mandatory in some countries in 1946 and manufacturers soon included large doses in many foods and drinks following on from a period of "Vitamania". Interactions of nutrition and infection were noted as "mutually aggravating" in a comprehensive monograph in 1968 (as remains true for more recent infections such as HIV). Thomas McKeown (1976) proposed the decline in infectious disease particularly TB was largely down to better diet; this was criticised but is now almost orthodox – the data was available then to link it with meat and nicotinamide (an anti-TB antibiotic) but that had to wait till 2013 [66]. Amongst other anecdotal observations a study with a control group from 1946 had shown that a meat and milk richer diet supplied by the Red Cross reduced the incidence of TB in prisoners of war by 15 fold. From the 1930's to recently repeated studies in many populations showed that milk increased linear growth as did meat and animal products and cognition along with increasing foetal weight relative to gestational age (SGA) and reduced mortality [45, 67–69]. Recommended daily intakes adjusted for physiological stresses, including for nicotinamide, have been in place for many years but are rarely monitored and screening programmes for NAD/Tryptophan deficiency have rarely been used even though eminently possible [70].

4. Poverty and subsistence: Austerity never works

Poverty has long been measured in terms of it's relationship to basic nutritional subsistence and the role of rulers responsibilities in providing calories at least for the proletariat long recognised in early empires. Landmark studies included Eden's "The state of the poor" (1797) that was explicit on the effects of rising food prices and was an empirical foundation for Poor Laws as were Booth and Rowntree's surveys and descriptions of "frugal lines" in the late 19th century. Poverty was often then thought inevitable or necessary to encourage work or was the fault of the poor and their habits and addictions. Poverty Enlightenment in the1800's saw the emergence of a new respect for the poor as people and no longer "shadows in a painting" with the economy seen as a tool for promoting human welfare – an important insight of Adam Smith. By 1890 Marshall in his "Principles of Economics" was writing of the "cumulative evil" of poorly fed children on economic potential and the need for progressive income taxes to help children rise out of poverty. Further enlightenments saw poverty as a severe constraint both on personal self-fulfilment and on aggregate economic growth and spawned many anti-poverty and better nutritional policies – even as opposition re-surfaced in the 1970's. More recently the dangers of poverty to the well-being of the rich have become better realised whether from war, migration or pandemics. Engel's law as modified by Bennett has been used as a measure of the escape from poverty line and traps and has been substantiated on many occasions whether at a house-hold, country (enabling a move from farming to industrialisation) or at an international level correlating with economic growth. Growth on its own rarely led to reductions in poverty unless it started in the agricultural and service sector enabling lower real food prices and better diets and higher meat intake before there is disposable income to support other aspects of an industrial and consumer society. Once economies are more established (or have used natural resources wisely) this becomes hard to appreciate as agriculture then plays a fairly insignificant role in gross GDP and financialization and "rentier capitalism" that conveniently forgets its roots and responsibilities. Progress is being made given that around 1820 80% of the world population was extremely poor but nutritional status as measured by amount of "wasting" or "stunting" (that includes cognition) is still high and the poor or poor national governance are still being blamed. The example of pellagra demonstrates that most lingering criticisms are however the effect not the cause of poverty in a vicious transgenerational cycle as we will demonstrate.

5. Preconception nutrition: precondition for equality of opportunity for unborn billions

The evidence for the importance of good diet not only early and throughout life but pre-conception and in previous generations continues to mount, building on the epidemiology of the Dutch winter and studies on low birth weight, and affects on both initial (brain) health but also the risk of later onset metabolic disease. This can account for much of the "missing heritability "of many diseases and demonstrates that very early investments (Heckman Curve) can be by far the most cost effective compared with end – of- life health costs- and emphasises the dangers of hunger and austerity measures [71–76] (**Figure 5**). Epigenetic mechanisms include DNA methylation and interactions with NAD/Nicotinamide metabolism with interventions favouring animal based foods and fairly low quantities of milk and eggs having major beneficial effects on cognitive performance and behaviour [77–84]. This also explains the high investment parents will place to allow long successful childhoods so important to our evolution but when impossible comes at the price of reducing social mobility for many but does allow for significant neurodiversity [85–90]. Other periods when brain development is sensitive to environmental and dietary influences are adolescence and "mother brain" with changes in white and grey matter and a pruning of synaptic connections that may be important for brain reserves and ageing well [91, 92]. Improvements of diet throughout life would have short-term and long-term future gains that compound so there is a high discount rate for the current cost for descendant generations fulfilling an obligation to unborn billions for their health and a more cosmopolitan attitude rather than more superficial concerns for local social justice [93, 94]. Better nutrition such as school "breakfast clubs" and free lunches (with dignity) works at all ages and even once in college (or prison, hospital or care homes) it can be shown that hunger on campus affects learning and results, let alone later in life [95, 96].

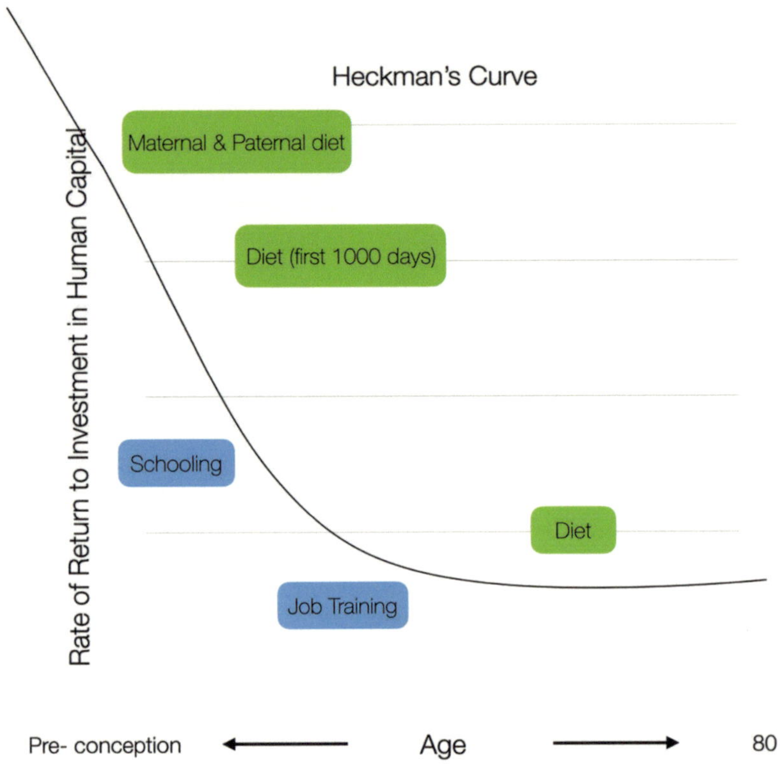

Figure 5.
Heckman's curves illustrate how the rate on return to investment for diet and education rises steeply if begun early - with reductions in many forms of brain and cardiovascular disease. In the case of diet this includes pre-conception parental and even grand-parental diet working through epigenetic and DNA-methylation paths that interact with nicotinamide metabolism. The graph would be steeper for poor countries with short-term benefits merging with long-term compound benefits for future generations. There are limited time windows where catch-up growth can occur with some developmental plasticity but this can create mismatches later in life where ample diets can precipitate metabolic disease and the full gamut encompassed in the developmental origins of adult disease (barker) hypothesis. Low nicotinamide early may cause a phenotype with less neurones more prone to the ravages of ageing but also leads to low induction of nicotinamide methylation that does not forecast a luxury diet and risks nicotinamide toxicity even if diet improves later.

6. NAD history of a key molecule

Harden (1906) discovered NAD as a cofactor whilst investigating fermentation and this was followed by demonstrations of NAD(P)(H) role in anabolism and calcium signalling and NAD(H) in hydride transfers and mitochondrial energetics. Mitochondrial failure is a major feature of many neurological degenerative conditions and often involves other organs. More recently NAD-synthases "feeders" (NMNATs) have been shown to have an important role in (Wallerian) degeneration [97, 98]. NAD-consumers as ADP-Ribose polymerases are known to be important in DNA repair (PARPs) and Sirtuins in ageing (NAD declines with age) with agonists or supplementation extending lifespans and rejuvenating stem cells and promoting tissue regeneration at times of stress, including food scarcity [99–101]. Sirtuin deacetylation (including histones) and NAD affect epigenetic mechanisms regulating genomic and epigenomic stability. NAD turnover is high and supply pool is competed over and is compartmentalised and fluctuates such as with circadian rhythms and is dependent on dietary nicotinamide. There is an endogenous pathway from the degradation of (dietary derived) tryptophan on the kynurenine pathway that also profoundly affects immune tolerance and is affected by many disease processes and pregnancy [102–104]. Hormetic anti-ageing influences from plant derived compounds, as part of their own defences, such as resveratrol also predominantly act through agonist actions in NAD-Consumer pathways, as does exercise and caloric restriction [105]. NAD connects with the social and sexual dyad relationships through oxytocin regulation [106, 107]. NAD fluxes through energetics and its consumers and signalling is a central master-controller of metabolism and we, in a very real sense, live in a NAD metabolome and microbiome with symbiotic and social breakdowns when the supply fails [108]. Satiety and avoidance of food, sex and gambling addiction and substance abuse from alcohol to cocaine has been linked with circadian NAD replete pathways affecting opioid, nicotine and cannabinoid receptor and dopamine, serotonin, oxytocin and adenosine reward circuits [109]. We may be addicted to these surrogates for NAD as we may be to NAD and its metabolites just like some extreme mutant cancer cells [110]. NAD replacement may work for a wide variety of diseases of ageing as may Nicotinamide N-Methyltransferase (NNMT) inhibition as this enzyme is induced, perhaps by high nicotinamide in diet, in a wide variety of "diseases of modernity" whether metabolic, cancerous or degenerative [109, 111, 112]. We are be in a "NAD World" with many of our senses, metabolism, physiology and actions attuned to obtaining and the optimal use of this supply in our "umwelt" [113, 114].

7. Pellagra's epidemics

Pellagra was first described in the medical literature by Gaspar Casal (1735) in a European epidemic amongst poor polenta eating peasants. He recognised its basis in a poor maize based and low meat diet as had the people affected (one offering to sell her house for a round of butter) [115]. Another famous epidemic was a century ago in the south-eastern "cotton" states of America triggered by economic hardship amongst sharecroppers and slaves. South Africa also had outbreaks triggered by rinderpest in cattle, poverty and apartheid. More recently outbreaks have occurred usually in war-torn African countries. Death rates are high and undiagnosed let alone untreated cases are common [116, 117].

8. Clinical features: 4 D dysfunction & degeneration across generations

Dermatitis was common and sometimes pathognomonic as a seasonal form of sunburn. Darker skins were somewhat protected limiting diagnosis and self-diagnosis that could otherwise lead to self-treatment and eating more meat before specific treatment was even available risking cognitive decline. (During our early history this may have favoured white skinned peoples in more northern climes that also had more meat in diet but less sun). Dementia and poor intellectual development were important features that included a raft of other neuropsychiatric syndromes, including a myoclonic encephalopathy, and poor and antisocial behaviour. Mimics of Parkinson's and Motor Neurone disease were noted. Diarrhoea was common due to inflammation and dysbiotic infections – other dysbiosis included TB – and there was susceptibility to acute infections. Death and premature ageing were common although rarely mentioned by the current ageing literature. Women were affected twice as often as men (and often have less access to meat in the home). Pigmented skin protects against the characteristic sunburn rash but may be a mixed blessing if then poor cognition is overlooked and not (self-) treated particularly as in Africa, Asia and the Americas meat was less available.

Pellagra was thought at the time as being a systems failure and degenerative in the atavistic reverse evolutionary sense of the term (even before the importance of meat in our evolution was realised) let alone being neurodegenerative in the modern sense. Pellagra is also an excellent example to remember in the nature via nurture and in Mind-Gut axis debates [118]. Pathology crossed many current classifications as there was evidence of degeneration, mitochondrial dysfunction, oxidative stress, dysbiosis, amyloidosis and manifestations with no discernible histopathology [41, 119].

9. Addiction: early opioid and other drug endemics and wars

Alcohol addiction and ritual drinking bouts "potationes" was well recognised but by interfering further with NAD status makes things worse and can be a cause of pellagra alongside thiamine deficiency and Wernicke's and Korsakoff syndrome both treated with multivitamin preparations. Addiction to food and feast days is more understandable but other addictions whether sex of gambling played in to developing pellagrins a poor reputation. Self or institutionalised and medicalised treatment with drugs even infants is more surprising. In the 17th century the poor cultivated on an industrial scale and ate poppy seed bread "pavrato" or "pane papaverino" and may well have caused the first "opioid" and "cannabinoid" epidemic. Other pain killing and hallucinatory herbs, including hemp, or spoiled rye with ergotism and St Vitus's dance were as a response to this "disease of wretchedness" benumbing an already stunned undernourished population who could only dream of a land of Cockaigne and a world to be turned upside down "mundus inversus" as often too weak to rebel. Opium has been used since the ancient world but not on this scale that subsequently got magnified with major effects in China, note 19th C "opium wars", and recently as is well documented causing incarcerations (though not of the Sackler's or other manufacturers of fentanyl and analogues) and life expectancy falls particularly in the USA [120–124]. This ultimate in unconscionable abuse of bio-power perhaps started with pellagra-related "deaths of despair" three centuries ago but is only recently recognised as having a link with NAD deficiency helped by NAD infusion [125].

10. Treatment

Joseph Goldberger in the USA during their epidemic beginning in 1902 is rightly credited for renewing the dietary link (proposed for the Nobel Prize three times) as then it was widely thought to be due to a poison in maize or infectious or genetic. He recommended high meat and fresh diets or the use of brewer's yeast to good effect for both prevention and cure and recognised the similarity with "black tongue" in dogs and the link to tryptophan. Conrad Elvehjem discovered nicotinic acid and the link to pellagra (1937) and Tom Spies and colleagues rapidly (Time magazine Team of the Year 1938) treated patients who even in advanced cases could respond.

11. Butterfly castes to butterfly effects: racial, class and other divides are not so black and white

Pellagrins whether white or black were derided as inferior human beings or even as a different race, with many epithets still used against "deplorables", and were known as the "Butterfly caste" (referring to the characteristic dermatitis) and "problem pellagrins". At various times those from Ireland and Italy have been treated much the same so colour may be a distraction from the real underlying cause of many "them and us" as really "have and have-not" divides and extreme reactions such as the Klu Klux Klan, segregation, apartheid and genocides and the occasional successful revolt such as on Haiti.

Racial and colour awareness can be date back to at least biblical times as in the Christian and Jewish myth of the "curse of black skinned Ham but the origin of whole populations having inferiority complexes has been dated to the 17th century German speaking countries (extraordinary given the later history of racial purity and Aryan supremacy) relative to the French who were dominant in all spheres of life whether material or cultural and was exacerbated by the thirty year war – when German livestock were rounded up and driven away - and these countries were in an early financial crisis with hyperinflation and a credit crunch [126].

Anthropologists travelling the world, such as the neurologist W.H.R.Rivers, dispelled some myths that "primitive" races had sharper perception as their minds used little energy for higher mental functions (the "Spencer hypothesis") although not many were convinced during strong eugenic movements [127–129]. Other anthropologists from the 1930's on have documented behaviour changes and dehumanisation during the "hungry season" (when meat is scarce and pellagra outbreaks start) or when peoples are removed from their hunting grounds noting societal and familial breakdown (with cannibalism and infanticide at the extremes) and include Turnbull's 1960's studies of the Ik tribe "Mountain People" and Scheper-Hughes 1990's Brazilian study "Death without weeping" of slow starvation often medicalised and treated with drugs [130–134]. Similar observations were made in the Ukraine 1930's famine, the Siege of Leningrad in 1941, and many Chinese famines as described for the 1960's Mao responsible famine in "Hungry Ghosts" [135]. Acquired constitutionally poor phenotypes secondary to financial hardship and a slide down Engel's curves precede societal stigma and institutional bias to "wastrels" and discrimination and medicalisation can hide the basic nutritional cause and responsibility even in the face of rebellion [63].

Bacon's 1676 rebellion is remembered as, for once, poor whites and poor blacks united to fight for their rights, yet still failed as did similar attempts later in America, well analysed (by Du Bois around 1900) and South Africa. The calculated response by well fed and nutri-genomically well-endowed northern European groups, who felt superior (and were taller), was that poor whites, often indentured servants or imported criminals, needed preferential material or status enhancing treatment so as not to spoil their own case for supremacy (upset by many shocks such as the Japanese attack on Pearl Harbour) [136, 137]. Hence Jim Crow laws, with apartheid menial jobs with and low incomes then prohibition "wars" first on alcohol and then drugs with incarceration all denying "inferior" groups any realistic chance of levelling up.

As Martin Luther King said in 1968 when about to launch the Poor People's Campaign just before he was assassinated *"in one America people have quality food and education, in the other the best minds can never come out"* [138]. Based on personal reversible experiences, the current mayor of New York has recognised that a diet derived from slavery to unhealthy soul food is a key front line in the fight for civil rights [139]: and, we say, was the original bottom line with a low meat intake affecting physiology and behaviour with a the willful drive to degeneration by further dietary restrictions and then blaming the victims [140–142]. Recent calls support this view that poor diet is the common denominator as in Louis Gates comments in *"America beyond the colour line" ."if King came back, he'd say we need another civil rights movement based on class not race"* reflecting class divides in people of all colours and the long-term effects of "segregation sequiturs" such as low wealth and income and poor diet - democratic socialism with strong economic rights underpinning diet and (electrical) energy rights is the way forward [143–146]. Pellagra's butterfly effects from origins in the southern states and south Africa bastions to this day of some of these frictions are still the most acute could worsen despite the importance of diet and pellagra prevention being well known for over a century [147].

12. Euthenics not eugenics

Ellen Swallow (1900) is credited with emphasising diet and ecology in human health and combating eugenics with euthenics and social welfare responses to inequality and the vagaries of birth and racial improvement; life not being self-authored [148]. Others despite their interest in eugenics emphasised diets with both proponents for vegetarianism improving the race, such as Graham and Kellogg (1914) and for meat, such as Woods Harrison (1911) - later taken seriously by countries such as Japan and China. McCarrison working in India and Boyd-Orr in Kenya understood that there was an interaction between tribes, races and diet with the Masai and Sikhs benefitting from blood and meat- based diets with taller heights and better health (including less TB) [149]. Even monarchies were at times pioneers of welfarism and sceptical of genealogy or meritocracy as were some social reformers - common prosperity as a mantra is now even championed in China is realised to have a dietary and euthenic habitat related basis – much based on the need for more meat [150].

13. Pellagra prevention

For most of our evolution and history as hunter-gatherers this was pre-solved as our diet was so meat based both as a source of micronutrients and for calories particularly in temperate zones so catastrophes would have caused localised or seasonal famines rather than chronic or widespread or lifelong malnutrition [151–155]. High intake of nicotinamide from hunting may have been important to the evolution of large cooperative brains with pellagra being an atavistic example of evolution in reverse gear. Furthermore there was a very strong sharing culture particularly for meat with anyone attempting to obtain the "lion's" share actively shunned [156]. Storing and preserving meat developed to overcome "seasonal" hungers or other interruptions to the usually ample supply helped by hunting parties leading the global diaspora. However, that led, along with climate change, to megafaunal extinctions on land and sea and a move to smaller prey and the beginnings of herding. After the agricultural neolithic revolution pastoralism developed to help boost the meat supply. The neolithic revolution was overall a net negative meat transition with adverse effects on health, height and brain size whilst populations exploded on a more pellagra-genic diet with malnutrition as hinted at by descriptions on "steles of famine" [157, 158].

Pastoralists and agriculturalists had trading arrangements allowing for a balanced diet (despite some taboos), even if these were at times fraught – much early raiding and warfare can be seen in this light as will be later empires acquiring pastureland. Pastoralists were not only providers of meat but also developed "secondary products", at multiple independent sites, butter/ghee and soured dairy products from milk curds

MACRO & MICRO-NUTRIGENIC ADAPTATIONS - NAD RELATED		
Amylase copy number	-	Starches
Alcohol dehydrogenase	-	Beer / Wine
Lactase persistence (Northern European)	-	Milk
Nicotinamide methylation (Carnivores)	-	Meat / Milk
NAD ENVIRONMENTAL - HOMEOSTATIC INTERFACES		
NAD / NADH (Redox state)	-	Mitochondrial ATP (Brain) power
NAD (P) H	-	Anabolism Xenobiotic detox
NAD - Consumers + sensors (Sirtuins)	-	Circadian rhythms Foraging - meals Master metabolic controls
Signalling / Chromatin making	-	Genetic memory Fetal programming / thrifty phenotype (DoHaD & transgenerational effects)
Nicotinamide / SAM metabolism	-	Ageing (Calorie restriction) Exercise

Figure 6.
There were several nutrigenomic adaptations during our recent evolution that enhanced our NAD diet most notably the development of lactose tolerance. Nicotinamide n-methyl transferase (NNMT) induction protected us from nicotinamide excesses and close control of NAD metabolism evolved through upgrades of NAD -consumers. Close links with environmental stimuli were forged affecting most metabolic and (neuro) physiological pathways.

(and whey) as fermented cheeses and yoghurts with low lactose levels but just as rich in nicotinamide riboside and more easily stored and transported.

In the best example of parallel cultural and genomic convergent evolution, suggesting extremely strong selection pressures lactase persistent genes emerged independently in several pastoralist populations, particularly in northern Europe, to overcome the natural decline in enzymic activity after infancy and avoiding the unpleasant symptoms of lactose intolerance such as flatulence and diarrhoea whilst cultural evolution developed cheeses [159] Other nutrigenomic developments allowed for greater ability to digest starches and alcohol that connect to NAD pathways and NNMT that, as with other carnivores may guard against nicotinamide overload as would many rearrangements in NAD-consumer pathways all of which are active in general and brain metabolism (**Figure 6**) [160].

Some population, such as those in Mesoamerica, had to adapt to low meat diets due to a lack of easy animal domesticates compared to the "big four" of current animal domesticates that were all herding pre-domesticates in the middle east's fertile crescent. A major staple maize co-evolved and achieved God-like status (as did cattle elsewhere as the "food of the Gods") as it is easy to grow and a good source of calories but not of tryptophan or nicotinamide [161–163]. In a cultural evolutionary approach, suggesting strong selection pressures, in the fields it led to the "3 Sisters" joined by beans and squash enabling growth through nitrogen fixation and both protection of the soil and as sunshade - and a broader diet with extra nicotinamide and amino-acids. As striking was the (subconscious) development of cooking with lime as "nix-tamalization" releasing bound nicotinamide - a practice not always exported to other continents leading to pellagra outbreaks in Europe and elsewhere. There are other examples of cooking practice that evolved to reduce toxicity from cyanogens and other toxins as did enhanced xenobiotic enzymes but this is the most striking example of releasing a micronutrient and reducing toxicity from its potential deficiency [162].

After the discovery of nicotinamide and its relationship to pellagra in the 1940's supplementation programmes were introduced in places starting with bread. Brown bread contains some nicotinamide in the germ but when processed to white bread is largely removed: although white bread was initially the preserve of the rich it became more popular, in a curious and unfortunate inversion, amongst the poor on low meat diets. (Later the pasteurisation of beer had the same effect by removing brewer's yeast an important source of vitamin B3 used as treatment in some epidemics). Manufacturers soon followed first of cereals, as "Vitamania" took hold and later widespread currently in "high energy" drinks and even teas. However, this usually targets the wrong global populations at risk of deficiency and has the disadvantage of being a potential drain on the methylome, unlike adequate meat. Along with effects on NAD-Consumer enzymic activity this could give mechanisms for a hypervitaminosis B3 as is suggested by excessive induction of NNMT in many diseases of "modernity". In other words, creating the worst of all possible "NAD Worlds" with missing treating cases of subclinical endemic pellagra at the same time as increasing the potential for nicotinamide overload and toxicity [17].

14. Endemic pellagra

During the pellagra epidemics, particularly the American outbreak in the southern cotton states a century ago, not only was it recognised that many cases went undiagnosed but that the rash was often not present and a condition of "pellagra sine pellagra" was endemic and probably had been since well before the Civil War – and may indeed have contributed to the confederate defeat. Low IQ in recruits and poor physical health, such as diarrhoea (now called "environmental enteropathy" or TB, were well recognised as a problem at the time: the finding of better IQ scores from black northerners compared with white southerners was an important if inconvenient truth for many. TB disappears once income and diets improve, presumably as nicotinamide is a TB antibiotic, and this has had a profound effect on population health although may lead to the emergence of allergic and auto-immune disease [164, 165] (**Figure 7**). At risk groups on low to very low meat diets may always have harboured sub-clinical cases and may still do whether seen through the lens of class or continent/country. It is worth a short historical detour to investigate this possibility accepting a lack of data due to no biochemical screening tool for NAD/Tryptophan deficiency in use, inexplicable and unacceptable for an easily treatable condition.

15. Class and NAD scales and ceilings: meat pecking orders

Since the Mesolithic and neolithic move down the food chain to a less meat based and more horticultural then cereal and calorific diet hierarchical class-based systems evolved in all sedentary civilisations [19, 152, 166]. If meat is less available it might make some sort of teleological sense to develop an upper ruling and

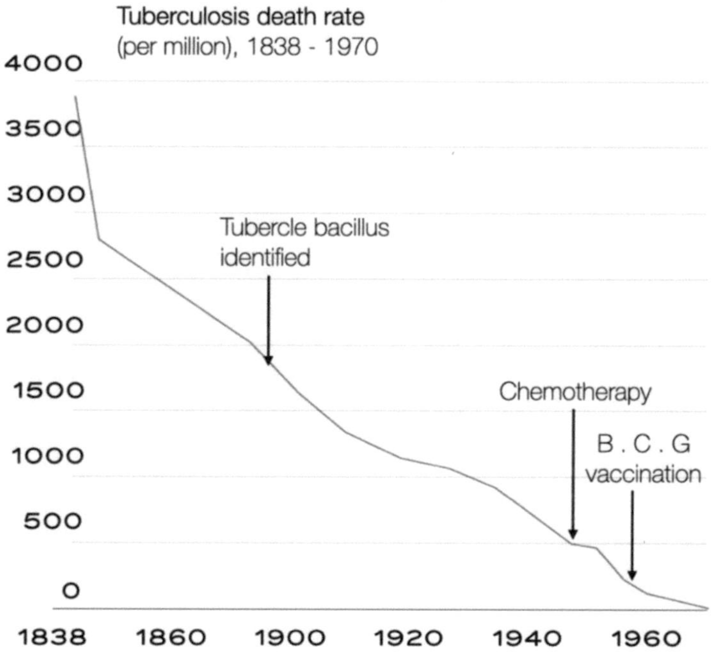

Figure 7.
TB rates fell dramatically before medicines became available and is a hallmark of modernity with a shift that increases life-span but comes with more allergic and auto-immune disease. Sanitation has little effect on this non-water or food-borne microbe and better housing increased rents reducing money for food and has, on documented occasions, led to more TB. Nicotinamide and it's analogues are TB antibiotics so there is a biological explanation and given TB's role in "educating" the immune system absence of this "old friend" may explain it's replacement with the emergence of allergic and auto-immune disease.

officer class on the better diet with high level cognition and a proletariat working and infantry class with, almost by definition, higher fertility. Higher populations allowed for greater divisions of labour and inventiveness. Much of human history can be explained in this light with trading (and wars) with pastoralist (mounted) "barbarians" to obtaining more pastureland by conquest ("Lebenstraum") – given class wars when the ruling classes did not honour their promises to feed the "common" people whose lands and hunting or farming rights they had taken as part of "enclosure" movements (**Figures 8** and **9**). Extreme versions such as wars to obtain captives for human sacrifice rather than land as state sponsored cannibalism for the upper classes in Mesoamerica [167]. Recent versions are evident the history of the working classes and their "snakes and ladders" of failures, whether white or black, if only education and not income or diet is addressed [168–171]. When the political will is there it is possible as examples include successful victualling often cited as being important for victory in all wars and well documented for the Napoleonic wars with significant meat intakes for the sailors compared with their enemies (and non-combatants at home) [172, 173].

High meat, milk, butter and cheese intake amongst the wealthy often amounting to gluttony and "orthorexia" is well documented everywhere that it has been studied for the last 10,000 years with documented effects on height, health, IQ and power as is the opposite effects of forced often monophagic vegetarianism on the working classes or low-income unemployed, especially women and including children [174–176]. Meat as the centrepiece (sometimes along with cheeses) has

Figure 8.
Gillray's 18thC cartoon may have been the first to illustrate the connection between class differences and cuts of the national "plum pudding" with meats to the king and his courtiers down to the "scrapings" for the labouring poor (the value of roast beef and the evils of alcohol were also well depicted by Hogarth). The carving up of empires between great powers and "state epicures" and" greedy-guts" divided the world to the great detriment of the poor in poor southern nations. Now rather than Palmerston and Napoleon leaders from Eurasian and Western blocs and food monopolies carve up the world – The dangers of this are profound and a common purpose needs to be established fast to correct this basic human right and create a world safer from pandemics.

long been important for the rich and for distribution in feast days - that used to be a lot commoner – or by monasteries (who had a beneficial effect on agriculture even after their dissolution). Peasant dishes can get gentrified, an old example being the addition of meat in medieval times to "frumenty" porridge, with many more examples when dishes were imported to richer climes as in the "hunger for America" diasporas from Europe [177, 178].

The rise of the middle class bourgeoisie on middling and more meaty diets honed in restaurants in countries who by happenstance got ahead on the meat curve is a likely basis for prosperity in a feed-forward cycle given effects on cognition and behaviour and a culture of truth and "matters of fact" that recognises talent from all classes as the history of the Royal Society and other legal and scientific institutions demonstrate [179–183].

Nevertheless a dangerous class divide persists that is known as the "Lockian proviso" and can be summarised as *"if the process of civilization has deprived the members of society of certain liberties (to gather, pasture, engage in the chase) compensation would be due to those persons for whom civilization was a net loss"* allowing a moral argument to steal under those circumstance offers taken up from "Robin Hoods" if there is inadequate pre-or re- distribution to the poor [184–186]. Rulers cannot rule by force alone so a cultural consensus and social contract is normally achieved even if periodic realignments have to be made after periods of austerity and "gilded" ages – we may be in one now exacerbated by Covid with monetary policy in the rich world risking a two-track divergent economic recovery between both classes and countries with their

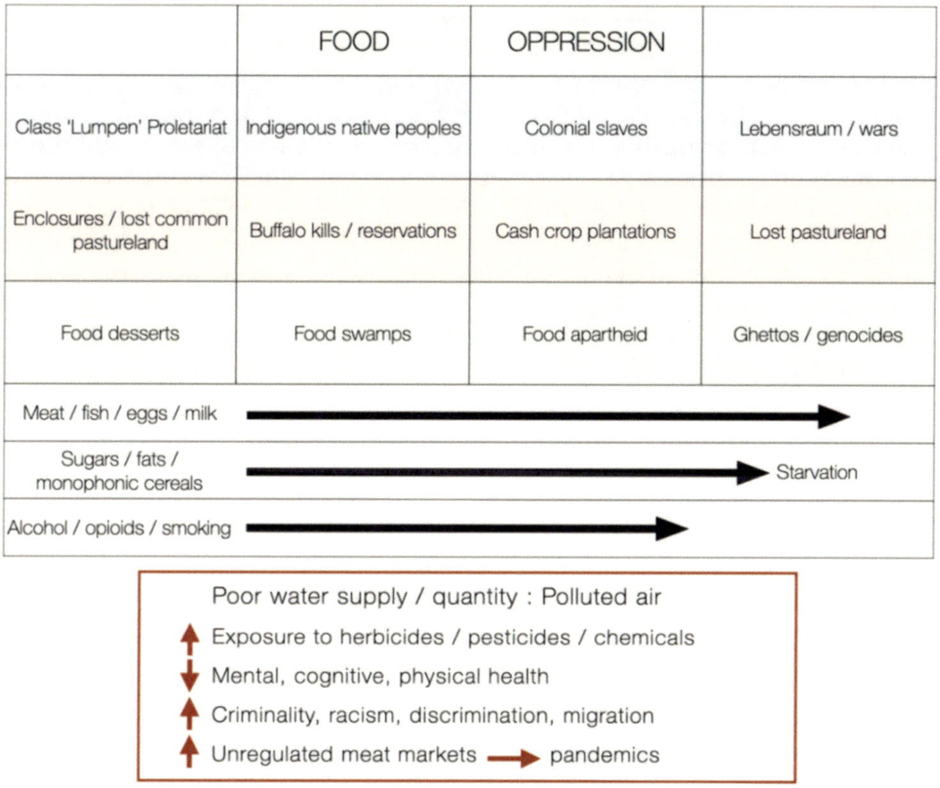

Figure 9.
Most forms of oppression have been crimes of commission (despite being resisted and criticised at the time) mediated by interrupting "swords v ploughshares" food and animal product chains often in (civil or class) and "guns or butter" conflicts, raids and wars. Sometimes these constraints targeted those who started from a poorer dietary base such that infectious disease collaborated in population destruction, such as the Americas, but not always as the bourgeoisie were attacked in the midst of collectivism as with the Russian kulaks and in Cambodia, China - and many Jewish pogroms. Crimes of omission have contributed to at least as much malnutrition and premature deaths or weaponised food supplies.

debt burdens and poor access to increasingly expensive food and agricultural investment funds [187–189]. Cries that we cannot afford more solidarity should remember what Keynes (1942) said *"anything we should actually do, we can afford.*

16. Discussion

Discussions about causation and the origins of class and racial inequalities typically consider only the last 3–400 years and when dysfunctional blame proximate factors such as the need for cheap or surplus labour or, various forms of (neo) slavery and related lack of (hereditary) capital or, "laissez-faire" neoliberalism or, weak democratic institutions and corruption or, property laws and loss of the "commons". However if one takes a longer view it is clear that our pre-history ancestors were far more egalitarian in general and particularly so over the sharing of meat. Later class gradients of meat eating emerged and perhaps allowed some increased diversity in a largely isogenic species with high human capital when meat intake was high but higher fertility amongst the proletariat that may have been of overall benefit. However, when taken to extremes with very low intakes "precariats" and "underclasses" let alone frank pellagra developed that became racialised and has caused much friction over the last few millennia. This underlying dietary motor has become obscured although poor diet as a transgenerational and within lives mechanism for exacerbating or causing racial, sexual and transgenerational inequalities is gaining traction and is, we believe, a constructive and practical way of dealing with reparations by eliminating food deserts and swamps with their "empty" calories and "amputation capitals" and other metabolic ghettoes – we should re-invent being" together at the table" and adequate sustenance with sustainability for all [190].

Reducing resource stress and food insecurity improves cognition and reduces loss of cognitive bandwidth by having to spend less time on earning to provide a basic diet over and above calories and giving more time to spend on education and family - with no need for breaking laws.

17. Conclusions

Engel's Law (1857), is one of the most robust regularities in economics, showing an uncanny fit both within and across countries with increasing incomes (or falls in prices) leading to a decline in the share of family income spent on food. Indeed, given other necessities such as heating, clothes, shelter and entertainment, it can be used as a measure of poverty. Bennetts (1941) modification demonstrated that as income rises less is spent on starches or cereals and more on milk and meat and fruit and vegetables up to a point where it then can fall off for meat. Indeed, we have previously proposed a hypervitaminosis B3 and there are other signals that too much meat may be toxic. Engel's curves are subject to major distortions if calories become very cheap but meat fruit and vegetables remain expensive as then this normal progression will become attenuated – especially if there is also competition from other "luxuries" with remarkable penetration such as mobile phones. These destroy the market for meat as demand seems to fall as there is not enough disposable income to support a safe market encouraging dangerous black markets in "bushmeat". No wonder there is no longer the same clear divide between the move from infectious to auto-immune disease and obesity now becoming common and now a feature of poverty due to the emphasis on cereals and oils and sugars and empty calories in food swamps creating a "double whammy" and sick societies [191].

We have given some background on NAD metabolism with pellagra being the "tip of a metabolic iceberg" and the importance of animal products as the source of nicotinamide and how it's deficiency can cause many disease, dysbiotic and social phenotypes that include poor cognition and anti-social behaviour with addictions creating classes that become dysfunctional. In *"Silencing the past: power and the production of history"* Trouillot (1995) insisted that *"the ultimate mark of power maybe it's invisibility; the ultimate challenge, the exposition of it's roots"* and this meat-power relationship needs to be made visible so that it can be corrected and allow multiple "Phoenixes" to rise.

Acknowledgements

This study was funded by QEHB Charity, Birmingham, UK.

Conflict of interest

The author(s) declared no potential conflicts of interest with respect to the research, authorship and/or publication of this article.

Author details

Adrian C. Williams[1]*, Christina Wood[1] and Lisa J. Hill[2]

1 Department of Neurology, University Hospitals Birmingham NHS Foundation Trust, Birmingham, UK

2 School of Biomedical Sciences, Institute of Clinical Sciences, University of Birmingham, UK

*Address all correspondence to: adrian.williams1949@gmail.com

References

[1] Arora N. The Lottery of Birth: On Inherited Social Inequalities. India: Three Essays Collective; 2017

[2] Smil V. Energy and Civilization: A History. MIT Press; 2017

[3] Marshall M. The Genesis Quest: The Geniuses and Eccentrics on a Journey to Uncover the Origin of Life on Earth. University of Chicago Press; 2020

[4] Pinson J. The rise and reign of the mammals: A new history, from the shadow of the dinosaurs to us. In: Amer Assoc Advancement Science. New York, Washington, DC; 2022

[5] Scheidel W. The Great Leveler. Princeton University Press; 2018

[6] Kent S. Farmers as Hunters: The Implications of Sedentism. Cambridge University Press; 1989

[7] Shepard P. The Others: How Animals Made us. Human: Island Press; 1997

[8] Maestripieri D. Games Primates Play: An Undercover Investigation of the Evolution and Economics of Human Relationships. Basic Books; 2012

[9] Kean H, Howell P. The Routledge Companion to Animal-Human History. Routledge; 2018

[10] Morris B. The Power of Animals: An Ethnography. Taylor & Francis; 2020

[11] von Hippel W. The social leap: The new evolutionary science of who we are. In: Where we Come from, and What Makes us Happy. Harper Collins; 2018

[12] Jebelli J. How the Mind Changed: A Human History of our Evolving Brain. Little, Brown; 2022

[13] Pollan M. This Is your Mind on Plants. Penguin Publishing Group; 2021

[14] Burton FD. Fire: The Spark That Ignited Human Evolution. University of New Mexico Press; 2011

[15] Berg AD, Muscat RJ. The Nutrition Factor; its Role in National Development [by] Alan Berg (Portions with Robert J). Muscat: Brookings Institution; 1973

[16] Pimentel D, Marcia H, Pimentel MS. Food, Energy, and Society. Third ed. CRC Press; 2007

[17] Williams AC, Hill LJ. Inequality: The dangers of meat haves and have-nots in a nicotinamide-adenine-dinucleotide world. In: Meat and Nutrition. Vol. 3. London, UK: Intech Open; 2021

[18] Bowles S et al. Poverty Traps. Princeton University Press; 2011

[19] Outram AK, Bogaard A. Subsistence and Society in Prehistory: New Directions in Economic Archaeology. Cambridge University Press; 2019

[20] Altieri MA. Agroecology: The Science of Sustainable Agriculture. CRC Press; 2018

[21] Weis T. The Ecological Hoofprint: The Global Burden of Industrial Livestock. Bloomsbury Publishing; 2013

[22] de Wit MM. Abolitionist Agroecology, Food Sovereignty and Pandemic Prevention. Daraja Press; 2021

[23] de Blij H. Why Geography Matters, more than Ever. USA: OUP; 2012

[24] Rieff D. The Reproach of Hunger: Food, Justice, and Money in the Twenty-First Century. Simon & Schuster; 2016

[25] Ungar PS, Teaford MF. Human Diet: Its Origin and Evolution. Bergin & Garvey; 2002

[26] Baccini P, Brunner PH. Metabolism of the Anthroposphere, Second Edition: Analysis, Evaluation, Design. MIT Press; 2012

[27] Bailey KD. Social Entropy Theory. State University of New York Press; 1990

[28] Clark RP. The Global Imperative: An Interpretive History of the Spread of Humankind. Taylor & Francis; 2018

[29] Frank AG. ReOrient: Global Economy in the Asian Age. Univ of California Press; 1998

[30] Ayres RU, Simonis UE. Industrial Metabolism: Restructuring for Sustainable Development. United Nations University Press; 1994

[31] Lessenich S. Living Well at Others' Expense: The Hidden Costs of Western Prosperity. John Wiley & Sons; 2019

[32] Parr T et al. Active Inference: The Free Energy Principle in Mind, Brain, and Behaviour. MIT Press; 2022

[33] Prigogine I. Order out of Chaos: Man's New Dialogue with Nature. Verso; 2017

[34] Thomas L. The Intersectional Environmentalist: How to Dismantle Systems of Oppression to Protect People +Planet. Little, Brown; 2022

[35] Williams AC, Hill LJ. Meat and Vitamin B3: Getting a Grip on Engel's Curve. Intech Open; 2021

[36] Clapp J, Moseley WG. This food crisis is different: COVID-19 and the fragility of the neoliberal food security order. The Journal of Peasant Studies. 2020;**47**(7):1393-1417

[37] Federico G. Feeding the World: An Economic History of Agriculture, 1800-2000. Princeton University Press; 2010

[38] Food M, C. Energy and the Creation of Industriousness: Work and Material Culture in Agrarian England, 1550-1780. Cambridge University Press; 2011

[39] Pilcher JM. Food in World History. Taylor & Francis; 2017

[40] Standage T. An Edible History of Humanity. Bloomsbury Publishing; 2009

[41] Williams AC, Hill LJ. Poverty and Pellagra's penumbras. In: B-complex Vitamins. London, UK: Intech Open; 2021

[42] Escribano AJ et al. Instant Insights: Integrated Crop-Livestock Systems. Burleigh Dodds Science Publishing Limited; 2021

[43] Headey DD, Alderman HH. The relative caloric prices of healthy and unhealthy foods differ systematically across income levels and continents. The Journal of Nutrition. 2019;**149**(11):2020-2033

[44] Forouzanfar MH et al. Global, regional, and national comparative risk assessment of 79 behavioural, environmental and occupational, and metabolic risks or clusters of risks, 1990-2015: A systematic analysis for the

global burden of disease study 2015. The Lancet. 2016;**388**(10053):1659-1724

[45] Health MC. Food and Social Inequality: Critical Perspectives on the Supply and Marketing of Food. Taylor & Francis; 2015

[46] Shaw I, Aldridge A. Consumerism, health and social order. Social Policy and Society. 2003;**2**(1):35-43

[47] Reckwitz A. The End of Illusions: Politics, Economy, and Culture in Late Modernity. John Wiley & Sons; 2021

[48] Tupy ML, Pooley GL. Superabundance: The Story of Population Growth, Innovation, and Human Flourishing on an Infinitely Bountiful Planet. Cato Institute; 2022

[49] Purkiss D. English Food: A Social History of England Told through the Food on its Tables. William Collins; 2022

[50] Fanzo J, Davis C. Global Food Systems, Diets, and Nutrition: Linking Science, Economics, and Policy. Springer International Publishing; 2021

[51] Unger PK, Press OU. Living High and Letting Die: Our Illusion of Innocence. Oxford University Press; 1996

[52] Stop LP. Thief!: The Commons, Enclosures, and Resistance. PM Press; 2014

[53] Sen A. Poverty and Famines: An Essay on Entitlement and Deprivation. Oxford: OUP; 1983

[54] Slocum R, Saldanha A. Geographies of Race and Food: Fields, Bodies, Markets. Taylor & Francis; 2016

[55] Piketty T. A Brief History of Equality. Harvard University Press; 2022

[56] Collier P. The Bottom Billion: Why the Poorest Countries Are Failing and What Can be Done about it. USA: Oxford University Press; 2008

[57] Smith D. Zygmunt Bauman: Prophet of Postmodernity. John Wiley & Sons; 2013

[58] Robertson C. What the future holds: Democracy, corruption, ESG and emigration. In: The Time-Travelling Economist. Springer; 2022. pp. 219-242

[59] Newman G. The Building of a Nation's Health. Garland; 1985

[60] de Pee S et al. Nutrition and Health in a Developing World. Springer International Publishing; 2017

[61] McMichael AJ. Planetary Overload: Global Environmental Change and the Health of the Human Species. Cambridge University Press; 1993

[62] Dubos RÃJ. So Human an Animal: How we Are Shaped by Surroundings and Events. Scribner; 1968

[63] Price WA, Nguyen T. Nutrition and Physical Degeneration: A Comparison of Primitive and Modern Diets and their Effects. EnCognitive.com; 2016

[64] Scrimshaw NS et al. Interactions of nutrition and infection. Monograph Series. World Health Organization. 1968;**57**:3-329

[65] Pammi M et al. Nutrition-Infection Interactions and Impacts on Human Health. Taylor & Francis; 2014

[66] Williams AC, Dunbar RI. Big brains, meat, tuberculosis, and the nicotinamide switches: co-evolutionary relationships with modern repercussions? International Journal of Tryptophan Research. 2013;**6**:S12838

[67] Patel R. Stuffed and Starved: The Hidden Battle for the Food System. Brooklyn: Melville House Publishing; 2007

[68] Schindler B. Eat like a Human: Nourishing Foods and Ancient Ways of Cooking to Revolutionise your Health. Hodder & Stoughton; 2022

[69] Simpson SJ, Raubenheimer D. The Nature of Nutrition: A Unifying Framework from Animal Adaptation to Human Obesity. Princeton University Press; 2012

[70] Liu Y et al. Quantitation of NAD+: Why do we need to measure it? Biochimica et Biophysica Acta (BBA)-General Subjects. 2018;**1862**(12):2527-2532

[71] Gluckman P, Hanson M. Mismatch: The Lifestyle Diseases Timebomb. Oxford University Press on Demand; 2008

[72] Austerity BM. The History of a Dangerous Idea. Oxford University Press; 2013

[73] Scott P. The wolf at the door: The trade union movement and overseas multinationals in Britain during the 1930s. Social History. 1998;**23**(2):195-210

[74] Titmuss RM. Birth, Poverty and Wealth. A Study of Infant Mortality. Read Books; 1943

[75] Russell SA. The hunger experiment. The Wilson Quarterly (1976). 2005;**29**(3):66-82

[76] Cattan S et al. Early Childhood Inequalities. Institute for Fiscal Studies; 2022. p. 1

[77] Leyton G. Effects of slow starvation. Lancet. 1946;**248**:73-79

[78] Allen LH, Dror DK. Effects of animal source foods, with emphasis on milk, in the diet of children in low-income countries. Milk and Milk Products in Human Nutrition. 2011;**67**:113-130

[79] Hulett JL et al. Animal source foods have a positive impact on the primary school test scores of Kenyan schoolchildren in a cluster-randomised, controlled feeding intervention trial. British Journal of Nutrition. 2014;**111**(5):875-886

[80] Pollitt E et al. Nutrition in early life and the fulfillment of intellectual potential. The Journal of Nutrition. 1995;**125**(suppl_4):1111S-1118S

[81] Sitwell W. Eggs or Anarchy: The Remarkable Story of the Man Tasked with the Impossible: To Feed a Nation at War. UK: Simon & Schuster; 2016

[82] Balehegn M et al. Animal-sourced foods for improved cognitive development. Animal Frontiers. 2019;**9**(4):50-57

[83] Black RE et al. Maternal and child undernutrition: Global and regional exposures and health consequences. The Lancet. 2008;**371**(9608):243-260

[84] Iannotti LL et al. Eggs in early complementary feeding and child growth: A randomized controlled trial. Pediatrics. 2017;**140**(1):e20163459

[85] Duffy B. Generations: Does When You're Born Shape Who you Are? Atlantic Books; 2021

[86] Hassett B. Growing up Human: The Evolution of Childhood. Bloomsbury USA; 2022

[87] Doepke M, Zilibotti F. Love, Money, and Parenting: How Economics Explains

the Way we Raise our Kids. Princeton University Press; 2020

[88] Bourne H. Am I Normal Yet? Usborne Publishing Limited; 2015

[89] Singer J. Neurodiversity: The Birth of an Idea. Judy Singer; 2017

[90] David W. The Pinch: How the Baby Boomers Took their Children's Future and Why they Should Give it Back. Atlantic Books; 2010

[91] Blakemore SJ. Inventing Ourselves: The Secret Life of the Teenage Brain. Transworld; 2018

[92] Conaboy C. Mother Brain: How Neuroscience Is Rewriting the Story of Parenthood. Henry Holt and Company; 2022

[93] MacAskill W. What we Owe the Future. Basic Books; 2022

[94] Doyle A. The New Puritans: How the Religion of Social Justice Captured the Western World. Little, Brown Book Group; 2022

[95] Goldrick-Rab S. Paying the Price: College Costs, Financial Aid, and the Betrayal of the American Dream. University of Chicago Press; 2016

[96] Mandery E. Poison ivy: How Elite Colleges Divide us. New Press; 2022

[97] Nikiforov A et al. The human NAD metabolome: Functions, metabolism and compartmentalization. Critical Reviews in Biochemistry and Molecular Biology. 2015;**50**(4):284-297

[98] Zhou M et al. Neuronal death induced by misfolded prion protein is due to NAD+ depletion and can be relieved in vitro and in vivo by NAD+ replenishment. Brain. 2015;**138**(Pt 4):992-1008

[99] Brenner C. Why is mom stressed: Homeorhesis as the potential problem and nicotinamide riboside as the potential solution. Journal of Experimental Neuroscience. 2019;**13**:1179069519869679

[100] Ear PH et al. Maternal nicotinamide riboside enhances postpartum weight loss, juvenile offspring development, and neurogenesis of adult offspring. Cell Reports. 2019;**26**(4):969-983.e964

[101] Platten M et al. Treatment of autoimmune neuroinflammation with a synthetic tryptophan metabolite. Science. 2005;**310**(5749):850-855

[102] Chang RQ et al. The role of indoleamine-2, 3-dioxygenase in normal and pathological pregnancies. American Journal of Reproductive Immunology. 2017;**79**:e12786

[103] Xu J et al. Brain SIRT1 mediates metabolic homeostasis and neuroprotection. Frontiers in Endocrinology (Lausanne). 2018;**9**:702

[104] Deng Q, Barbieri JT. Molecular mechanisms of the cytotoxicity of ADP-ribosylating toxins. Annual Review of Microbiology. 2008;**62**:271-288

[105] Calabrese EJ. Hormesis: Path and progression to significance. International Journal of Molecular Sciences. 2018;**19**(10):2871

[106] Gerasimenko M et al. Nicotinamide riboside supplementation corrects deficits in oxytocin, sociability and anxiety of CD157 mutants in a mouse model of autism spectrum disorder. Scientific Reports. 2020;**10**(1):1-12

[107] Salmina AB et al. Integrative neurochemistry and neurobiology of social recognition and behavior analyzed with respect to CD38-dependent brain

oxytocin secretion. Current Topics in Medicinal Chemistry. 2013;**13**:2965-2977

[108] Imai S-i. The NAD world 2.0: The importance of the inter-tissue communication mediated by NAMPT/NAD+/SIRT1 in mammalian aging and longevity control. NPJ Systems Biology and Applications. 2016;**2**(1):1-9

[109] Braidy N et al. Sobriety and satiety: Is NAD+ the answer? Antioxidants. 2020;**9**(5):425

[110] Tateishi K et al. Extreme vulnerability of IDH1 mutant cancers to NAD+ depletion. Cancer Cell. 2015;**28**(6):773-784

[111] Conlon N, Ford D. A systems-approach to NAD+ restoration. Biochemical Pharmacology. 2022;**198**:114946

[112] Gao Y et al. Nicotinamide N-methyl transferase (NNMT): An emerging therapeutic target. Drug Discovery Today. 2021;**26**:2699-2706

[113] Ball P. The Book of Minds: How to Understand Ourselves and Other Beings, from Animals to AI to Aliens. University of Chicago Press; 2022

[114] Yong E. An Immense World: How Animal Senses Reveal the Hidden Realms around us. (THE SUNDAY TIMES BESTSELLER): Random House; 2022

[115] Harris S. Clinical Pellagra. St. Louis: C. V. Mosby Company; 1941

[116] Williams AC, Ramsden DB. Pellagra: A clue as to why energy failure causes diseases? Medical Hypotheses. 2007;**69**(3):618-628

[117] Roe DA. A Plague of Corn: The Social History of Pellagra. Cornell University Press; 1973

[118] Deary IJ. Intelligence: A Very Short Introduction. Oxford University Press; 2020

[119] Mayer E. The Gut-Immune Connection: How Understanding the Connection between Food and Immunity Can Help us Regain our Health. HarperCollins; 2021

[120] Camporesi P. Bread of Dreams: Food and Fantasy in Early Modern Europe. Polity Press; 1996

[121] Inglis L. Milk of Paradise: A History of Opium. Pan Macmillan; 2018

[122] Jay M. High Society: Mind-Altering Drugs in History and Culture. Thames and Hudson Limited; 2013

[123] Pleij H, Webb D. Dreaming of Cockaigne: Medieval Fantasies of the Perfect Life. Columbia University Press; 2003

[124] Westhoff B. Fentanyl, Inc.: How Rogue Chemists Are Creating the Deadliest Wave of the Opioid Epidemic. Scribe Publications Pty Limited; 2019

[125] Higham S, Horwitz S. American Cartel: Inside the Battle to Bring down the Opioid Industry. Grand Central Publishing; 2022

[126] Bonney R. The Thirty Years' War 1618-1648. USA: Bloomsbury; 2002

[127] Chaplin JE. Subject Matter: Technology, the Body, and Science on the Anglo-American Frontier, 1500-1676. Harvard University Press; 2003

[128] Linstrum E. Ruling Minds: Psychology in the British Empire. Harvard University Press; 2016

[129] Taylor G. Buying Whiteness: Race, Culture, and Identity from Columbus to Hip Hop. Springer; 2005

[130] Carrasco D. City of Sacrifice: The Aztec Empire and the Role of Violence in Civilization. Beacon Press; 1999

[131] Hogarth RA. Medicalizing Blackness: Making Racial Difference in the Atlantic World, 1780-1840. University of North Carolina Press; 2017

[132] Russell SA. Hunger: An Unnatural History. Basic Books; 2006

[133] Scheper-Hughes N. Death without Weeping: The Violence of Everyday Life in Brazil. University of California Press; 1993

[134] Turnbull CM. The Mountain People. Triad/Paladin Books; 1984

[135] Snyder TD, Dillow SA. Digest of education statistics 2013. In: NCES 2015-011. National Center for Education Statistics; 2015

[136] Lake M, Reynolds H. Drawing the Global Colour Line: White men's Countries and the Question of Racial Equality. Melbourne Univ. Publishing; 2008

[137] Suranyi A. Indentured Servitude: Unfree Labour and Citizenship in the British Colonies. McGill-Queen's University Press; 2021

[138] Hägglund M. This Life: Why Mortality Makes us Free. Profile; 2019

[139] Adams E. Healthy at Last: A Plant-Based Approach to Preventing and Reversing Diabetes and Other Chronic Illnesses. Hay House, Inc.; 2021

[140] Eliav-Feldon M et al. The Origins of Racism in the West. Cambridge University Press; 2013

[141] Hinton E. From the War on Poverty to the War on Crime: The Making of Mass Incarceration in America. Harvard University Press; 2016

[142] Smith DL. Making Monsters: The Uncanny Power of Dehumanization. Harvard University Press; 2021

[143] Hatch AR. Blood Sugar: Racial Pharmacology and Food Justice in Black America. University of Minnesota Press; 2016

[144] Reed A. Class Notes: Posing as Politics and Other Thoughts on the American Scene. New Press; 2021

[145] Rogers JA. Nature Knows No Color-Line. Wesleyan University Press; 2012

[146] Reed AL. The South: Jim Crow and its Afterlives. Verso Books; 2022

[147] Bailey R, Wellesley L. Chokepoints and Vulnerabilities in Global Food Trade. London, UK: Chatham House; 2017

[148] Richards EH. Euthenics, the Science of Controllable Environment: A Plea for Better Living Conditions as a First Step toward Higher Human Efficiency. Good Press; 2019

[149] Hazareesingh S. Black Spartacus: The Epic Life of Toussaint Louverture. Penguin Books Limited; 2020

[150] Warren WJ. Meat Makes People Powerful: A Global History of the Modern Era. University of Iowa Press; 2018

[151] Gráda CÓ. Famine: A Short History. Princeton University Press; 2021

[152] Harris M. Cannibals and Kings: Origins of Cultures. Knopf Doubleday Publishing Group; 2011

[153] Kelly RL. The Foraging Spectrum. Washington, DC: Smithsonian Institution Press; 1995

[154] Le S. 100 Million Years of Food: What our Ancestors Ate and Why it Matters Today. Picador; 2016

[155] Stanford CB, Bunn HT. Meat-Eating and Human Evolution. Oxford University Press; 2001

[156] Widerquist K, McCall GS. Prehistoric Myths in Modern Political Philosophy: Challenging Stone Age Stories. Edinburgh University Press; 2017

[157] Bocquet-Appel JP, Bar-Yosef O. The Neolithic Demographic Transition and its Consequences. Netherlands: Springer; 2008

[158] Cohen MN. Health and the Rise of Civilization. Yale University Press; 1989

[159] Mendelson A. Milk: The Surprising Story of Milk through the Ages: Alfred a. Knopf; 2008

[160] Carlberg C et al. Nutrigenomics: How Science Works. Springer International Publishing; 2020

[161] Fussell BH. The Story of Corn. Albuquerque: University of New Mexico Press; 1992

[162] Johns T. With Bitter Herbs they Shall Eat it: Chemical Ecology and the Origins of Human Diet and Medicine. University of Arizona Press; 1990

[163] Jones MO. Corn: A Global History. Reaktion Books; 2017

[164] Krishnan V. Phantom Plague: How Tuberculosis Shaped History. Public Affairs; 2022

[165] Offer A et al. Insecurity, Inequality, and Obesity in Affluent Societies. Oxford University Press; 2012

[166] Kohler TA, Smith ME. Ten Thousand Years of Inequality: The Archaeology of Wealth Differences. University of Arizona Press; 2019

[167] Carrasco D. City of Sacrifice: The Aztec Empire and the Role of Violence in Civilization. Beacon Press; 2000

[168] Baradaran M. The Color of Money: Black Banks and the Racial Wealth. Gap: Harvard University Press; 2017

[169] Jones GS. Outcast London: A Study in the Relationship between Classes in Victorian Society. Verso Books; 2014

[170] Todd S. The People: The Rise and Fall of the Working Class, 1910-2010. John Murray Press; 2014

[171] Todd S. Snakes and Ladders: The Great British Social Mobility Myth. Random House; 2021

[172] Knight R. Britain against Napoleon: The Organization of Victory, 1793-1815. Penguin Books Limited; 2013

[173] Grainger I et al. The Royal Navy Victualling Yard, East Smithfield. London: Museum of London Archaeology; 2010

[174] Cederström C, Spicer A. The Wellness Syndrome. John Wiley & Sons; 2015

[175] Johnston J, Baumann S. Foodies. Taylor & Francis; 2015

[176] Poole S. You Aren't What you Eat: Fed up with Gastroculture. Aurum Press; 2012

[177] Colas A. Food, Politics, and Society: Social Theory and the Modern Food System. University of California Press; 2018

[178] Diner HR. Hungering for America. Harvard University Press; 2009

[179] McCloskey DN. Bourgeois Equality. University of Chicago Press; 2021

[180] Rauch J. The Constitution of Knowledge: A Defense of Truth. Brookings Institution Press; 2021

[181] Shapiro BJ. A Culture of Fact: England, 1550-1720. Cornell University Press; 2000

[182] Sitwell W. The Restaurant: A History of Eating out. UK: Simon & Schuster; 2020

[183] Stjernø S. Solidarity in Europe: The History of an Idea. Cambridge University Press; 2009

[184] Nelson PJ. Global Development and Human Rights: The Sustainable Development Goals and beyond. University of Toronto Press, Scholarly Publishing Division; 2021

[185] Nozick R. Anarchy, State, and Utopia. New York: Basic Books; 1974

[186] Holt JC. Robin Hood. Thames & Hudson; 2011

[187] Banerjee AV, Duflo E. Good Economics for Hard Times. Juggernaut Books; 2019

[188] Green T. The Covid Consensus: The New Politics of Global Inequality. Hurst Publishers; 2021

[189] Rosin C et al. Food Systems Failure: The Global Food Crisis and the Future of Agriculture. Taylor & Francis; 2013

[190] Allen P. Together at the Table: Sustainability and Sustenance in the American Agrifood System. Penn State University Press; 2015

[191] Stuckler D, Siegel K. Sick Societies: Responding to the Global Challenge of Chronic Disease. Oxford: OUP; 2011

Chapter 4

Pellagra: Down Not Out If Down and Out (and South) – Part 2

Adrian C. Williams, Christina Wood and Lisa J. Hill

Abstract

North-South variation in the supply of meat has always been present. Sharing of meat was the rule but in the multi-centric Neolithic revolution when domestication of animals and plants co-evolved class differences became pronounced-aristocrats and inferior proletariats and "lesser breeds and lower orders" started to form. The distribution of natural domesticates was uneven with the near-east and a temperate band across Europe well off compared with Africa and the Americas. The Columbian exchange changed this as meat became abundant in the New World who then exported to Europe. Wars, expropriations and genocides were over the meat supply and acquiring pastureland or water. Colonial plantation profits paid for meat imports from "settler colonies" indigenous or poor peoples on low meat pro-pellagrous diets were considered inferior whatever their colour and had poorer health and life expectancy. Attempts to correct hunger in the resultant ramshackle "Third world" concentrated on calories fuelling population booms and busts and delaying demographic, epidemiological and economic transitions. High meat variances are narrowing in China and Asia but need help elsewhere in the South. Dangers of not developing with a safe and sufficient meat supply include the emergence of zoonoses and mass migration. Reparations, rehabilitation and rejuvenation should concentrate on reconstituting a meat commons giving us a shot at redemption and survival.

Keywords: NAD, meat transitions, demographic transitions, epidemiological transitions, TB, poverty traps, metabolic syndromes, genocide

1. Introduction

Almost from time immemorial luminaries such as Cato, Cicero, de Quesney, Turbot, and Smith up until the present day have debated whether agriculture or industry or new energy sources or free trade created wealth and progress. Concepts on the "Biopower and Biopolitical" uses of diet can be dated back to Aristotle and Plato and elaborated on by Foucault and followers. Omnivorous diets for all as a prerequisite tailwind for firing on all body and brain cylinders neverthelessis rarely heralded (even by neurologists or psychiatrists) and even then the links between the "Grand Transitions" involving demographics, economics and epidemiology and the "Mega-threats" of pandemics and climate change are rarely made at a diet- metabolic level [1–7].

In Part 1 we discussed the consequences of being "down and out" or on "skid row" from meat and milk and therefore nicotinamide deprivation and mitochondrial energy within countries since the neolithic agricultural revolution and the domestication of crops and animals—here we shall discuss meat privilege on a broader canvas across time and space and the implications for human and planetary health. All agree that it's "inequality that kills" but how does this actually work? Sub-clinical pellagra stunts and slows individuals and fails to develop prosperity and a bourgeoisie creating bradykinetic macroeconomics and poverty from the pathology of ill health [8–10]. First, however, lets recap on some basic building blocks of civilisations starting with water but then majoring on meat and nicotinamide.

Splitting water by photosynthesis supplies oxygen and combined with CO_2 carbohydrates (CHO) and subsequently NADH to mitochondrial oxidative phosphorylation to produce energy as ATP. Water and water vapour, as a green-house gas, with a supply of photons makes the world habitable. Riparian "river cultures" such as by the Euphrates, Tigris, Nile, Indus and Yellow River with their floodplains and alluvial soils comprised the majority of early civilisations. "Hydraulic" empires, such as Rome, China or the desert kingdoms can be compared with those, some in Europe, with easier rainfall patterns or those with monsoon seasons, for their necessary adaptations for crops and the prodigious amounts of water needed to raise cattle and for their responses to climate changes-often inadequate causing collapses ("the Curse of Akkad"). Water is important for hydropower and trade routes (and Sea power states such as Athens, Carthage, Venice, Netherlands or Britain) and affects the politics of dams and canals and landscapes, such as terracing, and social structures, sanitation programmes, and wars when mutual interest fails [11–17]. A Gordian knot and nexus exists between water, air and diet and (for the last century) oil.

The role of diet, particularly grain in the rise, often with " ecological windfalls" or "ecological imperialism" and fall of empires or other collapses, often with "ecocides" from exploitation of the earth's resources (from Sumeria to Egypt to the Mayan and Roman empires and Easter Island) has been discussed: some enlarge on a role for meat and steppe nomads and, here we emphasise the importance of a sustainable balance between cereals and meat and, innovatively, discuss a metabolic mechanism suggested by the narrative of pellagra and NAD deficiency for both class and country effects [18–26]. Although air is free, unlike meat, but like water often not clean, asphyxiation has been a metaphor (and not always a metaphor thinking of lynching, Covid or George Floyd) for how to subjugate people—and their collective struggle for resources or to breathe as Franz Fanon (1959) said it is *"oxygen that invents a new humanity"*—but NAD deficiency (considered a "pseudo-hypoxia") is a more plausible mechanism [27–29].

2. Hypothesis: Short Story

Geographical gradients of meat intake have always existed but in the last few centuries they diverged and more recently have converged except in parts of the South where some billion people live on below $1.90 a day—the World Bank's definition of extreme poverty and so become enforced and often monophagic vegetarians with effects on health and longevity (**Figure 1**). Initially a more plant-based diet in the Neolithic may have increased fertility and a population growth that allowed for division of labour and high collective intelligence as the " Human Swarm" even as health and height deteriorated; but this may now be a developmental over-run producing "underclasses/ countries" [30, 31]. Here we attempt to describe how and why all this happened, hoping to illuminate a universal from the specific of pellagra, and that high meat intake is both

DOI: http://dx.doi.org/10.5772/intechopen.110061

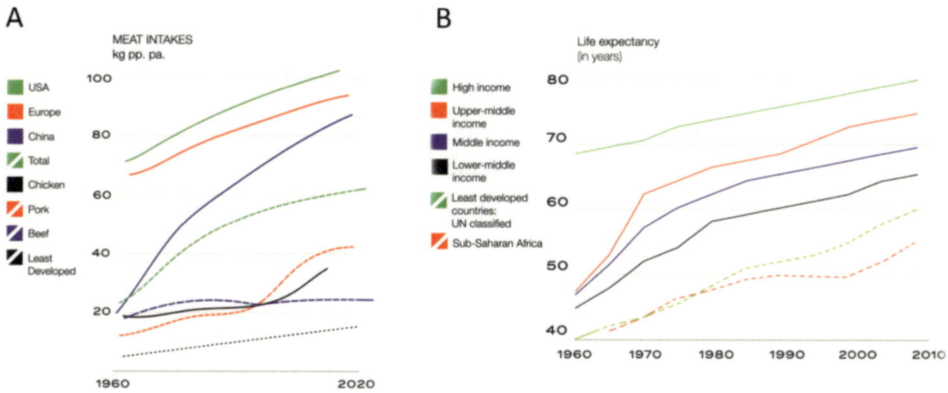

Figure 1.
(A) Meat by type & intakes by continent 1960-now. China shows a remarkable increase as have other "Asian Tigers". Europe and the USA long term leaders on intake are beginning to drop. India is in the middle on intake (better in the south), but Sub-Saharan and some other African countries are stalling as are their epidemiological and demographic transitions. Beef and ruminant intake worldwide have stabilised with pork (omnivorous scavengers) and industrially farmed chicken showing remarkable increases. (B) International Life expectancy by income. Remarkable increases shown here from 1960 but starting much earlier before medicine would have had much impact implying that public health measures such as clean water and a better diet were more important. Nicotinamide metabolism is important to most models of longevity in all species suggesting that high meat/milk intake could be causative.

the cause and the effect of economic progress (generally considered an unresolved "wicked problem"). Countries and continents have different epidemiological and demographic trajectories that may link to timing of safe meat transitions. If inequality is to be overcome its root cause and the source of biopower relationships, whether by class or by nation, needs to be understood and may lie with quality diets and quality populations. Sorting this may be the moral and safe way forward—and intersects with solutions for climate change and reducing civil unrest and pandemics [32–39].

3. More background and current relevance

Plows and petroleum have largely fuelled the food systems but despite successes has not served the poor or the planet that well [40–42] (**Figures 2** and **3**). Food insecurity underlies widespread micro- and macro-nutrient malnutrition, emigration and pandemics in this "nomadic" and "pandemic era" spread by commerce from zoonotic microbial cauldrons [43–46]. Such mega-challenges to feed with "sustainable intensification" increasing populations in the South rarely discuss a relationship between diet and fertility and the interactions with climate change [47–49]. There is a need for a mega-Marshall plan that as Bruntland's classic commission of 1987 said "*meets the needs of the present poor without compromising the needs of future generations*", that may include novel foods and fermented or cultured meat or exceeding "planetary boundaries" will cause a "ecological shock" with loss of biodiversity and further climate change [50–53]. Poor meat food-chains facilitate "bushcraft" and species-jumping zoonoses and decreases resistance to infectious pathology. Wars, including civil wars, and land enclosures and confiscations and land-grabs can be related to "lebenstraum" and the appropriation of well-watered and fertilised pastureland or the oil to support high intensity farming and a better diet-traditionally at others, usually in the global south's, expense. Geopolitical tensions and the follies of war include "slash and burn" tactics, unplanted or unharvested

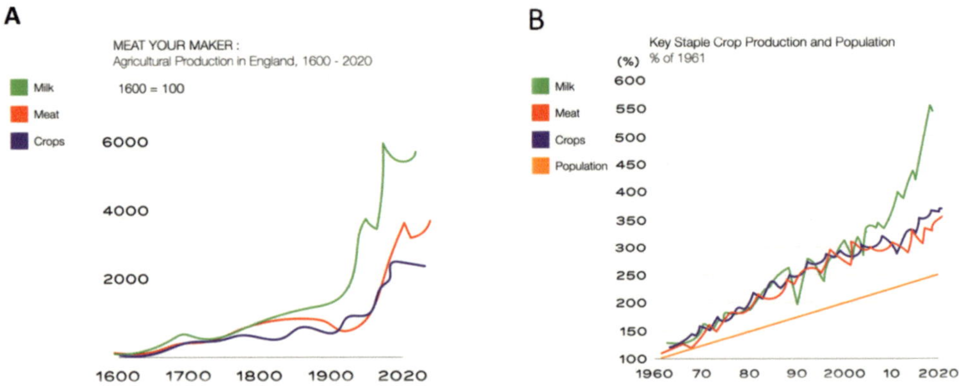

Figure 2.
(A) Meat production England 1600–2020. Meat production increased but barely kept up with population growth. Imports paid for by the profits from the cotton and sugar trade caused the real improvements helped by the repeal of the corn laws and better transportation and refrigeration allowing imports from the "settler" colonies and states. (B) Crop production yields have increased significantly over the last 50 years. Some GM crops contain vitamin A such as golden rice and recently vitamin D in tomatoes otherwise sourced from animal produce (or sunlight). Breads and cereals are often supplemented with vitamins, including nicotinamide, but less often in poor countries and are expensive. Oatmilk is now popular but does not contain as much nicotinamide-riboside as dairy milk.

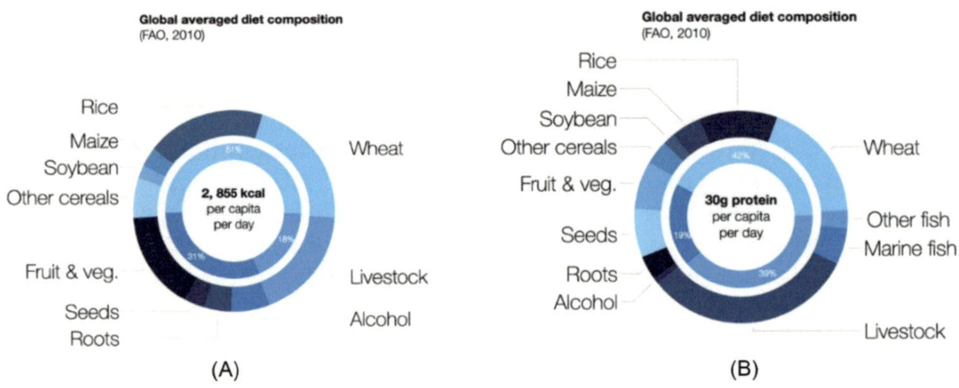

Figure 3.
(A) Global averaged diet composition. However, it is the extremes not the averages that are the problem with the danger of undiagnosed and untreated "pellagra sine pellagra" on the one hand and a possible hypervitaminosis B3 on the other. Switching from one extreme to the other within lifetimes can risk metabolic syndromes. Switching over generations to more meat and milk can lead to demographic, economic and epidemiological transitions. (B) Global sources of protein. All sources are not equal as far as Tryptophan is concerned as an alternative precursor to NAD where animal sources are better, as they are for nicotinamide. Other sources exist such as insects and larval forms—mealworms—that may play an increasing role in the future as may algae and macroalgae seaweeds such as kelps—and artificial laboratory grown meats from stem-cells.

crops, pillaging, stock-piling, blocked ports and sea channels, and sieges and often genocidal "hunger-plans" that along with price rises or tariff arrangements create a vicious cycle and more deprivation, again worst felt by the poor and in the South.

Existential threats such as climate change are both caused by and effect food systems (an initial boost to plant growth by high CO_2 is rapidly cancelled out by drought, floods and fires and locusts). Shorter-lived but potent methane as a greenhouse gas (80xCO_2) is largely derived from microbial synthesis in expanding wetlands and thawing permafrost "thermocrast" lakes as well as ruminants (and has increased markedly from 2007 in a vicious cycle that could form a "methane bomb"). Alongside the effect on CO_2 and nitrous oxide emissions this raises major concerns on meat intake and linked agricultural methods that increase emissions, such as transportation and excess use of fertilisers or tillage that destroy carbon sinks on land and acidified seas. Well maintained pastureland (of no use for crops) and perennial and tree silvacultures can however, like the sea, be massive carbon sinks [54].

High meat intake in the affluent west needs to drop but we will argue that for those on a low income, usually in the south in Asia or Africa it needs to increase (or its synthetic cell-based or safer non-ruminant/insect or plant/fungi and algae substitutes or at least with fortification of vitamins) to allow healthy demographic and epidemiological transitions. "Farmerian" structural violence has been attributed to poor food distribution networks and unfair property laws and metabolic and behavioural dysfunction in ghettoes as well as the incidence of infectious disease and early death [55–57]. A thought experiment imagining a "Dog-land" within the USA showed dogs to be at a high average global meat intake for humans [58].

High meat intake is not so much the problem, although the Delphic MHΔEN ATAN (Nothing in Excess) applies, as high global variances and lowering these drastically is the solution as a wholesale move to a plant-based diet often with ultra-processed high calorific but cheap foods is problematic [59–63]. The Lancet EAT commission suggested as much but was criticised for not explaining how the under-developed world was going to afford to eat a more varied vegetarian diet or for not fully recognising the need for some animal-sourced produce [64–66]. Despite living in a "Superabundant" world with a "breakfast bounty" whereby the price of meat and eggs has fallen for blue-collar workers by over 90% in real terms as measured by how long it takes to work to pay for them many of the poor within countries have not benefited and many countries in the south are no richer than the UK was in 1800 with per capita GDP still not exceeding 10$ a day [67, 68]. English breakfasts indeed became popular largely at the time when the international meat trade really took off with chilled American beef and deep frozen Argentinian and Australasian meat carried by refrigerated ships (such as the Circassia) in the 1880s such that by the eve of WW1 40% of the meat consumed by Britons came from abroad, a trade that shaped the world [69]. But let us start again toward the beginning as this all relates to how the high meat variances and quality of diets between nations as well as classes developed during our history, that were not part of our initial evolutionary trajectory. Not addressing this may de-rail our " slouch toward Utopia" instead directing us toward apocalyptic dystopias, if we do not develop climate smart food supplies and an "ethical omnivory" sensitive to animal rights and refrain from "milking" the poor [70–78].

4. Continental NAD divides

At the time of the Neolithic agricultural revolution all was not equal when it came to available animal domesticates and therefore the meat and milk supply even if sharing meat in our social leap was the early norm hierarchy, exploitations, enslavement and inequality rapidly developed [79, 80]. The middle east and then Europe and the Asian steppes were well off for amenable hoofed ruminants such as goats, sheep and cattle [24]. Africa the "cradle of mankind" having been a meat cornucopia with abundant wild herbivores in the savanna had few natural domesticates and more than their fair share of animal diseases (such as trypanosomiasis (sleeping sickness) in the wide tsetse belt and in the nineteenth century rinderpest introduced from Europe) and competing large carnivores. Nomads and pastoralists and relationships with other animals were important components of our history and culture [81, 82].

Cattle, Chattels and Capital all derive from the same etymology and in practices such as bride-wealth. More damaging was cattle rustling and the concept of "terra nullius" that allows for pastureland not obviously farmed by indigenous people to be appropriated and territorial gains through war or land-grabs and were all meat related as was cannibalism [83]. The New world after the original animal population was decimated by hunters had few natural domesticates other than turkeys and guinea pigs; China had foraging pigs, chickens and many ducks so was reasonably well supplied at times, but still largely cereal based and famine prone [84, 85].

5. Columbian exchange of NAD suppliers and attitudes to other peoples

The Columbian exchange changed the New World dramatically by developing settler states that improved their own and then others supply of meat (**Figure 4**). Cattle, sheep, goats and horses rapidly bred in this new ecology and on ranches [86–88]. This was usually to the detriment of indigenous peoples and the slaves, as illustrated by the manmade deplorable "Trail of Tears" and "Middle Passage", even although Africans had helped by smuggling (Black) rice across the Atlantic and contributed their considerable agricultural and even medicinal skills [89–93]. Some appreciation that native Americans were not, after all, monsters or savages having their own civilisations was there as a philosophy of "All Mankind is One" was supported by the Spanish monk Las Casas and even had royal support but did not unfortunately translate for long. As Montesquieu (1748) said (sarcastically) in his Spirit of the Laws

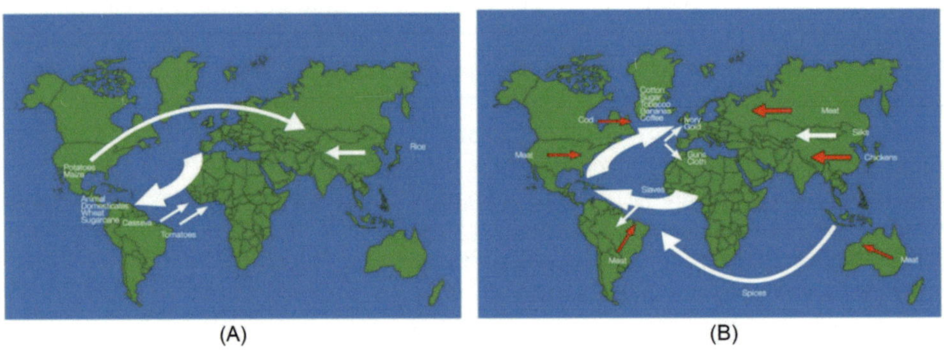

Figure 4.
(A) Columbian exchange. Animal Domesticates, such as sheep and cattle, were introduced to the New World that otherwise had very few such as turkeys and guinea pigs. Earlier megafauna had, as elsewhere, been killed off by hunters and climate change, and later bison were killed by (white)hunters to the detriment of indigenous tribes. Imported ungulates rapidly expanded swarming in the lush new pastures "ghost acres" tripling the pastureland for England and other European powers further aided by cowboys, barbed wire and ranching with the import of new breeds. Later feeds replaced grassland and the "livestock complex" was born. Maize and potatoes went East fuelling population booms and busts. (B) The slave trade by supporting sugar, tobacco and cotton indirectly supported the meat trade as the profits held by Europe paid for the meat imports as had silver and gold. Settler colonies exported meat, now frozen and easily transported, and usually had enough for their own benefit. Others with or without plantations had their farming systems damaged and their peasants taxed (salt and tea taxes provoked rebellions in India and America). Diets deteriorated (such as the Bengal famines). Indigenous peoples contributed much on farming technique and introduced crops such as rice in the West and medicinal "secret cures of slaves" (such as quinine) and commercial plants (such as rubber) of great value that like the silver, gold and (blood)diamonds were basically stolen as "primitive accumulation" to fund capitalist ventures and more profit for the centre. In the Irish famine and even quite recently examples of exporting meat whilst the supply to the local population declined markedly include Guatemala and Costa Rica in the 1960s [26].

"having extirpated the natives Europeans had no choice but to make slaves of Africans besides which sugar would be too dear" may provide the explanation along with other expropriations of useful materials and free or cheap labour from indentured servants.

Previously many racial differences had been attributed to "environmentalism" and climate or diet but the mood changed as Europeans, except when they inter-married (miscegenation may have even saved us from a speciation event) did not change their phenotype and so genetics became a more favoured explanation with whites genes being superior. The underclass and "cultures of poverty" were invented (with planters pitting working class whites and "hillbillies" against people of colour)—a self-fulfilling prophesy when the indigenous natural diets were destroyed in favour of producing cereals and sugar/molasses, and a pellagrogenic diet allowing for "class cluelessness and callousness" [94, 95]. Furthermore "divide and rule" policies by elites and "dog whistle" racial politics were encouraged on the plantations and thereafter stalled many worker class rebellions. More subtly on Haiti in 2008 food price increases were bailed out with strings attached that are damaging longer term for the nutritionally weakened poor—(even though they were the base of agriculture and capitalism in general)— all old and new methods that create superficial frictions that include pitching disadvantaged racial and tribal groups against each other [96–99]. This is now known as the "Southern Strategy" though moved North with the mass great migrations of African Americans after the Civil War (Underground Railroad) and is a "Pluto Populism" and "Racecraft" and often evangelical and nationalistic policy with vote gerrymandering and other suppressions much copied (including ironically by the party of Abraham Lincoln) to this day.

Some original abolitionists retaliated with criticism of eating slave-foods (such as sugar), and other forms of plant robbery such as rubber and cinchona (quinine anti-malarial) sentiments that continue with fair trade movements (but the connection with the meat trade was not realised) [100–103]. Gold and silver and later profits from the slave-based sugar and cotton industry paid for massive levels of meat and fish imports to Europe particularly after the invention of steam engines for ships and the railways, and refrigeration. Colonised settler states after the Americas included Australasia exporting meat; other colonies contributed indirectly such as India and Africa (inspired by Cecil Rhodes) were taxed or used to boost income from mining and "cash mono-crops" using cheap labour advantaging the European core at the expense of the "periphery." Outbreaks of pellagra, in the poor southern "cotton states" of the USA and later in "banana republics" were the tip of a malnutrition iceberg. Advances in farming beyond "hoe-culture" were discouraged as was the manufacture of goods, even traditional cotton clothing—all helped to create a now dependant "Third World" [104, 105].

Cash crops such as from peanuts, cacoa or tea took precedence over staples such as rice or yams often in the name of comparative advantage (that then had to be imported with no control over price) and destroyed mixed farming and the meat supply for all except surviving pastoralists. Ironically the Masai were even admired by their colonial masters for their height and health. As Daniel Defoe author and pamphleteer, said at the time *"No African trade, no negroes, no sugar, indicoes, cotton; no islands, no continent no (meat) trade"* and change or even insight has come slowly. This all allowed the "great subsistence crisis" in the north-western world triggered by war, volcanic eruption (Tambora 1815) and the "year without summer" and effects from El Nino to be the last crisis in the North but came at the expense of the South [106–108]. Modern equivalent practices include the export of Palm Oil and soybean (used as animal fodder in industrial farms) causing much local ecological damage—swaths of the Brazilian Cerrado region turned in to cropland and rainforests cleared for grazing cattle contributing to both local and global climate change and destroying much biodiversity in order to supply the west with more meat or biofuels [109, 110].

6. The West then the rest: meat as propellant. More circumstantial evidence

Ink (and blood) has been spilled over the issue of the reason for the rise of the west and we have added meat centrality to previous arguments relevant to all empire building and as advice to their builders [111–122]. Poor cooperation between agricultural empires, such as the Roman and the Ming empires (who decried milk and its products in an anti-Mongol gesture except in Zhangua province that fared better), and their "barbarian" pastoralist neighbours has been implicated in their falls (perhaps answering Needham's puzzle in the case of China as to why they were overtaken) [123, 124]. Similar issues have emerged for the Ottoman and Mughal empires with lack of animal fodder or exhausted irrigation systems and ecocides leading to "high equilibrium traps and inadvertent selection for quantity rather than quality population growth in low meat economies [125–127].

Julius Caesar noted of British tribes in 54BC "lacte et carne vivant" but most observers date Britain's success to the response to the 14th C Black Death that itself followed on from a great famine (and is thought to have originated in the East and spread along the Silk Road by Mongols as a zoonosis involving the fur trade in marmots and grain supplies eaten by both man and rats with their fleas all in close proximity with gruesome stories of early biological warfare catapulting the dead into a city (Caffa) under siege [128]). Here, the argument goes, the high death rate led to a shortage of peasants and yeomen so, despite early aristocratic resistance, wages rose and agrarian fields turned to pastureland innovations such as turnips for winter fodder avoiding autumnal culls and spring seasonal hardship (when pellagra often emerged). Meat intake increased markedly for several centuries—all reflected in increasing height and perhaps, IQ [129–134]. The need for labour saving devices became the "mother of inventions" for agriculture then industry. Studies have used estimates of affordable "baskets" at various times with international variations and fluctuations in the incidence of the more pellagra-genic basket support these claims [135–137] (**Figure 5**).

Elsewhere populations exploded on more cereal (and now maize) based diets and agricultural reforms lagged with many other obstacles in the East and South that involved ecological and unresolved water and irrigation issues that were not so difficult in rainy Europe [138]. The rise of the northern white protestants on a high dairy diet in the reformation compares with the poorer but more fertile Catholics which may have been another divergence in the making [139, 140]. In England population grew slowly for centuries but then populations boomed, at the time of Malthus, as there were dietary setbacks so by the 1800s many of the poor were enforced vegetarians culminating in the "hungry forties" before another meat transition took place.

Reforms of corn laws and deflation of food prices along with cotton and other profits allowing importation of meat on a grand scale meant that for once the poorest gained and could afford a better diet. This allowed for an expansion of the middle classes and the rise of "hawkers" of street meat and "fishwives", taverns and restaurants helped. The Chartist movement and the rise of trade's unions avoided the concurrent hunger inspired revolutions in Europe at least until the 1930's hunger

DOI: http://dx.doi.org/10.5772/intechopen.110061

	Basket	2-3000 calories
	Respectable: quantity pp, per year	Bare bones subsistence: pp, per year: Pellagrongenic
Beans / Peas	34 kg	20 kg
Beer	182 litres	
Bread	182 kg	
Oatmeal / grain		170 kg
Meat	26 kg	5 kg
Butter	5.2 kg	3 kg
Cheese	5.2 kg	
Eggs	5.2 kg	
Fuel	5.0 MDBT	2.0 MDBT
Linen / Cotton	5 metres	3 metres

Figure 5.
Subsistence baskets. Food baskets have been estimated for many countries over many centuries. Such baskets can be tracked—In England they deteriorated before the Black Death but then improved as income (after opposition) increased for centuries. The birth of the lead in agriculture and the industrial revolution has been attributed to this high meat intake and enough money for consumers to spend on luxuries developing commerce (once sumptuary laws were abandoned). Poor harvests and wars had led to economic failures widespread vegetarianism culminating in the "hungry" 1840s. Many then escaped from bare bones pellagrogenic diets and continued the lead in Europe but then were overtaken by the USA with China catching up more recently. These dietary divergences and convergences can be linked to the rise and fall of empires and civilisations.

marches. In England (and later America) and in Amsterdam and Antwerp's "glory years" butchers and their guilds prospered in the *luilrkkerland* (delicious-land) with stories depicting houses made of pancakes and bacon with pre-roasted pigs walking—butchers could not keep up with the nouveaux riche [141, 142].

Many observers have touched upon diet being important for talent, "industriousness" and progress: "Rosbifs" and "Roast Beef of Old England" were immortalised in paintings and song and Shakespeare's plays, such as Henry Vth [143]. Other dietary explanations have included a spice or "sugar rush" (although the dangers of sugar were pointed out by Thomas Willis around 1670 as diabetes "mellitus"). Importation of tea (that became part of the British identity despite no tea plantations in the UK) and coffee displaced beer, a safer source of liquid than water at the time, and distilled spirit addiction "drunk for a penny, dead drunk for tuppence" even feeding infants as depicted by Hogarth's Gin Alley, as preferable stimulants rather than depressants (whilst at the same time driving the opium trade depressing China may all have led to some international advantage). But back to meat important trades spices (curry becoming emblematic of the UK) relates to the preservation and taste of meat, or by camouflaging the added salt, along with other forms of charcuterie [144–148].

7. Nicotinamide not sugar rush: intellectual "savants and tinkerers"

We propose a meat and nicotinamide "rush" as a more likely explanation for moving the dial with non-dietary correlations such as coal and steam enabling improvements in the supply as would profits from sugar and cotton plantations [116, 149–152] (**Figure 6**). Long rotating global waves of prosperity allowing more (patentable) innovation noted by Kondratieff may have a basis in agricultural success and lower prices or higher wages and could relate to meat transitions that in turn are related to

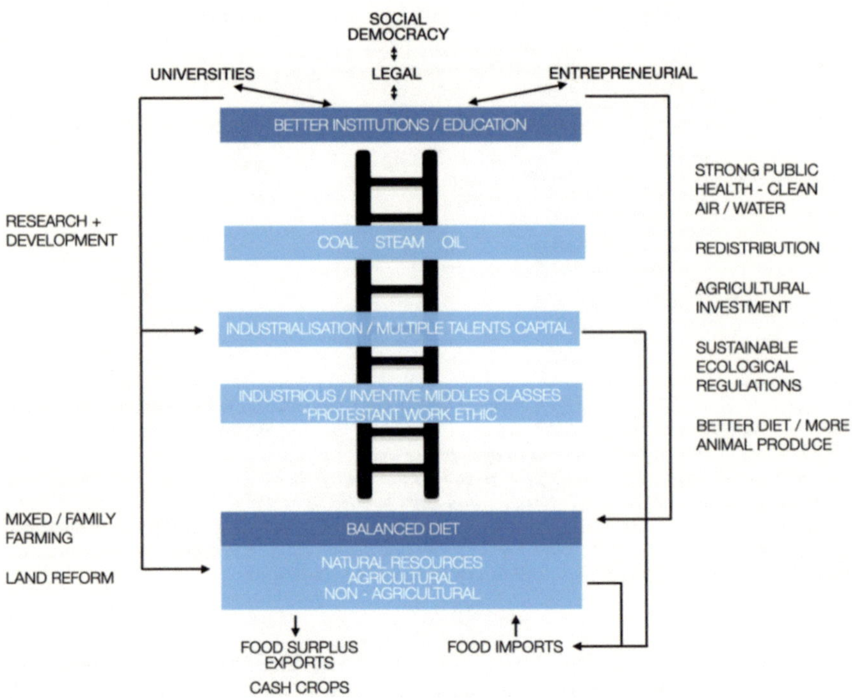

Figure 6.
Ladder of development. Better institutions and governance are the conventional wisdom and thought, after Weber, to be the major drivers of progress and may prove proximate causes. However, it is our contention that this develops only after a critical mass of the population have high human capital from an adequate and balanced diet. This became a positive feed-back loop in the nineteenth C helped by the rise of restaurants (and chefs no longer needed by the (French)aristocracy) further improving diet that gives the illusion that better diet is a consequence, but we would claim reverse causation. History does suggest this was learnt the hard way obscured with friction and peasant rebellions (such as Jack Kade's in 1450 over enclosures and privatisation of land usually for wool rather than meat that then became too expensive for the poor unless they worked)—and indeed has still not been learnt. Weber's example of protestants in northern Europe had high meat and dairy/butter intake and were lactose tolerant compared with southern Europe with less mixed farming and olive oil rather than butter that has a higher nicotinamide content. Southern Europe was the site of the first recorded pellagra outbreaks, by Casal, in the 18th century. Perhaps subconscious cultural lessons were learnt as Italy now has the most democratic of diets and this "Mediterranean" diet leads to relatively good health outcomes [153].

local agricultural and import opportunities all favouring good dinners for some; but let us look again at the example of nineteenth century Britain [153, 154].

8. Tectonic plates: meat, epidemiological and demographic transitions meet early in the UK

Agriculture evolved in Europe with fallows enriched by grasses and legumes "Ley farming" improving the soil and nitrogen content, and the rise and fall of tillage improving the soil microbiome, with alternation of crops and livestock with integrated systems demonstrated by Jethro Tull (1751) and "Turnip" Townsend for winter forage [155, 156]. Stock breeding and "High farming" were introduced by pioneers such as Robert Bakewell and John Webster as the UK became the world's "Stud Farm" and proponent of feedlots. England compared with many other countries including France had a high ratio of pasture to arable land supporting high meat intakes and horsepower along with ample manure for crops and was an important pre-requisite for the industrial revolution [157]. The fishing industry particularly for (dried and salted) herring, cod and whale took off supplying much meat as the "Northern Hunt Trades" and fourteenth C carp fish farming, helped by being allowed to eat fish on the many religious fasting/meat-free days, but soon was in trouble from over-fishing and later farms such as those for shrimp damaged the ecology of mangrove swamps as important carbon sinks even though done properly fish-farms may re-emerge as important sources of nicotinamide [158–160].

The astonishing rise of the chicken and egg market began to contribute [161]. However, with population increases numbers could not keep pace with demand. The well documented [60, 162–164] meat transition occurred early in the UK (1850–1900) with intake almost doubling for the general population although the idea that people and slaves need any more than subsistence died hard "scarcity promotes industry and less drunkenness" being a common view as was starvation as being a natural Malthusian population correction (as expressed by government in the Irish and Bengal famines). Better wages later became popularised by Henry Ford, as did conveyor belt manufacturing after observing the meat packing industry and is the basis of consumer society and luxury markets with all relating to Engel's curves.

The pattern of disease changed dramatically with far less infections and early deaths most notably for TB [165–169]. Sanitation or at least better housing and fresh air and clean water have been considered important in this pre-antibiotic era, but evidence is surprisingly unsupportive for TB perhaps because rents rose squeezing family budgets. There are many other correlations, such as increased sugar intake, but the only one, to our eyes, with a convincing biological explanation is more meat as meat is a source of nicotinamide. Nicotinamide has antibiotic activity against TB (and other organisms including Mycobacterium leprae that also "disappeared" in the western world in the earlier sixteenth century meat transition) and many TB antibiotics developed are nicotinamide analogues. Furthermore, TB excretes nicotinamide (used as a test for pathogenic strains) suggesting that it may be a nutritional symbiont that turns dysbiotic if relied upon too heavily [170]. As TB disappeared diseases of modernity particularly allergic and auto-immune diseases appeared [168, 171].

Biological explanations are to hand as the Tryptophan-NAD intrinsic pathway would not be activated on a NAD-precursor rich diet and is well known to be the immune tolerance pathway so over-reaction to otherwise harmless antigens might be expected (that might include the foetus); and absence of "Old Friends" could lead to loss of immune education again leading to immune over-reactions [171]. Relative infertility indeed rose once death rates had fallen as part of the demographic transitions that causes temporary population explosions and then stability—and then population declines as fertility falls below that needed for replacement. Such demographic patterns happen with modernity but the element of modernity responsible has never been identified (many happened before effective birth control was available or education improved.) Rather than talking of "African exceptionalism", for instance, diet should be reconsidered as there is a longer history of meat being held responsible for low fertility and cereal dependency or poor diet short of starvation for high fertility amongst the proletariat causing much alarm about "replacement" by "inferior" peoples, but little action. Higher quality of population rather than higher numbers should be the aim ("K" v. "r" selection) and that can be done starting with better diet [172].

9. Nineteenth-twentieth century meat exchanges: rural and third worlds

Observers ,such as de Sant Pierre (1769) noted that "*coffee and sugar were not essential to Europe but brought wretchedness and misery upon America and Africa*" including becoming reliant on local "bushmeat" eating their own, now rare, wild animals from elephants to monkeys and losing some indigenous cereals, that often have lower calories and more micronutrients let alone being a source of cultural and biodiversity that may become important with climate change. During the later colonial period meat and tropical goods flowed from the periphery in to the European core aided by refrigeration and canning and "Free Trade" using cotton and sugar profits to pay for the imports and essentially living well at others expense creating "Victorian Holocausts" in the tropics often exacerbated or even capitalising on climate change [106, 173–177] (**Figure 7**). Current telling examples can be seen in Madagascar currently and occasionally in the centre as in the 1930s Dust Bowl on the American prairies that are worrying signs for a future of climate change in coastal and natural deserts, even California [178].

Figure 7.
Sugar in Europe and the settler colonies was balanced by higher meat intake. This was not true for poor people particularly those in the Caribbean and "cotton" (and pellagra) southern states of the USA fed with high quantities of molasses (and rum) and little meat. This was a recipe for pellagra and poor nutrition throughout life and transgenerational epigenetic effects including metabolic syndromes. Reparations might be well directed to preventing this happening in the future. The same argument is true for poor peoples in rich economies where cheap calories with 2 for 1 deals and junk food do not help people ascend Engel's curves.

Within the core there were extraordinary concentrations of cattle and meat flowing from rural areas to magnets large towns and cities that majored in animal husbandry and use of pigs as well as horses, hens and cows contributing to "metabolic rifts". As Jenner rhymed (1772) *"Fat droves of sheep and oxen, consigned from Lincoln fens, that swearing drovers beat to Smithfield pens."* This need spread and the Scottish clearances and the "colonisation" of Ireland caused much hardship and emigration in order to supply meat centrally at the expense of the "crofters". Smithfield market is an exemplar in its Georgian day of a dominant 800-year-old industry of live animals "meat on the hoof" driven into the city for food and transport (and dung not easily re-cycled) dwarfing other better-known industries, such as sugar and coffee, and helped by advances in stockbreeding for meat and dairy. Abattoirs then moved out of town and out of view of the consumers with new tensions for animal rights by making the "cognitive dissonance" over killing animals easier.

10. Pellagra: plus Kwashiorkor

Pellagra epidemics re-emerged in the periphery, namely South Africa, where it had been unknown until 1914, as did poverty, racism and disease—chiefly infectious or nutritional such as Kwashiorkor, well- illustrated by Oxfam, and felt by many to be infantile pellagra particularly after the protein hypothesis was discarded [179–182]. Post WW2 local initiatives and cereal exports and some meat from the developed world supported some feed-lot cattle in the underdeveloped world allowing for their own meat transitions that drove demographic and epidemiological transitions [105, 183–188]. However, this was highly heterogeneous (often aid was linked with political and anti-communist motives or a way of exporting subsidised surpluses that did nothing for local

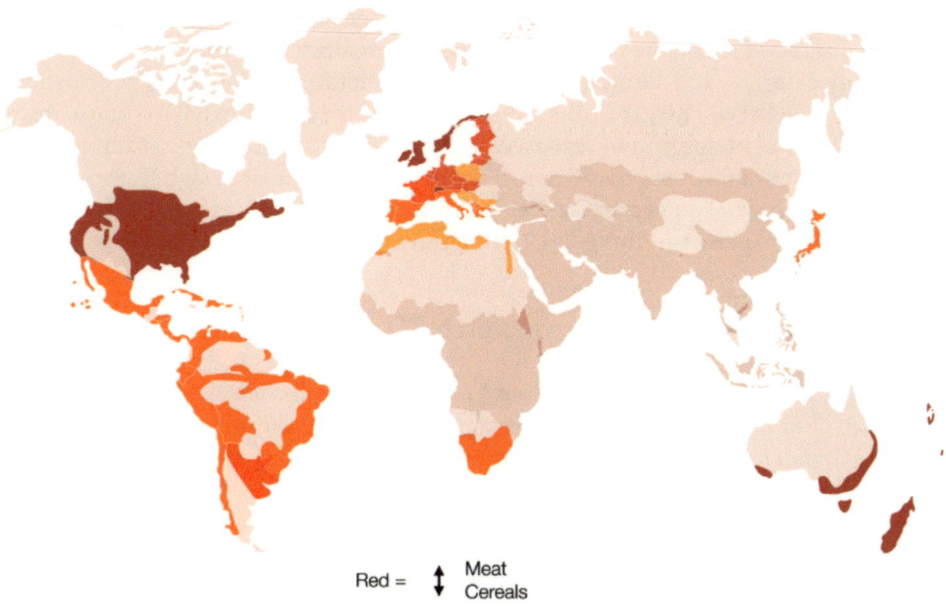

Figure 8.
Countries in shades of red whose cereal consumption fell as meat consumption rose correlate with high or very high (darker red) economic activity. This is Engel's/Bennett's law at an international level. Poor countries are trapped by high food and energy prices where they also make up a larger share of the national inflation basket.

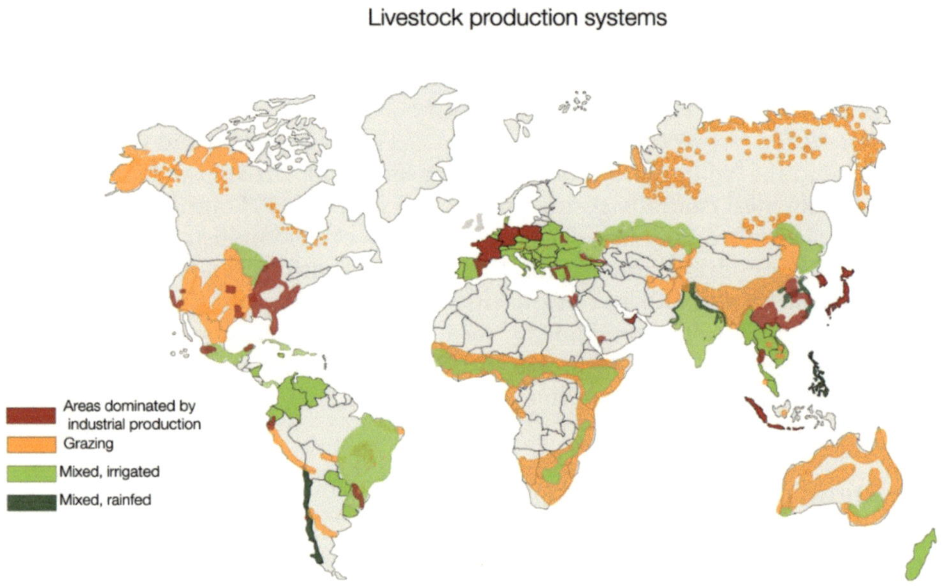

Figure 9.
Meat and wheat international trades are high as is maize and soyabean with much used to support meat industrialisation (and biofuels). Sources and industrialisation of meat is geographically very heterogeneous. More local production with mixed farming or pastoralism would help reduce variances across the globe and reduce the need for artificial fertilisers.

farmers) and more commonly being cereal based may have given rise to population explosions as much as alleviating hunger. Transitions are still incomplete in southern Asia and most of Africa where poverty, stunting and chronic infectious disease with population explosions remain common and, we suspect subclinical pellagra is common.

Such countries may be the victims of excessive cerealization then calorificization with "empty calories" and "junk foods" and other temptations by-passing the normal function of Engel's law [189, 190]. The so called "resource curse" may be when states get lucky with natural underground resources whether oil or minerals, or cash crops they do not spend it on improving diet where it is poor—by chance or design Europe did not do that with their wool and cotton trades and may have narrowly escaped this trap—probably "a damm close run thing" as Wellington said of Waterloo (**Figures 8** and **9**).

11. Poverty pathogenesis converges on NAD mechanisms

Stephen J Gould's comment channels as "*I am, somehow, less interested in the weight and convolutions of Einstein's brain than in the near certainty that people of equal talent have lived and died in cotton fields and sweat shops*"—much talent and brain power is indeed being wasted for lack of NAD-as William Blake said (in 1793) "Reason is the bound circumference of Energy". The poor (and particularly women as men commonly got the "Lion's share" of meat) have a NAD deficient headwind to contend with and an NAD-related degenerative pathology and functional neuropsychiatric effects ("feebleminded" and ripe at times for enforced sterilization or extirpation) from both low dietary intake and increased consumption as they are exposed to more than their fair share of environmental genotoxic and other toxins and infections [191, 192]. Chemical revolutions have shaped our world from chlorine (releasing bleaching fields for pastureland) (1785) to soap and

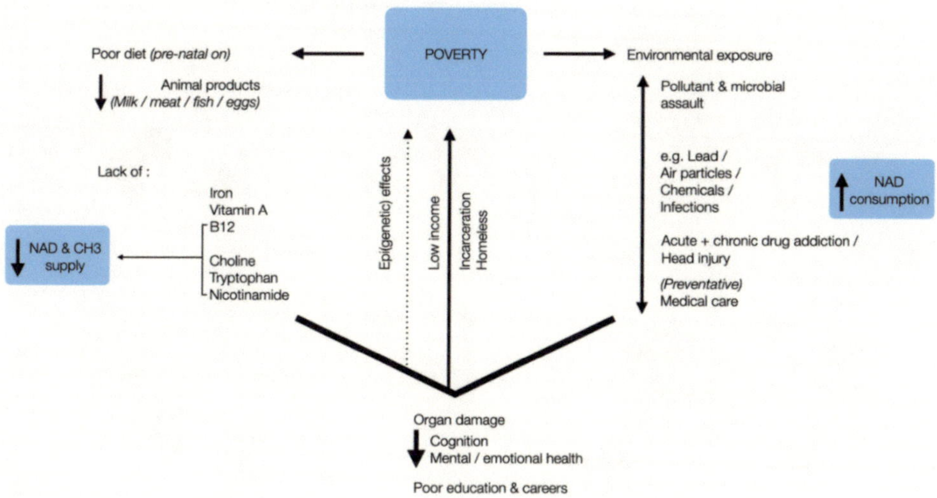

Figure 10.
A balanced diet has to include adequate nicotinamide supply along with other micronutrients that, until supplementation was introduced, had to come from animal produce. Supplementation and "nutritionism" may be an emergency procedure of value but is a technological fix (as is Golden Rice) to try to fix an economic problem and ignores broader cultural and health issues that rarely work and in the case of nicotinamide can drain the methylome [196]. Poor diet and poverty are interlinked in cycles of despair and increased environmental hazards adding insult to injury that are trans-generational.

cleaning water (1850) to refining sugar to artificial fertilizers to DDT and organophosphates and neonicotinoids to leaded gasoline and poison gases and safe refrigeration, (that enabled the meat trade) but all have come at a (neurotoxic)price and unforeseen consequences particularly in poor neighbourhoods [193–195] (**Figure 10**). Incentives for "credit invisible" disadvantaged people to set up business in the USA (as part of Nixon's "black capitalism" even concentrated on fast food outlets and franchises "force feeding junk" (although admittedly hamburgers, kebabs and "cod fish sandwiches" like fish and chips in Britain increase meat intake)rather than grocery stores as did targeted advertising and "Foodopolies" and all not helped by traditions of "soul/slave" food [197, 198]. Poor neurobiology has previously been blamed on genetics, but this plays a secondary role perpetuating the problem through epigenetic means well after the environmental assault [199]. Epigenesis may have direct effects through chromatin marks or in a wider sense of the term lead to poorer education both within and outside the family or access to (preventive) medical care or exercise. A large volume of evidence supports the idea that animal products such as milk and meat are important in child development in good studies in many countries over many years: given the evidence that increased income is spent on ascending Engel's curves this explains the beneficial effects of universal income and services being cost effective with improved labour outcomes and productivity and the failure of austerity programmes [129, 133, 200, 201].

12. Living high: letting die

The moral position on this *"thou shalt not kill"* is clear as many ethicists and philosophers from Aristotle on have pointed out that all should be allowed to flourish [202–205] (**Figure 11**). Many diseases of poverty can be related to meat deprivation and the "western diet" linked with allergy and auto-immune disease and (food) allergies often attributed

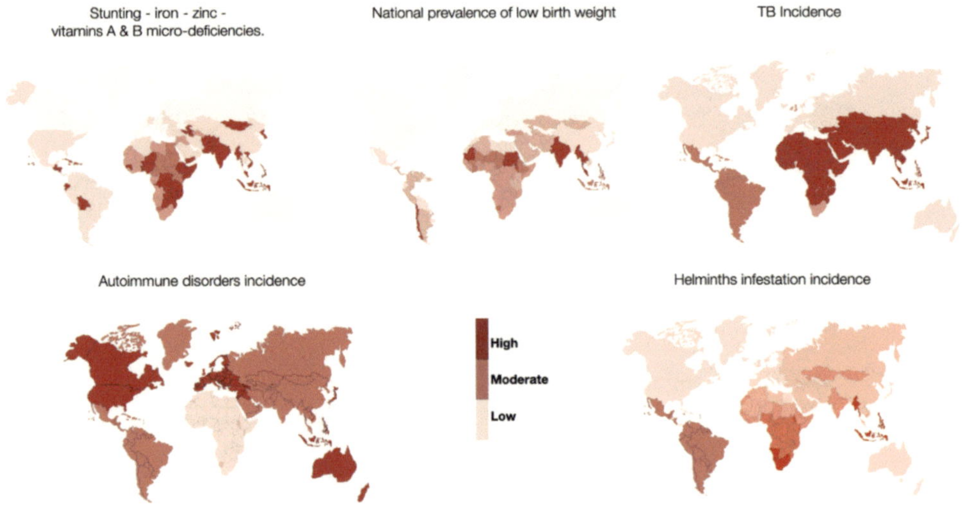

Figure 11.
Stunting whether measured by height or IQ tests is very geographic and correlates with poor diets. So do low birth weights that then even if diet improves runs the risk of metabolic syndromes. TB decreases over time if meat intake increases as it did in Victorian England but is still common in countries where meat intake is low. Nicotinamide has antibiotic actions against TB and many helminths. As TB and Helminth incidence decrease allergic and autoimmune disease quickly and consistently rise in incidence as does longevity and some age- related inflammatory, degenerative and cancerous conditions.

to lack of fibre and ultra-processed foods could be a direct effect of higher nicotinamide doses and that the theories merge as less fibre with changes in the gut microbiome and less butyrate and other short-chain fatty acids (SCFAs) affects the niacin receptors such as GPR43 and GPR109A and both T cell function and NAD metabolism [206–209].

We did not emerge from Hobbesian state of "solitary, poor, nasty, brutish and short" but from a (meat)affluent society. As Rousseau, Locke, Paine, John Stuart Mill and many since have pointed out the essence of the social contract and property rights with enclosures is that there is then an obligation to feed those whose land, pasture and hunting rights have been expropriated. Indeed that has often been taken as the hallmark of civilisation even if codes are regularly broken [210]. Many revolts, revolutions and even world wars have been based on the right to subsistence, or used food as a weapon such as the Nazi Hungerplan (see "The taste of war"-Collingham 2012 – for many relevant examples emphasising the need for meat). Some nearly succeeded such as the French revolution's "le droit a la subsistence" supervised by the "Comite de Mendicite" of 1790 that after long years of malnutrition under the Ancien Regime was working well until adverse finances due to the Napoleonic wars intervened [211–215].

The basis of a good diet is well known even if detail changes with fashions rather than evidence: Hippocratic and Aristotle's biopower recommendations for individuals and states on dietary and exercise regiments are not that different to modern recommendations or as observed in healthy "Blue Zones" [64, 216]. The problem is that for millions within even wealthy countries and billions globally this advice is not affordable or even available—and not because the poor cannot manage their domestic budget or cook properly but because high calorific foods are cheap and there is less waste and they keep you alive even if "starving on a full stomach" [66]. Cost of thriving indices (Coti) suggest that even the (crucial) middle classes in rich countries are over the last 50 years struggling for the basics including the nutritious food baskets – at lease on one income. Statesman like actions to alleviate intergenerational friction from "baby boomer"

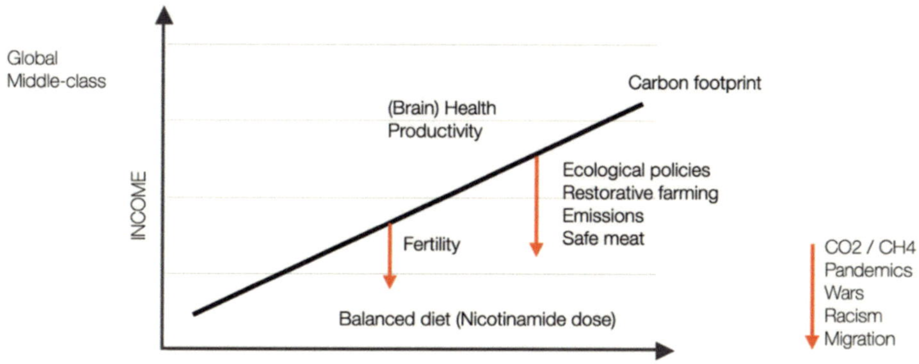

Figure 12.
A balanced diet leads to better demographics with high quality but lower numbers of offspring that raise productivity and inventiveness and eventually lowers emissions after an initial sharp rise at least in the past. This is different to the cereal related boom-bust demographics of the neolithic and recent green revolutions with their emphasis on calories that become demographic and health "time bombs".

pinches on subsequent generations prosperity and diet need accepting as breaks in the contract between generations (after Edmund Burke). A move back to "kitchen table" issues rather than the distraction of post-materialist cultural concerns treating lower status non-elites better may counteract populism and waves of migration along a metabolic gradient unstoppable by "Canute-like" border controls and walls [217].

13. Poverty traps

Previous worries about alleviating poverty and encouraging laziness or high fertility can now be shown, and the pellagra example helps, to be wrong—better diet cures laziness and poor or even criminal behaviour patterns and may decrease fertility (to the point it becomes a concern as the demographic dividend from more young people dissipates and immigration can become a necessity) [218] (**Figure 12**). The original Neolithic move to a more cereal based diet may have been a "Faustian" bargain increasing fertility and populations that could develop divisions of labour but at the expense of poorer health and human capital for some—a bargain we should recognise and take active steps to avoid.

"Tragedy of the commons" [219] claims that peasant cooperatives overgraze pastureland are rare whilst capitalist corporations frequently have short term non-sustainable profit orientated approaches that damage the soil and use artificial fertilisers and pesticides to such an extent that they damage ecosystems and diminish biodiversity. Mixed farming cooperatives usually do the more sustainable reverse unless driven out of the market when if close to starvation "seed corn" may get eaten or corn sold cheaply that has to be bought back early in the next season at greater expense or, not eating (or even destroying) cattle as it is the only source of wealth—classic poverty double traps that happen [220].

14. Politics and economics

No political or economic ideology has had unalloyed success in feeding their own populations (or of consistently helping others)—a key measure of the legitimacy of any government. Neoliberal policies emphasise industrial agribusiness' efficiency with monopolies and profits to shareholders, compared with earlier "Fordist" capitalist systems that encouraged higher pay leading to better diets and consumerism for workers and gains for subordinate classes, have not served meat equality well [221, 222]. Neither has the Ricardian economics of "comparative advantage" that makes some sense when exporting tropical goods (or wine from Portugal versus cloth from England) and free trade but taken too far can destroy local mixed farming and even staples that then have to be "offshored" to unreliable or expensive sources leading to food insecurity [223].

State involvement as socialist and collectivist experiments also often fail such as in China or in the Soviet Union as witnessed by the Ukrainian Holdamor with starvation even in the "breadbasket" of Russia [224]. Other widespread famines and Stalin's Russia "Terror-Famine" in the 1930s and China's and Mao's "Secret Famine" in the 1960s—neither put nutrition first but industrialisation. Meat issues and differentially successful agrarian policies may have been behind both the rise and fall of the Iron Curtain—McDonald's in Moscow was reduced to ersatz meat and meat queues were deeply unpopular in east Germany and Poland [225]. The need for square meals was re-discovered in the 1930s great depression in the capitalist USA and in Britain stimulated by hunger and job marches and both had to take a more socialist stance as in Roosevelt's "New Deal" [226, 227]. This deal incorporated anti-monopolist laws that dissolved the "Beef Trust" (alongside the more remembered Standard Oil) and supported the earlier populist movement of farmer cooperatives and the "Granger" movement notably and unusually from all "races" in the 1870s although "Big Ag" later has regained power [228–231].

The more capitalist and scientific focus (as with the "green" scientific hybrid and GM approaches) have been mainly on yields and calories rather than on nutrition and improving human capital—paradoxically reducing overt starvation but perhaps fuelling population booms on a "Sisyphean task" as far as keeping everyone well fed is concerned [26, 232–235]. There have been attempts to prioritise economies to basic human needs [236–240]. These sometimes put diet and meat via either jobs or direct cash ahead of other human or political or even international legal values (that often favour the rich) and could "unshackle" some economies (such as in India relative to China's undemocratic policies) as a (hopefully) temporary trade-off [241–243].

15. Big food—small holders

A basic challenge is to decouple growth from its reliance on fossil fuels and fertilisers (from bones to guano to potash, phosphorus and artificial oil fuelled nitrogen fixation) in a way that does not make things worse for the planet or its inhabitants: compound "green growth" is likely to require a combination of some reversion to traditional crop-pastureland recycling practices and technological innovation from vertical farming to artificial meat to methane-eating bacteria in dairy farms and measuring progress using Human Development indices not GDP [244–247].

Agribusiness has been guilty of short-term emission heavy and other "Silent Spring" approaches that contribute to climate change and will eventually lead to less suitable agricultural land being available. Somehow a way forward with "local for local" (as was true for almost the whole of our evolution) "10 mile diets" insourcing more sustainable farming and ranching needs to be found [248–251]. Loss of biodiversity includes loss of the rarer foods that might be needed later and include some sources of meat [252]. Goldschmidt's 1978 "As you Sow" research showed profound effects on

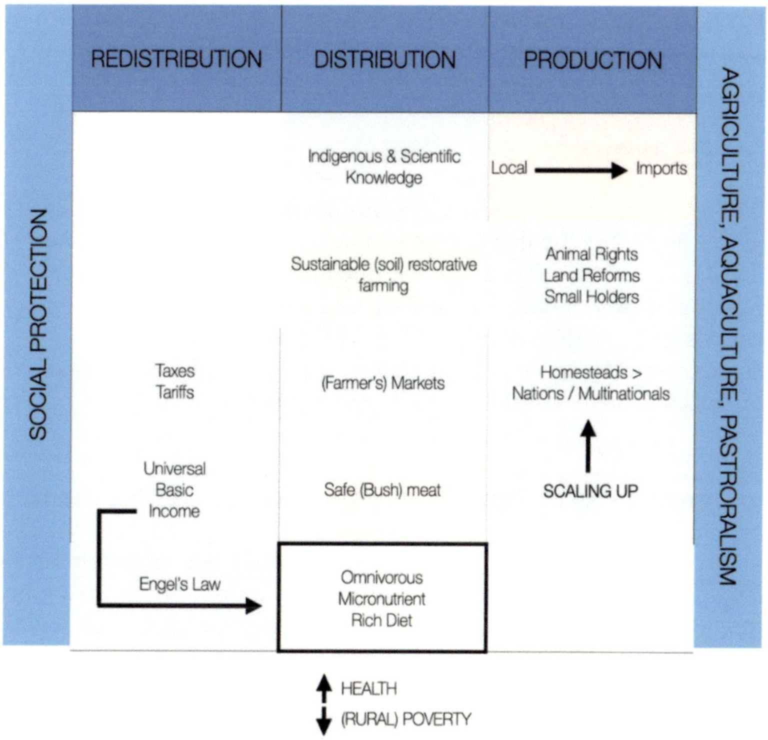

Figure 13.
Answers are likely to come from a combination of social protection of income ideally through well-paid jobs and better educational infrastructure in deprived areas as "pre-distribution" or, redistribution with universal lifelines for meat (and energy) consumption up to a certain point with food subsidies or cash payments. Local production of foods and the best of the scientific and indigenous knowledge approaches with restorative farming and better and safer distribution networks will help.

the community affecting poverty and health of industrial agriculture adversely versus family owned farms [253].

There are many suggestions and experiments ongoing that could solve these problems that integrate crop—livestock systems—if these decentralised systems can satisfy be scaled up creating a world system rather than globalisation with its high food miles and waste [254–267]. Local economies with food sovereignty going beyond food security with international peasant, slow food, cooperative solidarity movements like *La Via Campesina and Gruppi di Acquisto Solidale* with the many versions of food hubs may cut out middle-men and monopolies or speculators making produce cheaper, safer and closer products suitable to diverse dietary needs [244, 268]. Regenerative farming and agroecology, sometimes less strict than organic farming that does not allow weed-killers, with less tillage or chemicals and "soil doctors" may heal poor soil health and make them both more productive and more resilient to climate change (whether temperature, drought or floods) as well as habitat for a rich ecosystem for micro-organisms and a storage sink for carbon - but this is often only solvable by local measures as there is no "one size fits all" solution [269–272] (**Figure 13**). Care needs to be taken as radical overhauls such as banning all fertilisers and pesticides can cause a low yield crisis, as suggested by the experience of Sri Lanka. Laboratory grown meat has come a long way since the first burger (2013 and

$330,000) as have shifts from beef and lamb toward chicken and tofu but may remain too expensive for the poor and unacceptable "Frankenfoods" to the rich [273].

16. Is more data needed?

Cartesian logic says that when we have enough knowledge of a balanced diet, we do not need the "logic" of the marketplace to help us valorise or price meat. Neither do we need more data to act. The long overdue measurement and screening of NAD pathways in populations should still be done and would act as a lever to target deficient individuals and to adjust dosage at times of stress.

A simple goal reclaiming a meat and diet commons should be set that would quite possibly pay for itself several-fold in productivity, resilience and safety [274–277].

17. Pro-pandemic meat variances: poor plagues can plague the rich world

If altruistic and moral arguments do not suffice, then fear of pandemics and climate change might tip the balance toward further action. The meat supply has long been a source for concern from nineteenth century meat packers in Americas "Red meat Republic" and Chicago's "Porkopolis" as depicted by Cronon and Sinclair in "the Jungle" with dangers to the packers and the consumers triggering legal constraints on labour and consumer safety (the birth of the FDA). This did not stop the variant

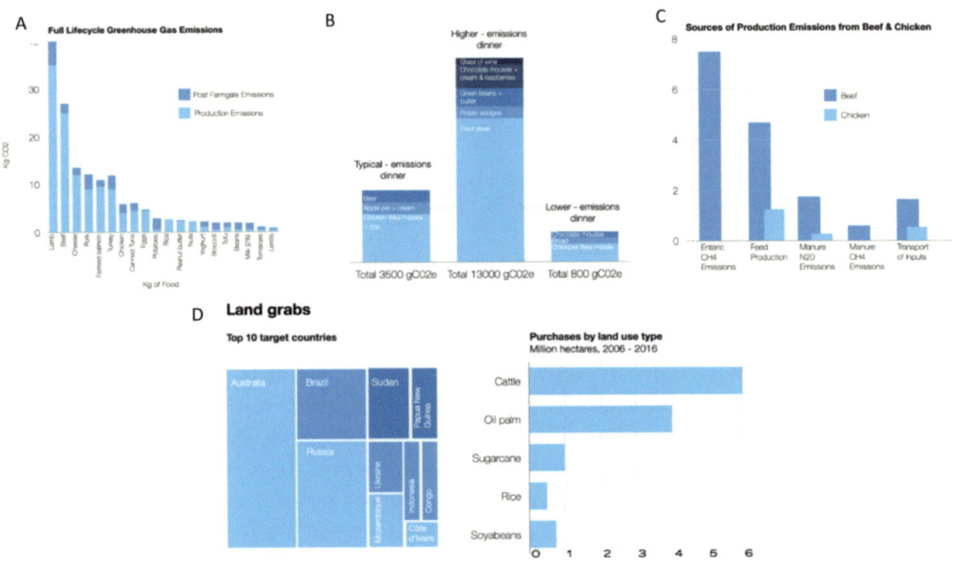

Figure 14.
(A) Greenhouse emissions by food type. (B) Emissions from beef and chicken. (C) High to low emission meals. (D) Land grabs and their main uses. Ruthless pastureland and cropland appropriations and exploitations have been important drivers of our history. Territorial gains (Ukraine excepted) are now rare but the goal of most deals remains to feed the western centre or create profits from cash crops, oil or mining increasingly of rare metals. All benefitting from information technology and "offshoring" with cheaper labour rather than improving the farms or the diet of the local inhabitants [294–296]. When local governments are astute to avoid "resource curse" or the "Dutch disease" they use profit to farm and feed their population properly and do not become too dependent on others for aid or their food surpluses or manufactured goods.

CJD outbreak in the UK from unnatural farming and feeding practices associated with "industrial meat regimes" or in China "dragonhead enterprises"—COVID-19 highlights the need for better and safer meat supplies as did other earlier emergent zoonoses; pathogen spillover is more likely when there are rapid changes in bat ecology including climate change and deforestation and when they are hungry and nomadic themselves bringing them closer to intermediate viral hosts and the human food-chain [62, 278–284].

The quest for meat is strong enough that risks get taken both by the poor in desperation, such as widespread poaching endangering species, and equally by the bored uber-rich desiring exotic foods usually meat. Bush meat and other wildlife hunting or farming particularly in deforested areas can bring species together so closely that it allows for microbial/viral species-jumping an important source of new (and old) zoonoses and food poisoning. Animal culls from emergent veterinary infection or known pathogens that could have been avoided by better vaccination programmes can impair the meat supply and cause price spikes. Regulating the market with a visible hand banning illegal wildlife, improving livestock densities and revamping with less cruelty to animals and fish, tidier slaughterhouse conditions combined with early warning systems are urgent to correct this "biological experiment" as are fresh looks at conservation in zoos and parks with public health in mind [285–288]. Restricting the rich in a "meat retreat" but supplying the poor is the crucial point that must be kept central to policies and not compatible with calls to abolish livestock farming and replace with veganism and microbial ferments (reminiscent of the dystopian 1973 film Soylent Green) [45, 289–293] (**Figure 14**).

NAD depleted populations are prone to infections and more likely to infect others with the infections further depleting NAD from tissue damage and perhaps impaired uptake of tryptophan in a metabolic trap. Indeed some manifestations of COVID and other infections (and long lasting side-effects such as chronic fatigue) may be new versions of pellagra with its documented disturbances of nicotinamide metabolism [297].

18. Answers: nutritional repairs as reparation and route to prosperity

Well-meaning attempts, and utopian ideals, have been made to alleviate nutritional poverty from feasts and potlatches in feudal times to poor laws and charity including from churches and monasteries and later foundling hospitals, school meals and food banks—and the far less well-meaning depictions of the "undeserving poor" and establishment of workhouses [298]. The usual reason for support was to provide enough basic subsistence to socially "produce and reproduce" a surplus workforce in case needed for industry or for "cannon-fodder". More enlightened policies exist but often rush to education or electricity or fertility control or good institutions without much mention of diet the gift that underpins these worthy goals [299–303] (**Figure 15**). In other words a meat and fish ("Blue") Commons and a "Glocalisation" foodscape needs to be recreated counteracting the "Lauderdale paradox" (as the "Charter of the Forest" a part of the Magna Carta had attempted in 1217 as later did the 17th C Leveller and Digger upheavals in the English Civil War, with "warning tears of the oppressed", and later the Quakers and Chartists). However, the rich still enriched themselves "contrived scarcity" of what had been abundant and was to become super-abundant. The worries of "Grim reaper" Malthusians such as Paul Ehrlich (as in his 1980 bet with the cornucopian Julian Simon) are perhaps resolvable by high "K" quality populations on good diets not "r" "population bombs" [304]. On the whole up to now the optimists have been right with Human

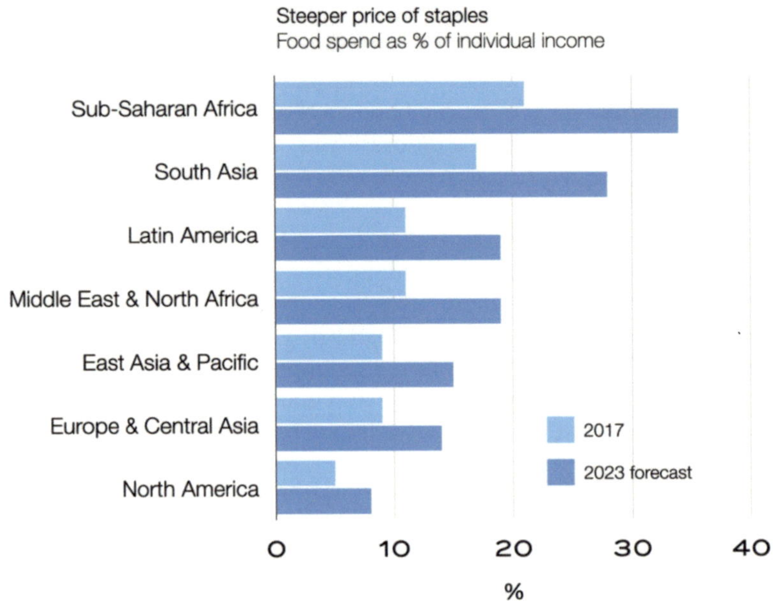

Figure 15.
Food spend as % of income remains high in many countries and will be a lot worse for the poorest families. Engel's/Bennett's law is clear cheap empty calories will be the first priority and meat/milk and vegetables will be squeezed out. By contrast rich people in rich countries spend a negligible proportion of their income on food that is in effect free as it barely affects their disposable income. Overindulgence can however be exacerbated by excessive fortification in foods and drinks running the risk of hypervitaminoses B3 for the affluent.

Development reports showing reductions from 60% to 10% of peoples in extreme poverty (less than 2$ a day) since 1950 with increases in schooling, life expectancy and happiness but this has been unequally spread; the pandemic has reversed progress in a detour that may last given an "uncertainty complex" in a "polycrisis" or ticking "cluster bombs" best avoided by returning to basics such as wholesome diets.

Global problems, we have illustrated can be dated to unlucky geographies exacerbated by colonialism and post-colonial self-determination and the long shadow of slavery or neo-slavery, such as indentured servants and child labour, frequently resulting in pro-pellagrous diets [305–311]. Indeed, some good data is on catch-up growth after the stunting of American slave children that gives hope, as do many other studies across the world on migrants that diet with more animal products and protein can improve height cognition and behaviour and life chances within and across generations. Massive reparations were made to the slave owners not the slaves and reconstruction deals such as " Forty acres and a mule" that would have helped were soon reneged on [28]. If we set our minds we could disallow a range of black-white disparities in the prevalence of chronic disease including (Caribbean) "amputation capitals" caused by metabolic and "blood sugar" syndromes as well as excessive deaths from infection. Targeting the vulnerable may be more effective, more workable and more affordable than other proposals or repatriating looted cultural artefacts and could be off-set by other stolen assets from corrupt dictators laundered in the West as more recent "sins of the fathers" [57, 312–319].

Reparations, it will be remembered were poorly managed in Versailles in 1919 with fatal consequences (paving the way to WW2), despite help from the American Red Cross and "Save the Children" (who against opposition also helped starving Russians).

DOI: http://dx.doi.org/10.5772/intechopen.110061

By common assent the German famines led to fascism, further *lebenstraum (that was a factor initiating WW1 seeking "a place in the sun" for both Germany and Japan)* and WW2: The "Hungerplan" to starve Soviets and Jews mixed up such nutritional policies, racism and later genocide and "Holocausts" (whilst the world watched "*a grotesque dream is forming*"—Joseph Roth 1924) [212, 320, 321]. Half-starving people on a low meat diet turns them into versions of pellagrins, underlined by Primo Levi's description of the "Musselman" who close to death "*with head dropped and shoulders curved whose face and eyes not a trace of thought was to be seen*"—that then "proves" them inferior.

In the WW2 case this chain of events was defined (by Lemkin and Lauterpacht -1945/6) as "genocide" and "crimes against humanity" and their "Final Solution" of euthanized death [302–304, 322–324]. Many genocides can also be seen in the light of dividing pastoralist and farmer groups and deciding they are ethnically different as evidenced in the history of the San (hunted as bushmen) or frictions over dairy farming in Kenya (the Mau Mau uprising in 1952) or in Rwanda or Darfur or inventing national borders that encourage ethnolinguistic diversity and friction as in Africa and India/Pakistan with similar colonial-settler like frictions and "indiginocides" in Gaza, Sri Lanka and Myanmar [323].

In these self-fulfilling policies half-starving people on a low meat diet turns them into versions of pellagrins 'legitimising' denationalisation followed by dehumanization and should be incorporated into Stanton's "10 stages of genocide" under ecological and economic headings. We should re-think development emphasising the opposite outcome from a better diet turning the investment of reducing famine and malnutrition and circumvent various vested interests to avoid future criticisms and our own "autogenocide" [245, 325–331]. This is particularly important for Africa and some parts of Asia allowing them to complete their demographic and epidemiological transitions to everyone's advantage [274, 332–336]. The remarkable success of countries such as South Korea, a war-torn "basket case" in the 1950s concentrating on agriculture, land reform, exportable goods and culture shows that "Shrimp to Whale" progress can be done [337].

19. Better policies: nutrition front and centre "to each according to their needs"

Food stamps and maternal support and attempts to maximise employment or provide universal basic incomes have had some luck [338]. Multiple studies over nearly a century in many countries have shown the benefits of animal sourced foods the tall Masai relate to pastoralism and the tall Dutch to high milk intake in the Netherlands—international IQ comparisons may similarly relate to diet [339–353]. Universal basic income first suggested in Thomas More's old Utopian dream and enacted in Speenhamland (1795) almost became American policy in 1969 as part of the "War on Poverty" and many recent studies have shown large cost-effective gains in many measures across the board—this may well reflect Engel's curves in improving the amount of milk and meat in diet [354]. Universal basic services is a closely related idea that would have similar results if it included diet as a priority rather than jump to welfare objectives such as free education and health by-passing diet [355, 356]. A reinvention of "gas and water" municipal socialism (as instigated by Chamberlain in Birmingham) and "New Jerusalem's" with communes driven then by labour unions and strikes to include diet and international subsistence revolts often led by women ("Rebecca food riots") andhelp to local farmers and their markets may be a the exemplar narrative and pre-requisite for success [357–362].

Policies that encourage rather than marginalise pastoralism (as a protected local industry as important as other forms of security such as energy, computer, ships or military technology) and invest in "precision" mixed farming and sustainable regenerative agriculture and aquaponics combining local indigenous knowledge with the best of the scientific approaches—and learning from well—constructed trials sensitive to local opportunities and cultural needs have to make sense [53, 76, 256, 266, 363–370]. Pastoralists are particularly at risk of climate change and its effects on forestation such as the advance of Boreal forests and reindeer populations or, their loss from fires, and is in the Sahal a flashpoint for outbreaks of violence [371].

Furthermore, agriculture and silviculture improve biodiversity reflecting the co-evolution between man, birds, bees and other insects, and symbiotic microbes in our gut and in the soil [372–374]. Neonicotinoids and other pesticides have damaged insect populations and the use of antibiotics to encourage growth in animals contributes to changes in our microbiome with antibiotic resistance and the emergence of "superbugs" now a major health hazard [375].

20. Humanitarianism: re-igniting a concern for distant others

Humanitarianism and its institutions have a history that some date from the consequences of decisions in Paris (1919) and Biafra (1970), where pellagra was well described, was another historical stimulus—that is still relevant to poor nutrition and population explosions in Nigeria [364, 376–379]. Charity organisations, such as Medicins Sans Frontières and the NGO Oxfam [380] with local development is the way forward including supporting pastoralists. The need for migration should then be manageable with less friction—after all behaviourally modern *Homo Sapiens* has been on a transcontinental global quest for meat or fish along eastern and western coastal "kelp and salmon highways" (and remember the Polynesians and Thule Inuit and the more commercial Basques) for 200,000 years or so and only migrates away from family when it is forced to [381, 382].

Calls for internationalisation of the meat, milk and egg supply pointing out the related failures of water privatisation and to the overall success of national meat rationing in time of war may be in order [11, 383, 384]. Such combined national and international approaches with a broader view of "A Duty of Care" [385–387] should lessen the loaded dice of being born in the wrong place with little chance up upward mobility (on the Gatsby curve) and may even be helped by technology and increased ability to virtually see the world from other perspectives, although runs the risk of flaming inequality, with gamers even playing farming games (originally "Harvest Moon") rather than shooting "baddies" and rewards for being kind not vengeful [388, 389].

21. NAD ups and downs over time and place—"Only one Earth"

Pellagra, often familial, is an outstanding example, alongside eco-genetic opposite side of the genetic coin examples, such as phenylketonuria also cured by dietary intervention, of good nature via good nurture [390]. Cotton market failures led to pellagra outbreaks just as loss of ship building led to a dietary and health collapse in Glasgow [391, 392]. Although the rest of "Bad Food Britain" or the USA is far from ideal (with many food-banks necessary) hope comes from Mediterranean countries where the class food divide is not so marked, and neither are health inequalities. Perhaps as they were in the eye of the original pellagra storm lessons were learnt Elsewhere the "fallacy of composition" states that because many have access to luxuries one must not presume that necessities are catered for as heavy marketing of new technologies distort Engel's law even in slums and "shanty towns" [393].

Politics should lower its sights and deal with Masferrer's (1929) "El minimum vital" regardless of the hand dealt by fate and as Berthold Brecht said in his Three Penny Opera "*You may proclaim, good sirs, your fine philosophy but till you feed us right and wrong can wait*" or Scott Fitzgerald (1925 -Jazz age) in the Great Gatsby " *No human difference is so profound as between the sick and the well*" and this failure to allow ascension up the food chain is the real source of (brain)health inequality [382, 394–397]. A "Constitution of knowledge" or "Adaptive intelligence" or "Metacognition" depend upon diet and if these do not function well neither science or the arts or political and democratic institutions can emerge capable of solving the major problems facing the planet and our own survival [263, 296, 380, 398–404].

Poor nutrition commonly gets medicalised shifting the blame and responsibility from the state to the individual and then blaming willpower or incarcerating or driving to drug and food addictive behaviour with opioid epidemics and mappable deaths of despair as was even documented three centuries ago in the pellagra epidemics [78, 405]. Failure to progress nutrition and human capital may explain the "Great economic disappointment" of the last 20 years [406]. Poor classes and countries particularly in the South face a NAD headwind that results in a developmental and degenerative pathology of poverty and inferiority, as judged by many in richer habitats who choose to forget the quote from Corinthians "*What do you have that you did not receive? And if you received it, why do you boast as if it were not a gift.*" Or the meritocrats conveniently forget "*Der Mensh was iBt*" (Feuerbach 1850) that can now be shown by stable isotope research, that has tracked our meat: plant dietary history, to be literally true that we are what (and where) we eat [407–409]. Rich gastronomy and exotic diets look decadent at best if the south is included in "Only One Earth" (Ward 1972) and contributes to a dangerous syndemic and as said to the rich after Belshazzars feast in the Book of Daniel "*thou art weighed in the balances, and found wanting*". Meat is not a luxury for billions and should be democratised or as Benjamin Franklin (1758) stated "For want of a nail, the kingdom was lost". Metabolic rifts allow for dangerous metabolic ghettoes with poor brain development crucial for abstract thought, language writing and arithmetic and the many Neuro- prefixed disciplines that "supersize the mind" [312, 316, 410, 411]. A more moral economy should focus on this need as did the "physiocrats" and "levellers" several centuries ago—and was Confucian philosophy long before that [186, 397, 412]. Meat markets are (Pareto) inefficient as it would be possible to make many people better off without making anyone worse off (indeed they may well be better off too; giving money would pay for the meat and would not be wasted as predicted by Engel and has been shown in several randomised trials (see Banerjee and Duflo's Nobel prize winning studies).

22. Postcript 1: current concerns

Much of the heated controversies over the rights and wrongs of meat based or plant-based diets can be resolved if seen in the light of needing homeostasis and NAD(H) based metabolism [284, 413–420] that may yet rhyme with a green hydrogen energy revolution [421]. The writing is on the wall over the ethics and safety of allowing billions to be poorly fed so we need to sort this out or any future generations will obloquy us in the same way as we criticise the diet-mediated slave trade and most genocides [269, 422].

Stagflation with increasing prices and no growth was triggered in the 1970s by energy and then food price spikes and may happen again turbocharged by current shocks—for every percentage point increase in food prices another 10mn people are pushed into extreme poverty and developed countries are far from immune to this effect (prices are up 30% this year) with households paying over 50% of income on food. Heavy current subsidies of foods may become unaffordable for some countries and already concentrate on calories encouraging junk food and "buy one get one free" deals. Repeating dated mantras such as the poor needing help with cooking and domestic budgeting or, protectionism with export bans over foods or fertilisers are not an answer.

Recent events in Ukraine bring to a head longstanding problem in the food-chain affecting everyone but particularly those teetering in countries affected by conflict—including Yemen, Ethiopia Sudan and Egypt—risking another "Arab uprising" that began symbolically with the self-immolation of a Tunisian vegetable vendor and were followed by Sudanese and other protests over the price of bread [423].

Individuals may be best served by free money but some states need "Special Drawing Rights" and support via World Food Programmes and combined with investment to improve agricultural yields such as in Africa and incentives to cut down

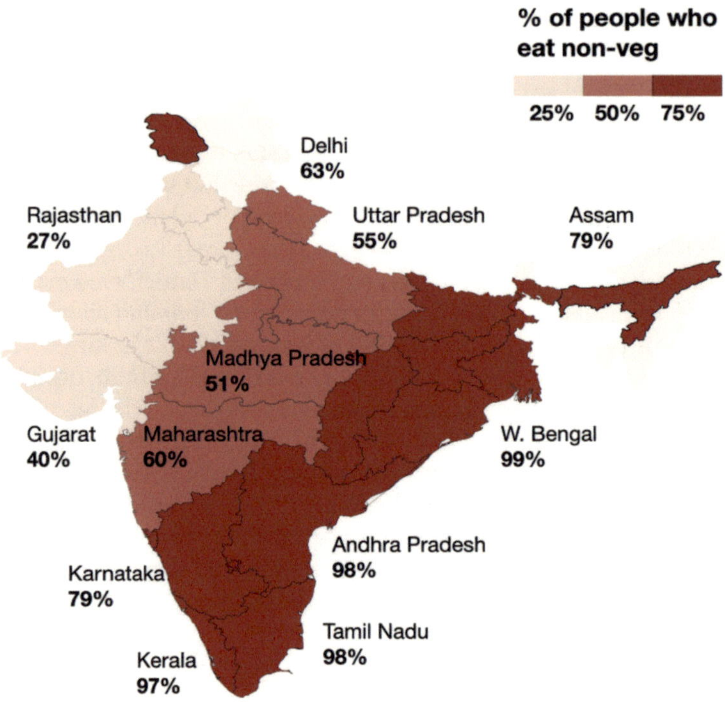

Figure 16.
Southern v, Northern India may be as close to a controlled experiment as it is possible to get compared to the natural experiments we have described. A child born in the South is far less likely to die in their first year of life or lose her mother in childbirth. She will receive better nutrition with free midday meals and go to school and stay in school longer then attend college and secure employment with better pay. She will have fewer children who in turn will be healthier and more educated than her. Southern India, and Kerala in particular, over the last 50 years has developed meat and especially beef as a secular dish cutting across class, caste and religion—the latter of course is complex historically with taboos that remain influential in the North. Initially the extra expense may have been kick-started by Gulf migration and remittance monies sent home but is now self-sustaining due to economic success and high meat production as well as consumption.

on scandalous food waste (as in European "butter mountains" and "milk lakes") and overindulgence in the developed world [424–426].

23. Postcript 2: QED?

These are man-made disasters, as was overt pellagra. Hope should come from realising that darkest London of 200 years ago was not that different to the poorest regions of the world now. Strong correlations with progress such as fossil-fuel dependency that may have tempted emerging economies, such as in Africa, to believe that they cannot leap-frog to green economies can be shown that the legacy carbon path, or at least an extreme form of it, is not necessary if a balanced diet is prioritised.

Concrete optimism could spring from an almost randomised series of experiments in India where the southern states have overtaken the Northern states in educational and economic terms with better health (such as infant mortality or TB incidence) with enlightened policies on meat/milk/eggs eating and midday school meals [427] (**Figure 16**).

Improved diet would increase resistance to infection more than the most powerful antibiotics and avoid "superbugs" and improve gene maintenance, expression, regenesis (including of neurones) and alter our inheritance in a way that glitzy gene therapies—for the rare and rich—could only dream of (although it has been suggested that such technology could by using gene-drives reduce the desire for meat and make us more altruistic) [278, 428–431]. At the same time the nightmare of more inequality would be avoided as so much unmet need would be addressed. The fairly high hereditability of traits such as height and IQ and many diseases are much less if poor suggesting that diet is the crucial, cheaper and fairer intervention [432].

24. Discussion

It does not take a paranoid or an anxious mind-set to sense trouble ahead if food-security and nutrition is not taken more seriously—the Maputo declaration of 2003 committed African leaders to devote 10% of budgetary allocations to agriculture; few have come close but placate populations with food imports that they increasingly cannot afford and should rely less on free trade but produce more at home or "friend-shore" rather than deal with international monopolies with few value based rules [433–435]. Commerce and market forces along with some optimism over new technology could, if it prioritised diet, create its long- promised potential for peace, (as first suggested by Montesquieu (1748) and Angell (1910)) and " freedom" and democracy between interdependent classes and countries that currently are made dependant on those with (non-seasonal and varied) affluent dietary and other (decadent) appetites—and could stop us all hurtling toward disaster [436–438].

A post liberal cosmopolitan economics that considers others more, and agrees with Joesph Roth (1938) " People aren't pigments and the world is not a palette!" and is not about the survival of the richest (even planning escapes to Mars (!) or eternal lives) or even a "golden billion" and other "great replacement" conspiracy theories by remembering Benjamin Franklin's 1776 dictum "we must hang together or surely we shall hang separately" is in order [439–441]. Originally race was closely linked to class and poverty (as are immigration and other white laws) exemplified by parochial commentaries such as "the Bethnal Green poor are a caste apart, a race of whom we know nothing" force-feeding inferiority that then needed to be despised or civilised in hostile environments cloaked in superficial markers such as colour or religion, as noted by WEB. Du Bois when visiting a Warsaw ghetto in 1949 [442].

Better diet for all should speed up our "slouch toward Utopia" rather than "molecular utopianism" for the few. We finally begin to answer Keynes' 1924 criticism that "*we lack a coherent scheme of progress*" as that means meat for metabolic progress, not meat as a sign of progress or for "showing-off". Poverty alongside affluence definable as "stuffed and starved" metabolic diseases with pathologies not due to "status anxiety". A flatter not fatter earth with reduced meat variances is required and need not wreck the world and may save it [188, 443–446]. As (Sir) Bob Geldof, influenced by the history of *An Gorta Mor*—the Irish Great Hunger of the 1850s when governments also stood-by—recently said "*Development is not some feel-good virtue-signalling function, it is a critical necessity of humanity or the spores of poverty find fertile ground in the poisonous fields of ethnicity, populism and nationalism- ask Putin*". Giving in to our altruistic urge with something of a moral revolution may help when combined with a "fear factor" of pandemics and climate change and a concern for our kids and future generations and generational injustice [447–452].

Bernard Kouchner, the originator of Medecins Sans Frontiers, (1993) having observed starvation and pellagra in the Biafran war wrote in his memoir "*Le malheur des autres*"—"A quoi servaient medicins s'ils n'alertaint pas le monde du blocus alimentaire.....silencieux nous etions complices du massacre systematique". Preventative and moral medicine needs to avoid being accomplices amongst the geopolitical chaos and speak and act up over malnutrition and pandemic avoidance with safer meat supplies—and in keeping with Ernest Hemmingway's "*if the story is so simple it can be said in 6 words*" and to avoid the weight of circumstantial but (we hope) percipient material dragging our argument down we say loud and clear Square Meals—Poverty and Pandemics Gone. The lesson of the Black Death or of imports from "settler colonies" is that doubling meat intake for the peasants creates "Golden Ages" with more equality of early developmental outcomes and opportunities but this could be done without more plagues or creating more third worlds in the post-colonial age that could "boomerang" [453, 454]. As a critique of development policies, states: the silence on unequal (metabolic) rights for rich and poor made unequal by dietary not genetic or cultural "tangles of pathology",—and the lack of trust to make their own subsistence and other decisions has to end with less emphasis on blaming the victims and punishing laws and orders [455]. "Races" may correlate with skin colour but this oppressive social construct disconnects from causation as it includes Jews, Gypsies, Romas and many other poor economically disempowered white classes on poor diets (in the past including the Irish and Italians or indeed anyone non-Nordic), subject to (immigration) discrimination and genocide, that profoundly affects plastic phenotypes [456–459].

*His story relevant to child malnutrition was "*Baby shoes for Sale. Never used*"

Acknowledgements

This study was funded by QEHB Charity, Birmingham, UK.

Author contributions

ACW and LJH drafted review, CW produced figures/tables, all authors proofed and authorised submission.

DOI: http://dx.doi.org/10.5772/intechopen.110061

Funding

This study was funded by QEHB Charity, Birmingham, UK.

Conflicts of interest

The author(s) declared no potential conflicts of interest with respect to the research, authorship and/or publication of this article.

Author details

Adrian C. Williams[1]*, Christina Wood[1] and Lisa J. Hill[2]

1 Department of Neurology, University Hospitals Birmingham NHS Foundation Trust, Birmingham, UK

2 School of Biomedical Sciences, Institute of Clinical Sciences, University of Birmingham, UK

*Address all correspondence to: adrian.williams1949@gmail.com

References

[1] Hyman M. Food Fix: How to Save Our Health, Our Economy, Our Communities and Our Planet–One Bite at a Time. Hachette UK; 2020

[2] Smil V. Grand Transitions: How the Modern World Was Made. Oxford University Press; 2021

[3] Thunberg G. The Climate Book. Penguin Books Limited; 2022

[4] Waters A. We are What We Eat: A Slow Food Manifesto. Penguin; 2022

[5] Vardi L. The Physiocrats and the World of the Enlightenment. Cambridge University Press; 2012

[6] Carolan M. The Sociology of Food and Agriculture. Taylor & Francis; 2016

[7] Misra UK, Kalita J. Neurological Consequences of Nutritional Disorders: CRC Press; 2021.

[8] Bernstein WJ. The Birth of Plenty: How the Prosperity of the Modern Work was Created. McGraw Hill LLC; 2010

[9] McCloskey DN, Carden A. Leave Me Alone and I'll Make You Rich: How the Bourgeois Deal Enriched the World. University of Chicago Press; 2020

[10] Miles D et al. Macroeconomics: Understanding the Global Economy. Wiley; 2012

[11] Water SS. The Epic Struggle for Wealth, Power, and Civilization. HarperCollins; 2010

[12] Wilson-Lee E. A History of Water: Being an Account of a Murder, an Epic and Two Visions of Global History. HarperCollins Publishers Limited; 2022

[13] Jones TC. Desert Kingdom: How Oil and Water Forged Modern Saudi Arabia. Harvard University Press; 2010

[14] Halliday S. Water: A Turbulent History. Sutton Gloucestershire; 2004

[15] Boccaletti G. Water: A Biography. Knopf Doubleday Publishing Group; 2021

[16] Lambert A. Seapower States: Maritime Culture, Continental Empires and the Conflict That Made the Modern World: Yale University Press; 2018.

[17] Howarth DA. British Sea Power. Basic Books; 2003

[18] Davidson A et al. The Oxford Companion to Food. Oxford University Press; 2014

[19] Fraser EDG, Rimas A. Empires of Food: Feast, Famine and the Rise and Fall of Civilizations. Arrow Books; 2011

[20] Rimas A, Fraser E. Beef: The Untold Story of How Milk, Meat, and Muscle Shaped the World. Harper Collins; 2008

[21] Jacob HE, Winston RC. Six Thousand Years of Bread: Its Holy and Unholy History. Hauraki Publishing; 2016

[22] Tannahill R. Food in History. Paw Prints; 2008

[23] Khosrova E. Butter: A Rich History. Algonquin Books; 2017

[24] Scott JC. Against the Grain: A Deep History of the Earliest States. Yale University Press; 2017

[25] Crosby AW. Ecological Imperialism. Cambridge University Press; 2015

[26] Ponting C. A New Green History of the World: The Environment and the Collapse of Great Civilizations. Random House; 2011

[27] Fanon F. Studies in a Dying Colonialism. Earthscan; 1989

[28] Foner E. A Short History of Reconstruction, Updated Edition. HarperCollins; 2015

[29] Gomes AP et al. Declining NAD+ induces a pseudohypoxic state disrupting nuclear-mitochondrial communication during aging. Cell. 2013;**155**(7):1624-1638

[30] Bocquet-Appel JP, Bar-Yosef O. The Neolithic Demographic Transition and its Consequences. Springer Netherlands; 2008

[31] Kohler TA, Smith ME. Ten Thousand Years of Inequality: The Archaeology of Wealth Differences. University of Arizona Press; 2019

[32] Huber MT. Climate Change as Class War: Building Socialism on a Warming Planet. Verso Books; 2022

[33] Lieven A. Climate Change and the Nation State: The Case for Nationalism in a Warming World. Oxford University Press; 2020

[34] Parfit D. Reasons and Persons. OUP Oxford; 1986

[35] MacAskill W. What We Owe the Future. Basic Books; 2022

[36] Davis M. The Monster Enters: COVID-19, Avian Flu, and the Plagues of Capitalism. Verso Books; 2022

[37] McAlevey J. No Shortcuts: Organizing for Power in the New Gilded Age. Oxford University Press; 2016

[38] Zaretsky R. Victories Never Last: Reading and Caregiving in a Time of Plague. University of Chicago Press; 2022

[39] Medicine NAo. Global Roadmap for Healthy Longevity. Washington, DC: The National Academies Press; 2022. 320 p

[40] Ruddiman WF. Plows, Plagues, and Petroleum: How Humans Took Control of Climate. Princeton University Press; 2010

[41] Pfeiffer DA. Eating Fossil Fuels: Oil, Food, and the Coming Crisis in Agriculture. New Society Publishers; 2006

[42] Porritt J. Hope in Hell: A Decade to Confront the Climate Emergency. Simon & Schuster UK; 2020

[43] Vince G. Nomad Century: How Climate Migration Will Reshape Our World. Flatiron Books; 2022

[44] Harper K. Plagues Upon the Earth: Disease and the Course of Human History. Princeton University Press; 2021

[45] Harrison M. Contagion: How Commerce Has Spread Disease. Yale University Press; 2012

[46] Chan A, Ridley M. Viral: The Search for the Origin of Covid-19. HarperCollins Publishers; 2021

[47] Bess MD. Planet in Peril: Humanity's Four Greatest Challenges and How We Can Overcome Them. Cambridge University Press; 2022

[48] Roubini N. Megathreats: The Ten Trends that Imperil Our Future, and How to Survive Them. John Murray; 2022

[49] Yergin D. The New Map: Energy, Climate, and the Clash of Nations. Penguin Publishing Group; 2020

[50] Rockström J et al. Planetary boundaries: Exploring the safe operating space for humanity. Ecology and Society. 2009;**14**(2)

[51] Ninan KN. The Economics of Biodiversity Conservation: Valuation in Tropical Forest Ecosystems. Taylor & Francis Group; 2012

[52] Pretty J et al. Sustainable Intensification: Increasing Productivity in African Food and Agricultural Systems. Earthscan; 2011

[53] Rees M. If Science is to Save Us. Polity Press; 2022

[54] Toensmeier E, Herren H. The Carbon Farming Solution: A Global Toolkit of Perennial Crops and Regenerative Agriculture Practices for Climate Change Mitigation and Food Security. Chelsea Green Publishing; 2016

[55] Malm A. The Progress of This Storm: Nature and Society in a Warming World. Verso Books; 2020

[56] Baccini P, Brunner PH. Metabolism of the Anthroposphere, Second Edition: Analysis, Evaluation, Design. MIT Press; 2012

[57] Wells JCK. The Metabolic Ghetto: An Evolutionary Perspective on Nutrition, Power Relations and Chronic Disease. Cambridge University Press; 2016

[58] Korzeniewicz RP, Moran TP. Unveiling Inequality: A World-Historical Perspective. Russell Sage Foundation; 2009

[59] Fanzo J. Can Fixing Dinner Fix the Planet? Johns Hopkins University Press; 2021

[60] Otter C. Diet for a Large Planet: Industrial Britain, Food Systems, and World Ecology. University of Chicago Press; 2020

[61] Steinfeld H et al. Livestock's Long Shadow: Environmental Issues and Options. Food and Agriculture Organization of the United Nations; 2006

[62] Cronon W. Nature's Metropolis: Chicago and the Great West. W.W. Norton; 2009

[63] Monbiot G. Regenesis: Feeding the World Without Devouring the Planet. Penguin Books Limited; 2022

[64] Willett W et al. Food in the Anthropocene: The EAT–Lancet Commission on healthy diets from sustainable food systems. The Lancet. 2019;**393**(10170):447-492

[65] Hirvonen K et al. Affordability of the EAT–Lancet reference diet: A global analysis. The Lancet Global Health. 2020;**8**(1):e59-e66

[66] Wylie D, Societies ACoL. Starving on a Full Stomach: Hunger and the Triumph of Cultural Racism in Modern South Africa. University Press of Virginia; 2001

[67] Tupy ML, Pooley GL. Superabundance: The Story of Population Growth, Innovation, and Human Flourishing on an Infinitely Bountiful Planet. Cato Institute; 2022

[68] Koyama M, Rubin J. How the World Became Rich: The Historical Origins of Economic Growth. Wiley; 2022

[69] Bernstein WJ. A Splendid Exchange: How Trade Shaped the World. Grove Atlantic; 2009

[70] Scialabba G. Slouching Toward Utopia: Essays and Reviews. Pressed Wafer; 2018

[71] Godin S. The Carbon Almanac. Penguin; 2022

[72] Gibson M. The Feeding of Nations: Redefining Food Security for the 21st Century. CRC Press; 2016

[73] Reay D. Climate-Smart Food. Springer International Publishing; 2019

[74] Bridle S. Food and Climate Change Without the Hot Air. UIT Cambridge; 2020

[75] Randers J. A Global Forecast for the Next Forty Years. Chelsea Green Publishing; 2052. p. 2012

[76] Steel C. Sitopia: How Food Can Save the World. Random House; 2020

[77] Hunter J. The Making of the Crofting Community. Birlinn; 2018

[78] Percival R. The Meat Paradox: 'Brilliantly Provocative, Original, Electrifying' Bee Wilson, Financial Times. Little, Brown Book Group; 2022

[79] Kaplan RD. The Revenge of Geography: What the Map Tells Us About Coming Conflicts and the Battle Against Fate. Random House Publishing Group; 2012

[80] von Hippel W. The Social Leap: The New Evolutionary Science of Who We Are, Where We Come From, and What Makes Us Happy. HarperCollins; 2018

[81] Sattin A. Nomads: The Wanderers Who Shaped Our World. John Murray Press; 2022

[82] Gray J. Straw Dogs: Thoughts On Humans And Other Animals. Granta Publications; 2015

[83] Harvey D. Spaces of Global Capitalism: A Theory of Uneven Geographical Development. Verso Books; 2019

[84] Wolch JR, Emel J. Animal Geographies: Place, Politics, and Identity in the Nature-culture Borderlands. Verso; 1998

[85] Wilcox S, Rutherford S. Historical Animal Geographies. Taylor & Francis Group; 2018

[86] Crosby AW. The Columbian Exchange: Biological and Cultural Consequences of 1492. Greenwood Publishing Group; 2003

[87] Crosby AW. Ecological Imperialism: The Biological Expansion of Europe, 900-1900. Cambridge University Press; 1993

[88] Suranyi A. The Atlantic Connection: A History of the Atlantic World, 1450-1900. Taylor & Francis; 2015

[89] Schiebinger L. Secret Cures of Slaves: People, Plants, and Medicine in the Eighteenth-Century Atlantic World. Stanford University Press; 2017

[90] Elliott JH, Societies ACoL. The Old World and the New: 1492-1650. Cambridge University Press; 1992

[91] Boorstin DJ. The Discoverers. Knopf Doubleday Publishing Group; 2011

[92] Clark RP. The Global Imperative: An Interpretive History of the Spread of Humankind. Taylor & Francis; 2018

[93] Penniman L, Washington K. Farming While Black: Soul Fire Farm's Practical Guide to Liberation on the Land. Chelsea Green Publishing; 2018

[94] Williams JC, Cuban M. White Working Class, With a New Foreword by Mark Cuban and a New Preface by the Author: Overcoming Class Cluelessness in America. Harvard Business Review Press; 2019

[95] Vance JD. Hillbilly Elegy: A Memoir of a Family and Culture in Crisis. HarperCollins Publishers; 2016

[96] Baptist EE. The Half has Never Been Told: Slavery and the Making of American Capitalism. Basic Books; 2016

[97] Fryer P. Staying Power: The History of Black People in Britain. PlutoPress; 2018

[98] Sbicca J. Food Justice Now!: Deepening the Roots of Social Struggle. University of Minnesota Press; 2018

[99] Strings S. Fearing the Black Body: The Racial Origins of Fat Phobia. NYU Press; 2019

[100] Everill B. Not Made by Slaves: Ethical Capitalism in the Age of Abolition. Harvard University Press; 2020

[101] Stiglitz JE, Charlton A. Fair Trade for All: How Trade Can Promote Development. Oxford University Press; 2007

[102] Ind N, Horlings S. Brands with a Conscience: How to Build a Successful and Responsible Brand. Kogan Page; 2016

[103] Jackson J. The Thief at the End of the World: Rubber, Power, and the Seeds of Empire. Penguin Publishing Group; 2008

[104] Belich J. Replenishing the Earth: The Settler Revolution and the Rise of the Angloworld. OUP Oxford; 2011

[105] Rodney W. How Europe Underdeveloped Africa. Bogle-L'Ouverture Publications; 1972

[106] Linden E. The Winds of Change: Climate, Weather, and the Destruction of Civilizations. Simon & Schuster; 2006

[107] Keys D. Catastrophe: An Investigation Into the Origins of the Modern World. Ballantine Pub.; 2000

[108] Miller EM. The Last Great Subsistence Crisis in the Western World. JSTOR; 1978

[109] Robins JE. Oil Palm: A Global History. University of North Carolina Press; 2021

[110] Turzi M. The Political Economy of Agricultural Booms: Managing Soybean Production in Argentina, Brazil, and Paraguay. Springer International Publishing; 2016

[111] Edgerton D. The Rise and Fall of the British Nation: A twentieth-Century History. Penguin UK; 2018

[112] Goldstone JA. Why Europe? The Rise of the West in World History 1500-1850. McGraw-Hill Higher Education; 2009

[113] Goody J. The Eurasian Miracle. Polity; 2010

[114] Jones E. The European Miracle: Environments, Economies and Geopolitics in the History of Europe and Asia. Cambridge University Press; 2003

[115] Tainter J. The Collapse of Complex Societies. Cambridge University Press; 1988

[116] Vries P. State, Economy and the Great Divergence: Great Britain and China, 1680s-1850s. Bloomsbury Publishing; 2015

[117] Cannadine D. Victorious Century: The United Kingdom, 1800-1906. Penguin Books Limited; 2017

[118] Turchin P. Ages of Discord: A Structural-demographic Analysis of American History. Beresta Books; 2016

[119] Slobodian Q. Globalists: The End of Empire and the Birth of Neoliberalism. Harvard University Press; 2020

[120] Parthasarathi P. Why Europe Grew Rich and Asia Did Not: Global Economic Divergence, 1600-1850. Cambridge University Press; 2011

[121] Ravallion M. The Economics of Poverty: History, Measurement, and Policy. Oxford University Press; 2016

[122] Smith N. Uneven Development: Nature, Capital, and the Production of Space. University of Georgia Press; 2010

[123] Elvin M. The Retreat of the Elephants. The Retreat of the Elephants. Yale University Press; 2008

[124] Frank AG. ReOrient: Global Economy in the Asian Age. Univ of California Press; 1998

[125] De Vries J. The Economy of Europe in an Age of Crisis, 1600-1750. Cambridge University Press; 1976

[126] Kaplan H et al. A theory of human life history evolution: Diet, intelligence, and longevity. Evolutionary Anthropology: Issues, News, and Reviews. 2000;**9**(4):156-185

[127] O'Rourke KH. The European grain invasion, 1870-1913. The Journal of Economic History. 1997;**57**(4):775-801

[128] Welford M. Geographies of Plague Pandemics: The Spatial-Temporal Behavior of Plague to the Modern Day. Taylor & Francis; 2018

[129] Case A, Paxson CH. Stature and Status: Height, Ability, and Labor Market Outcomes. SSRN; 2010

[130] Deaton A. The Great Escape: Health, Wealth, and the Origins of Inequality. Princeton University Press; 2013

[131] Dyer C. Standards of Living in the Later Middle Ages: Social Change in England C.1200-1520. Cambridge University Press; 1989

[132] Floud R et al. Height, Health and History: Nutritional Status in the United Kingdom, 1750-1980. Cambridge University Press; 2006

[133] Manary M et al. Protein quality and growth in malnourished children. Food and Nutrition Bulletin. 2016;**37**(1_suppl):S29-S36

[134] Steckel RH, Research NBoE. Heights and Human Welfare: Recent Developments and New Directions. National Bureau of Economic Research; 2008

[135] Allen RC. The British Industrial Revolution in Global Perspective: How Commerce Created the Industrial Revolution and Modern Economic Growth. Oxford: Oxford University Press; 2006

[136] Findlay R, O'Rourke KH. Power and Plenty: Trade, War, and the World Economy in the Second Millennium. Princeton University Press; 2009

[137] Van Zanden JL. The Long Road to the Industrial Revolution: The European Economy in a Global Perspective, 1000-1800. Brill; 2009

[138] Daly J. The Rise of Western Power: A Comparative History of Western Civilization. A&C Black; 2013

[139] Gregory BS. The Unintended Reformation: How a Religious Revolution Secularized Society. Harvard University Press; 2015

[140] Ryrie A. Protestants: The Radicals Who Made the Modern World. William Collins; 2017

[141] Diner HR. Hungering for America. Harvard University Press; 2009

[142] Pye M. Antwerp: The Glory Years. Penguin Books, Limited; 2022

[143] Muldrew C. Food, Energy and the Creation of Industriousness: Work and Material Culture in Agrarian England, 1550-1780. Cambridge University Press; 2011

[144] Turner J. Spice: The History of a Temptation. Vintage; 2008

[145] Allen SL. The Devil's Cup: Coffee, the Driving Force in History. Canongate; 2001

[146] Schivelbusch W. Tastes of Paradise: A Social History of Spices, Stimulants, and Intoxicants. Vintage Books; 1993

[147] Keay J. The Spice Route: A History. John Murray; 2006

[148] Moss M. Salt, Sugar, Fat: How the Food Giants Hooked Us. Ebury Publishing; 2013

[149] Mintz SW. Sweetness and Power: The Place of Sugar in Modern History. Penguin; 1986

[150] O'Brien PK. The Economies of Imperial China and Western Europe: Debating the Great Divergence. Springer International Publishing; 2020

[151] Pomeranz K. The Great Divergence: China, Europe, and the Making of the Modern World Economy. Princeton University Press; 2009

[152] Walvin J. Sugar: The World Corrupted, From Slavery to Obesity. Little, Brown Book Group; 2017

[153] Viestad A. Dinner in Rome: A History of the World in One Meal. Reaktion Books; 2022

[154] Kondratieff ND. The long waves in economic life. Review (Fernand Braudel Center). 1979:519-562

[155] Overton M. Agricultural Revolution in England: The Transformation of the Agrarian Economy 1500-1850. Cambridge University Press; 1996

[156] Sheldrake M. Entangled Life: How Fungi Make Our Worlds, Change Our Minds & Shape Our Futures. Random House Publishing Group; 2021

[157] Wrigley EA. Continuity, Chance and Change: The Character of the Industrial Revolution in England. Cambridge University Press; 1990

[158] Kurlansky M. Cod. Random House; 2011

[159] Deaton A, Muellbauer J. Economics and Consumer Behavior. Cambridge University Press; 1980

[160] Clover C. Rewilding the Sea: How to Save our Oceans. Ebury Publishing; 2022

[161] Barber J. The Chicken: A Natural History. Ivy Press; 2018

[162] Kurlansky M. Cod: A Biography of the Fish that Changed the World. Vintage; 1999

[163] Oddy DJ, Miller DS. Making of the Modern British Diet. 1976

[164] Perren R. The Meat Trade in Britain, 1840-1914. Routledge and Kegan Paul; 1978

[165] Caballero B. The Nutrition Transition: Diet and Disease in the Developing World. Elsevier Science; 2002

[166] Lindeberg S. Food and Western Disease: Health and Nutrition from an Evolutionary Perspective. Wiley; 2010

[167] Pajuelo D et al. NAD(+) depletion triggers macrophage necroptosis, a cell death pathway exploited by Mycobacterium tuberculosis. Cell Reports. 2018;**24**(2):429-440

[168] Smith FB. The Retreat of Tuberculosis, 1850-1950. Croom Helm; 1988

[169] Tak U et al. The tuberculosis necrotizing toxin is an NAD+ and NADP+ glycohydrolase with distinct enzymatic properties. Journal of Biological Chemistry. 2019;**294**(9):3024-3036

[170] Williams AC, Dunbar RI. Big brains, meat, tuberculosis, and the nicotinamide switches: Co-evolutionary relationships with modern repercussions? International Journal of Tryptophan Research. 2013;**6**:73-88

[171] Rook G. The Hygiene Hypothesis and Darwinian Medicine. Birkhäuser Basel; 2009

[172] Meij J et al. Quality–quantity trade-off of human offspring under adverse environmental conditions. Journal of Evolutionary Biology. 2009;**22**(5):1014-1023

[173] McMichael P. Development and Social Change: A Global Perspective. Thousand Oaks, US: Pine Forge; 2008

[174] McMichael P. Food Regimes and Agrarian Questions. Practical Action Publishing; 2014

[175] Williamson JG. Trade and Poverty: When the Third World Fell Behind. MIT Press; 2013

[176] Lessenich S. Living Well at Others' Expense: The Hidden Costs of Western Prosperity. John Wiley & Sons; 2019

[177] Davis M. Late Victorian Holocausts: El Niño Famines and the Making of the Third World. Verso; 2001

[178] Richard A. The Sloth Lemur's Song: Madagascar from the Deep Past to the Uncertain Present. University of Chicago Press; 2022

[179] Collingham EM. Imperial Bodies: The Physical Experience of the Raj, C. 1800-1947. Wiley; 2001

[180] Downs J. Maladies of Empire: How Colonialism, Slavery, and War Transformed Medicine. Harvard University Press; 2021

[181] Stamp LD. The Geography of Life and Death. Collins; 1964

[182] de Castro J. The Geopolitics of Hunger. Monthly Review Press; 1977

[183] Davis M. Late Victorian Holocausts: El Niño Famines and the Making of the Third World. Verso Books; 2002

[184] Dixon J. The Changing Chicken: Chooks, Cooks and Culinary Culture. UNSW Press; 2002

[185] Hickel J. The Divide: A Brief Guide to Global Inequality and its Solutions. Random House; 2017

[186] Hopkins AG, Bruchey SW. An Economic History of West Africa. Columbia University Press; 1973

[187] Josephson PR. Chicken: A History from Farmyard to Factory. Polity Press; 2020

[188] Patel R. Stuffed and Starved: The hidden Battle for the Food System. Brooklyn: Melville House Publishing; 2007

[189] Laudan R. Cuisine and Empire: Cooking in World History. University of California Press; 2015

[190] Leyton G. Effects of slow starvation. Lancet. 1946:73-79

[191] Osmani SR. Nutrition and Poverty. Oxford University Press; 1992

[192] Perlmutter D. Grain Brain: The Surprising Truth about Wheat, Carbs, and Sugar--Your Brain's Silent Killers. Hachette UK; 2018

[193] McGrayne SB. Prometheans in the Lab: Chemistry and the Making of the Modern World. McGraw-Hill; 2001

[194] Phillips K. An Ethnography of Hunger: Politics, Subsistence, and the Unpredictable Grace of the Sun. Indiana University Press; 2018

[195] Washington HA. A Terrible Thing to Waste: Environmental Racism and Its Assault on the American Mind. Little, Brown; 2019

[196] Scrinis G. Nutritionism: The Science and Politics of Dietary Advice. Columbia University Press; 2013

[197] Bittman M. Animal, Vegetable, Junk: A History of Food, from Sustainable to Suicidal. Houghton Mifflin Harcourt; 2021

[198] Jou C. Supersizing Urban America: How Inner Cities Got Fast Food with Government Help. University of Chicago Press; 2017

[199] Hogarth RA. Medicalizing Blackness: Making Racial Difference in the Atlantic World, 1780-1840. University of North Carolina Press; 2017

[200] Haagh L. The Case for Universal Basic Income. Polity Press; 2019

[201] Van Parijs P, Vanderborght Y. Basic Income: A Radical Proposal for a Free Society and a Sane Economy. Harvard University Press; 2017

[202] Aiken W, LaFollette H. World Hunger and Morality. Prentice Hall; 1996

[203] Seligman MEP. Flourish: A Visionary New Understanding of Happiness and Well-being. Atria Books; 2012

[204] Singer P. Practical Ethics. Cambridge University Press; 2011

[205] Unger PK, Press OU. Living High and Letting Die: Our Illusion of Innocence. Oxford University Press; 1996

[206] Singh N et al. Activation of Gpr109a, receptor for niacin and the commensal metabolite butyrate, suppresses colonic inflammation and carcinogenesis. Immunity. 2014;**40**(1):128-139

[207] Macia L et al. Metabolite-sensing receptors GPR43 and GPR109A facilitate dietary fibre-induced gut homeostasis through regulation of the inflammasome. Nature Communications. 2015;**6**(1):1-15

[208] Tan J et al. Dietary fiber and bacterial SCFA enhance oral tolerance and protect against food allergy through diverse cellular pathways. Cell Reports. 2016;**15**(12):2809-2824

[209] Langley MR et al. Critical role of astrocyte NAD+ glycohydrolase in myelin injury and regeneration. Journal of Neuroscience. 2021;**41**(41):8644-8667

[210] McCabe H. John Stuart Mill, Socialist. McGill-Queen's University Press; 2021

[211] Offer A, Offer REHA. The First World War: An Agrarian Interpretation. Clarendon Press; 1989

[212] Gerhard G. Nazi Hunger Politics: A History of Food in the Third Reich. Rowman & Littlefield Publishers; 2015

[213] Forrest AI. The French Revolution and the Poor. B. Blackwell; 1981

[214] Hufton OH. The Poor of Eighteenth-century France 1750-1789. Clarendon Press; 1974

[215] McMeekin S. The Russian Origins of the First World War. Harvard University Press; 2013

[216] Buettner D. The Blue Zones Solution: Eating and Living Like the World's Healthiest People. National Geographic; 2015

[217] Harding J. Border Vigils: Keeping Migrants Out of the Rich World. Verso Books; 2012

[218] Soss J et al. Disciplining the Poor: Neoliberal Paternalism and the Persistent Power of Race. University of Chicago Press; 2011

[219] Linebaugh P. Stop, Thief!: The Commons, Enclosures, and Resistance. PM Press; 2014

[220] Bowles S et al. Poverty Traps. Princeton University Press; 2011

[221] Hobsbawm EJ, Wrigley C. Industry and Empire: From 1750 to the Present Day. The New Press; 1999

[222] Wolf S, Bonanno A. The neoliberal regime in the agri-food sector. Crisis, resilience. 2014

[223] Emmanuel A et al. Unequal Exchange: A Study of the Imperialism of Trade. Monthly Review Press; 1969

[224] Conquest R. The Harvest of Sorrow: Soviet Collectivisation and the Terror-Famine. Random House; 2018

[225] Conradi P. Who Lost Russia?: How the World Entered a New Cold War. Oneworld Publications; 2018

[226] Ziegelman J, Coe A. A Square Meal: A Culinary History of the Great Depression. HarperCollins; 2016

[227] Stevenson J, Cook C. The Slump: Britain in the Great Depression. Taylor & Francis; 2013

[228] Goodwyn L. Democratic Promise: The Populist Moment in America. Oxford University Press; 1976

[229] Leonard C. The Meat Racket: The Secret Takeover of America's Food Business. Simon & Schuster; 2014

[230] Lynn BC. Liberty from All Masters: The New American Autocracy vs. the Will of the People. St. Martin's Publishing Group; 2020

[231] Teachout Z, Sanders B. Break 'Em Up: Recovering Our Freedom from Big Ag, Big Tech, and Big Money. St. Martin's Publishing Group; 2020

[232] Barrett CB, Maxwell D. Food Aid After Fifty Years: Recasting its Role. Taylor & Francis; 2007

[233] Hayter T. Aid as imperialism. 1971

[234] Jun F et al. China's Agriculture and Rural Development in the Post-reform Era. ACA Publishing Limited; 2017

[235] Moskoff W. Perestroika in the Countryside: Agricultural Reform in the Gorbachev Era. Taylor & Francis; 2019

[236] Zadek S. The Civil Corporation. Earthscan; 2007

[237] Buller A. The Value of a Whale: On the Illusions of Green Capitalism. Manchester University Press; 2022

[238] Henderson R. Reimagining Capitalism in a World on Fire: Shortlisted for the FT & McKinsey Business Book of the Year Award 2020. Penguin Books Limited; 2021

[239] Mackey J et al. Conscious Capitalism, With a New Preface by the Authors:

Liberating the Heroic Spirit of Business. Harvard Business Review Press; 2014

[240] Esping-Andersen G. The Three Worlds of Welfare Capitalism. Princeton University Press; 1990

[241] Chhibber A, Soz SA. Unshackling India: Hard Truths and Clear Choices for Economic Revival. HarperCollins; 2021

[242] Tcherneva PR. The Case for a Job Guarantee. Polity Press; 2020

[243] Pistor K. The Code of Capital: How the Law Creates Wealth and Inequality. Princeton University Press; 2020

[244] Stoknes PE, Hawken P. Tomorrow's Economy: A Guide to Creating Healthy Green Growth. MIT Press; 2022

[245] Hawken P et al. Natural Capitalism. Little, Brown; 2007

[246] Moore JW. Capitalism in the Web of Life: Ecology and the Accumulation of Capital. Verso Books; 2015

[247] Handfield RB, Linton TK. Flow: How the Best Supply Chains Thrive. University of Toronto Press; 2022

[248] Robin V. Blessing the Hands that Feed Us: What Eating Closer to Home Can Teach Us about Food, Community, and Our Place on Earth. Thorndike Press; 2014

[249] Lent J. The Web of Meaning: Integrating Science and Traditional Wisdom to Find Our Place in the Universe. Profile; 2021

[250] Yunkaporta T. Sand Talk: How Indigenous Thinking Can Save the World. Text Publishing Company; 2019

[251] Howard PH. Concentration and Power in the Food System: Who Controls What We Eat? Revised ed. Bloomsbury Publishing; 2021

[252] Saladino D. Eating to Extinction: The World's Rarest Foods and Why We Need to Save Them. Random House; 2021

[253] Lobao LM. Locality and Inequality: Farm and Industry Structure and Socioeconomic Conditions. State University of New York Press; 1990

[254] Dixon JA et al. Farming Systems and Poverty: Improving Farmers' Livelihoods in a Changing World. FAO; 2001

[255] Shepard M. Restoration Agriculture: Real-world Permaculture for Farmers. Acres U.S.A.; 2013

[256] De Luna KM. Collecting Food, Cultivating People: Subsistence and Society in Central Africa. Yale University Press; 2016

[257] Antle JM, Ray S. Sustainable Agricultural Development: An Economic Perspective. Springer International Publishing; 2020

[258] Shaw HJ, Shaw JJA. Corporate Social Responsibility, Social Justice and the Global Food Supply Chain: Towards an Ethical Food Policy for Sustainable Supermarkets. Taylor & Francis; 2019

[259] Zhou ZY. Global Food Security: What Matters? Taylor & Francis; 2019

[260] Walia SS, Walia US. Farming System and Sustainable Agriculture. Scientific Publishers; 2020

[261] Yao S. Agricultural Reforms and Grain Production in China. Palgrave Macmillan UK; 2016

[262] Wolf SA, Bonanno A. The Neoliberal Regime in the Agri-Food Sector: Crisis, Resilience, and Restructuring. Taylor & Francis; 2013

[263] Rosin C et al. Food Systems Failure: The Global Food Crisis and the Future of Agriculture. Taylor & Francis; 2013

[264] Plague HJ. Population and the English Economy 1348-1530. Macmillan; 1977

[265] Cole DH, Ostrom E. Property in Land and Other Resources. Lincoln Institute of Land Policy; 2012

[266] Escribano AJ et al. Instant Insights: Integrated Crop-Livestock Systems. Burleigh Dodds Science Publishing Limited; 2021

[267] Dixon J et al. Farming Systems and Food Security in Africa: Priorities for Science and Policy Under Global Change. Taylor & Francis Limited; 2021

[268] Judge K. Direct: The Rise of the Middleman Economy and the Power of Going to the Source. HarperCollins; 2022

[269] Fiennes J. Land Healer: How Farming Can Save Britain's Countryside. Ebury Publishing; 2022

[270] Kassam A et al. Successful Experiences and Lessons from Conservation Agriculture Worldwide. Agronomy. 2022;**12**(4):769

[271] Kassam A, Kassam L. Rethinking Food and Agriculture: New Ways Forward. Elsevier Science; 2020

[272] Rooted SL. Stories of Life, Land and a Farming revolution. Viking; 2022. 368 p

[273] Hong TK et al. Current issues and technical advances in cultured meat production: A review. Food Science of Animal Resources. 2021;**41**(3):355-372

[274] Canning D et al. Africa's Demographic Transition: Dividend or Disaster? World Bank Publications; 2015

[275] Goodhart C, Pradhan M. The Great Demographic Reversal: Ageing Societies, Waning Inequality, and an Inflation Revival. Springer International Publishing; 2020

[276] Harrison M. Contagion: How Commerce Has Spread Disease. Yale University Press; 2013

[277] Paice E. Youthquake: Why African Demography Should Matter to the World. Head of Zeus; 2021

[278] de Wit MM. Abolitionist Agroecology, Food Sovereignty and Pandemic Prevention. Daraja Press; 2021

[279] Will RG et al. A new variant of Creutzfeldt-Jakob disease in the UK. The Lancet. 1996;**347**(9006):921-925

[280] Ritchie DL et al. Variant CJD: Reflections a Quarter of a Century on. Pathogens. 2021;**10**(11):1413

[281] Sinclair U. The Jungle. OUP Oxford; 2010

[282] Greger M. How to Survive a Pandemic. Pan Macmillan; 2020

[283] Davis M. The Monster at Our Door: The Global Threat of Avian Flu. Henry Holt and Company; 2006

[284] Specht J. Red Meat Republic. Princeton University Press; 2019

[285] Ogle M. In Meat We Trust: An Unexpected History of Carnivore America. Houghton Mifflin Harcourt; 2013

[286] Lien ME. Becoming Salmon: Aquaculture and the Domestication of a Fish. University of California Press; 2015

[287] Stead V, Hinkson M. Beyond Global Food Supply Chains: Crisis, Disruption,

Regeneration. Springer Nature Singapore; 2022

[288] Gliessman SR. Agroecology: The Ecology of Sustainable Food Systems. 3rd ed. CRC Press; 2014

[289] Green T. The Covid Consensus: The New Politics of Global Inequality. Hurst Publishers; 2021

[290] Tooze A. Shutdown: How Covid Shook the World's Economy. Penguin Books Limited; 2021

[291] Andiman WA. Animals Viruses and Humans, A Narrow Divide: How Lethal Zoonotic Viruses Spill Over and Threaten Us. Paul Dry Books; 2018

[292] Honigsbaum M. The Pandemic Century: A History of Global Contagion from the Spanish Flu to Covid-19. Ebury Publishing; 2020

[293] Thomas M, Green P. The return of cosmopolitan capital: Globalization, the state and war. Historical Materialism. 2006;**14**(4):203-232

[294] Coulson M. The History of Mining: The Events, Technology and People Involved in the Industry that Forged the Modern World. Harriman House; 2012

[295] Pitron G, Jacobsohn B. The Rare Metals War: The Dark Side of Clean Energy and Digital Technologies. Scribe Publications Pty Limited; 2020

[296] Baldwin R. The Great Convergence: Information Technology and the New Globalization. Harvard University Press; 2016

[297] Williams AC, Hill LJ. Inequality: The dangers of meat haves and have-nots in a nicotinamide-adenine-dinucleotide world. Meat and Nutrition. 2021;**3**

[298] Levine S. School Lunch Politics. Princeton University Press; 2011

[299] Robertson C. The Time-Travelling Economist: Why Education, Electricity and Fertility Are Key to Escaping Poverty. Springer International Publishing; 2022

[300] Hyde L. The Gift: Imagination and the Erotic Life of Property. Vintage; 1999

[301] Brand U, Wissen M. The Imperial Mode of Living: Everyday Life and the Ecological Crisis of Capitalism. Verso Books; 2021

[302] Evans H. Good Times, Bad Times. BSB Publishers; 2013

[303] Bregman R. Utopia for Realists: How We Can Build the Ideal World. Little, Brown; 2018

[304] Standing G. The Blue Commons: Rescuing the Economy of the Sea. Penguin Books Limited; 2022

[305] Burnard T. Planters, Merchants, and Slaves: Plantation Societies in British America, 1650-1820. University of Chicago Press; 2015

[306] Chancel L, DeBevoise M. Unsustainable Inequalities: Social Justice and the Environment. Harvard University Press; 2020

[307] Health MC. Food and Social Inequality: Critical Perspectives on the Supply and Marketing of Food. Taylor & Francis; 2015

[308] Patterson O. Slavery and Social Death: A Comparative Study, With a New Preface. Harvard University Press; 2018

[309] Philcox R et al. The Wretched of the Earth. Grove Press; 2004

[310] Winchester S. Land: How the Hunger for Ownership Shaped the Modern World. HarperCollins Publishers; 2021

[311] Winchester S. Land: The Ownership of Everywhere. HarperCollins Publishers Limited; 2021

[312] Borgstrom G. The Hungry Planet: The Modern World at the Edge of Famine. Macmillan; 1972

[313] George S. Feeding the few: Corporate Control of Food. Washington DC: Institute for Policy Studies; 1979

[314] Hayter T, First TW. The Creation of World Poverty. Pluto; 1981

[315] Patel P. The Tyranny of Nations: How the Last 500 Years Shaped Today's Global Economy. Bifocal; 2021

[316] Stuckler D, Siegel K. Sick Societies: Responding to the Global Challenge of Chronic Disease. OUP Oxford; 2011

[317] Salzberger RP et al. Reparations for Slavery: A Reader. Rowman & Littlefield Publishers; 2004

[318] Darity WA, Mullen AK. From Here to Equality: Reparations for Black Americans in the Twenty-First Century. University of North Carolina Press; 2020

[319] Herman A. Restitution: The Return of Cultural Artefacts. Lund Humphries Publishers, Limited; 2021

[320] MacMillan M. Paris 1919: Six Months that Changed the World. Random House. p. 2007

[321] Snyder T. Black Earth: The Holocaust as History and Warning. Random House; 2015

[322] Adhikari M. Genocide on Settler Frontiers: When Hunter-Gatherers and Commercial Stock Farmers Clash. Berghahn Books; 2015

[323] Naimark NM. Genocide: A World History. Oxford University Press; 2017

[324] Kiernan B. Blood and Soil: A World History of Genocide and Extermination from Sparta to Darfur. Yale University Press; 2008

[325] Hancock G. Lords of Poverty: The Power, Prestige, and Corruption of the International Aid Business. Atlantic Monthly Press; 1989

[326] Keen D. The Benefits of Famine: A Political Economy of Famine and Relief in Southwestern Sudan, 1983-1989. James Currey Publishers. p. 2008

[327] Goldin I. Development: A Very Short Introduction. Oxford University Press; 2018

[328] Dercon S. Gambling on Development. Hurst Publishers; 2022

[329] Adorno TW, Hoban W. Aspects of the New Right-Wing Extremism. Wiley; 2020

[330] Rifkin J. The Age of Resilience: Reimagining Existence on a Rewilding Earth. St. Martin's Publishing Group; 2022

[331] Slaughter AM. The Chessboard and the Web: Strategies of Connection in a Networked World. Yale University Press; 2017

[332] Zheng Y, Qian J. Development and Poverty Reduction: A Global Comparative Perspective. Taylor & Francis; 2019

[333] Faloyin D. Africa Is Not A Country: Breaking Stereotypes of Modern Africa. Random House; 2022

[334] Taiwo O. Against Decolonisation: Taking African Agency Seriously. C Hurst & Company Publishers Limited; 2022

[335] Goldin I. Rescue: From Global Crisis to a Better World. Hodder & Stoughton; 2021

[336] Shapiro BJ. A Culture of Fact: England, 1550-1720. Cornell University Press; 2000

[337] Pardo RP. Shrimp to Whale: South Korea from the Forgotten War to K-Pop. Hurst Publishers; 2022

[338] Ball W. The Gates of Europe: The Eurasian Steppe and Europe's Border. East & West Pub.; 2015

[339] Allen LH, Dror DK. Effects of animal source foods, with emphasis on milk, in the diet of children in low-income countries. Milk and Milk Products in Human Nutrition. 2011;**67**:113-130

[340] Bhutta ZA et al. Evidence-based interventions for improvement of maternal and child nutrition: What can be done and at what cost? The Lancet. 2013;**382**(9890):452-477

[341] Black RE et al. Maternal and child undernutrition and overweight in low-income and middle-income countries. The Lancet. 2013;**382**(9890):427-451

[342] De Beer H. Dairy products and physical stature: A systematic review and meta-analysis of controlled trials. Economics & Human Biology. 2012;**10**(3):299-309

[343] Fredriks AM et al. Continuing positive secular growth change in The Netherlands 1955-1997. Pediatric Research. 2000;**47**(3):316-323

[344] Galler JR et al. The influence of early malnutrition on subsequent behavioral development: II. Classroom behavior. Journal of the American Academy of Child Psychiatry. 1983;**22**(1):16-22

[345] Golden M. Trace elements in human nutrition. Human Nutrition Clinical Nutrition. 1982;**36**(3):185-202

[346] Grantham-McGregor SM, Cumper G. Jamaican studies in nutrition and child development, and their implications for national development. Proceedings of the Nutrition Society. 1992;**51**(1):71-79

[347] Grillenberger M et al. Food supplements have a positive impact on weight gain and the addition of animal source foods increases lean body mass of Kenyan schoolchildren. The Journal of Nutrition. 2003;**133**(11):3957S-3964S

[348] Hulett JL et al. Animal source foods have a positive impact on the primary school test scores of Kenyan schoolchildren in a cluster-randomised, controlled feeding intervention trial. British Journal of Nutrition. 2014;**111**(5):875-886

[349] Levitsky DA, Strupp BJ. Malnutrition and the brain: Changing concepts, changing concerns. The Journal of Nutrition. 1995;**125**(suppl_8):2212S-2220S

[350] Neisser U, Association AP. The Rising Curve: Long-term Gains in IQ and Related Measures. American Psychological Association; 1998

[351] Pollitt E et al. Nutrition in early life and the fulfillment of intellectual potential. The Journal of Nutrition. 1995;**125**(suppl_4):1111S-1118S

[352] Scotland GBDoHf et al. Milk Consumption and the Growth of School Children: Report on an Investigation in

DOI: http://dx.doi.org/10.5772/intechopen.110061

Lanarkshire Schools. H.M. Stationery Office; 1930

[353] Winick M. Nutrition: Pre- and Postnatal Development. Springer US; 2012

[354] Bregman R. Utopia for Realists: And How We Can Get There. Bloomsbury Publishing Plc; 2018

[355] Coote A, Percy A. The Case for Universal Basic Services. Polity Press; 2020

[356] Fan S. Public Expenditures, Growth, and Poverty: Lessons from Developing Countries. International Food Policy Research Institute; 2008

[357] Lawrence M, Buller A. Owning the Future: Power and Property in an Age of Crisis. Verso Books; 2022

[358] Vinen R. Second City: Birmingham and the Forging of Modern Britain. Penguin Books Limited; 2022

[359] Stromquist S. Claiming the City: A Global History of Workers' Fight for Municipal Socialism. Verso Books; 2023

[360] Thompson H. Disorder: Hard Times in the 21st Century. Oxford University Press; 2022

[361] Silver N. Finance, Society and Sustainability: How to Make the Financial System Work for the Economy, People and Planet. Palgrave Macmillan UK; 2017

[362] Soper K. Post-Growth Living: For an Alternative Hedonism. Verso Books; 2020

[363] Delgado CL et al. Determinants and Implications of the Growing Scale of Livestock Farms in Four Fast-growing Developing Countries. International Food Policy Research Institute; 2008

[364] Fan S. Halving Hunger: Meeting the First Millennium Development Goal through" Business as Unusual". Intl Food Policy Res Inst; 2010

[365] Fanzo J et al. Diversifying Food and Diets: Using Agricultural Biodiversity to Improve Nutrition and Health. Taylor & Francis; 2013

[366] Mellor JW. Agricultural Development and Economic Transformation: Promoting Growth with Poverty Reduction. Springer; 2017

[367] Ohlson K. The Soil Will Save Us: How Scientists, Farmers, and Foodies Are Healing the Soil to Save the Planet. Rodale Books; 2014

[368] Perfecto I et al. Nature's Matrix: Linking Agriculture, Conservation and Food Sovereignty. Earthscan; 2009

[369] Sánchez PA et al. Halving Hunger: It Can be Done. Earthscan; 2005

[370] Vogler P. Scoff: A History of Food and Class in Britain. Atlantic Books; 2020

[371] Rawlence B. The Treeline: The Last Forest and the Future of Life on Earth. Random House; 2022

[372] Buscher B, Fletcher R. The Conservation Revolution: Radical Ideas for Saving Nature Beyond the Anthropocene. Verso Books; 2020

[373] Foster JB. The Return of Nature: Socialism and Ecology. Monthly Review Press; 2020

[374] Mardon A et al. Antibiotic Resistance: An Emerging Pandemic. Golden Meteorite Press; 2021

[375] Blaikie P, Brookfield H. Land Degradation and Society. Taylor & Francis; 2015

[376] Forsyth F. The Biafra Story: The Making of an African Legend. Pen & Sword Books; 2015

[377] Heerten L. The Biafran War and Postcolonial Humanitarianism: Spectacles of Suffering. Cambridge University Press; 2017

[378] Ignatieff M. Human Rights as Politics and Idolatry. Human Rights as Politics and Idolatry. Princeton University Press; 2011

[379] Wright W, Middendorf G. The Fight Over Food: Producers, Consumers, and Activists Challenge the Global Food System. Penn State Press; 2010

[380] Dunbar RI. The social brain hypothesis and its implications for social evolution. Annals of Human Biology. 2009;**36**(5):562-572

[381] Boltanski L et al. Distant Suffering: Morality, Media and Politics. Cambridge University Press; 1999

[382] Marmot M. The Health Gap: The Challenge of an Unequal World. Bloomsbury Publishing; 2015

[383] Chellaney B. Water, Peace, and War: Confronting the Global Water Crisis. Oxford University Press; 2014

[384] Sitwell W. Eggs or Anarchy: The Remarkable Story of the Man Tasked with the Impossible: To Feed a Nation at War. Simon & Schuster UK; 2016

[385] Hennessy P. A Duty of Care: Britain Before and After Covid. Penguin Books Limited; 2022

[386] Herbert B. Losing Our Way: An Intimate Portrait of a Troubled America. Knopf Doubleday Publishing Group; 2015

[387] Winders W, Scott JC. The Politics of Food Supply: U.S. Agricultural Policy in the World Economy. Yale University Press; 2012

[388] Narula H. Virtual Society: The Metaverse and the New Frontiers of Human Experience. Penguin Books Limited; 2022

[389] Lobel O. The Equality Machine: Harnessing Digital Technology for a Brighter, More Inclusive Future. PublicAffairs; 2022

[390] Harden KP. The Genetic Lottery: Why DNA Matters for Social Equality. Princeton University Press; 2021

[391] Blythman J. Bad Food Britain: [How a Nation Ruined Its Appetite]. Fourth Estate London; 2006

[392] Okun AM. Equality and Efficiency: The Big Tradeoff. Brookings Institution Press; 2015

[393] McKendrick N, Plumb JH. The Birth of a Consumer Society: The Commercialization of Eighteenth-Century England. Edward Everett Root; 2018

[394] Atkinson AB. Inequality. Harvard University Press; 2015

[395] Dorling D, Dorling D. Injustice: Why Social Inequality Still Persists. Policy Press; 2015

[396] Wilkinson R, Pickett K. The Spirit Level: Why Equality is Better for Everyone. Penguin Books Limited; 2010

[397] Wilkinson RG. Poverty and Progress: An Ecological Model of Economic Development. Methuen; 1973

[398] Parr T et al. Active Inference: The Free Energy Principle in Mind, Brain, and Behavior. MIT Press; 2022

[399] Rauch J. The Constitution of Knowledge: A Defense of Truth. Brookings Institution Press; 2021

[400] Shea N, Frith CD. The global workspace needs metacognition. Trends in Cognitive Sciences. 2019;**23**(7):560-571

[401] Sternberg RJ. Adaptive Intelligence: Surviving and Thriving in Times of Uncertainty. Cambridge University Press; 2021

[402] Van Schaik CP et al. Explaining brain size variation: From social to cultural brain. Trends in Cognitive Sciences. 2012;**16**(5):277-284

[403] Verschelden C, Pasquerella L. Bandwidth Recovery: Helping Students Reclaim Cognitive Resources Lost to Poverty, Racism, and Social Marginalization. Stylus Publishing; 2017

[404] Ward A. The Social Lives of Animals. Basic Books; 2022

[405] Warren WJ. Meat Makes People Powerful: A Global History of the Modern Era. University of Iowa Press; 2018

[406] Haskel J, Westlake S. Restarting the Future: How to Fix the Intangible Economy. Princeton University Press; 2022

[407] Outram AK, Bogaard A. Subsistence and Society in Prehistory: New Directions in Economic Archaeology. Cambridge University Press; 2019

[408] Poole S. You Aren't What You Eat: Fed Up with Gastroculture. Aurum Press; 2012

[409] Gabaccia DR, Gabaccia DR. We Are What We Eat: Ethnic Food and the Making of Americans. Harvard University Press; 2009

[410] Baughan E. Saving the Children: Humanitarianism, Internationalism, and Empire. University of California Press; 2021

[411] Tallis R. Aping Mankind. Routledge; 2016

[412] Rogan T. The Moral Economists: R. H. Tawney, Karl Polanyi, E. P. Thompson, and the Critique of Capitalism. Princeton University Press; 2019

[413] Cordain L. The Paleo Diet: Lose Weight and Get Healthy by Eating the Foods You Were Designed to Eat. Wiley; 2010

[414] Hinman RB, Harris RB. The Story of Meat. Swift; 1942

[415] Marcus E. Meat Market: Animals, Ethics, & Money. Brio Press; 2005

[416] Rimas A, Fraser E. Beef: The Untold Story of How Milk, Meat, and Muscle Shaped the World. HarperCollins; 2009

[417] Sykes N. Beastly Questions: Animal Answers to Archaeological Issues. Bloomsbury Publishing; 2014

[418] Weis T. The Ecological Hoofprint: The Global Burden of Industrial Livestock. Bloomsbury Publishing; 2013

[419] Winders B, Ransom E. Global Meat: Social and Environmental Consequences of the Expanding Meat Industry. MIT Press; 2019

[420] Zaraska M. Meathooked: The History and Science of Our 2.5-Million-Year Obsession with Meat. Basic Books; 2016

[421] Alverà M. The Hydrogen Revolution: A Blueprint for the Future of Clean Energy. Hachette UK; 2021

[422] George S. Another World Is Possible If. Verso Books; 2004

[423] Morland P. Tomorrow's People: The Future of Humanity in Ten Numbers. Pan Macmillan; 2022

[424] Edkins J. Whose Hunger?: Concepts of Famine, Practices of Aid. University of Minnesota Press; 2000

[425] George H. Progress and Poverty: An Inquiry Into the Cause of Industrial Depressions, and of Increase of Want with Increase of Wealth—the Remedy. Kegan Paul, Trench; 1883

[426] Nordman E. The Uncommon Knowledge of Elinor Ostrom: Essential Lessons for Collective Action. Island Press; 2021

[427] Nilakantan RS. South vs North: India's Great Divide. Juggernaut Publication; 2002. p. 280

[428] Cobb M. The Genetic Age: Our Perilous Quest To Edit Life. Profile; 2022

[429] Baylis F. Altered Inheritance: CRISPR and the Ethics of Human Genome Editing. Harvard University Press; 2019

[430] Metzl J. Hacking Darwin: Genetic Engineering and the Future of Humanity. Sourcebooks; 2019

[431] Church GM, Regis E. Regenesis: How Synthetic Biology Will Reinvent Nature and Ourselves. Basic Books; 2014

[432] Turkheimer E et al. Socioeconomic status modifies heritability of IQ in young children. Psychological Science. 2003;**14**(6):623-628

[433] Ferguson N. The Square and the Tower: Networks, Hierarchies and the Struggle for Global Power. Penguin Books Limited; 2017

[434] Foroohar R. Homecoming: The Path to Prosperity in a Post-Global World. Crown; 2022

[435] Stoller M. Goliath: The 100-Year War Between Monopoly Power and Democracy. Simon & Schuster; 2020

[436] Edwards J. The Great Fox Illusion. Walker Books Limited; 2022

[437] Siddons AR. Fault Lines. HarperCollins; 1996

[438] Pilger J. Freedom Next Time. Transworld; 2011

[439] Pim K. Endless Flight: The Life of Joseph Roth. Granta Books; 2022

[440] Gordon LR. Fear of Black Consciousness. Penguin Books Limited; 2022

[441] Rushkoff D. Survival of the richest. Medium. 2018;(July):5

[442] Malik K. The Meaning of Race: Race, History and Culture in Western Society. Macmillan Education UK; 1996

[443] Mansbridge JJ. Beyond Self-Interest. University of Chicago Press; 1990

[444] Warren S. Survival of the Richest. Book Beautiful; 2018

[445] DeLong JB. Slouching towards Utopia: An Economic History of the Twentieth Century. Basic Books; 2022

[446] Thackray A. Private Science: Biotechnology and the Rise of the Molecular Sciences. University of Pennsylvania Press; 1998

[447] Preston SD. The Altruistic Urge: Why We're Driven to Help Others. Columbia University Press; 2022

[448] McAfee A. More From Less: The Surprising Story of How We Learned to Prosper Using Fewer Resources—and What Happens Next. Simon & Schuster UK; 2019

DOI: http://dx.doi.org/10.5772/intechopen.110061

[449] Appiah KA. The Honor Code: How Moral Revolutions Happen. W.W. Norton; 2011

[450] Putnam RD. Our Kids: The American Dream in Crisis. Simon & Schuster; 2016

[451] Mounk Y. The Great Experiment: How to Make Diverse Democracies Work. Bloomsbury Publishing Plc; 2022

[452] Klinenberg E. Palaces for the People: How To Build a More Equal and United Society. Random House; 2018

[453] Belich J. The World the Plague Made: The Black Death and the Rise of Europe. Princeton University Press; 2022

[454] Koram K. Uncommon Wealth: Britain and the Aftermath of Empire. John Murray Press; 2022

[455] Easterly W. The Tyranny of Experts: Economists, Dictators, and the Forgotten Rights of the Poor. Basic Books; 2014

[456] Coates TN. Between the World and Me. Text Publishing Company; 2015

[457] Malik K. Not So Black and White: A History of Race from White Supremacy to Identity Politics. Hurst Publishers; 2023

[458] Rich D. Everyday Hate How antisemitism is built into our world – and how you can change it. 2023

[459] Rosenberg AS. Undesirable Immigrants: Why Racism Persists in International Migration. Princeton University Press; 2022

Section 3

Vitamin E and Its Medicinal Values

Chapter 5

Medicinal Significance and Complications of Vitamin E

Naresh Podila, Sathish Kumar Konidala, Jithendra Chimakurthy, Srilatha Muddisetti, Suryaprabha Matangi, Natesh Gunturu, Yamarthi Venkateswara Rao and Mithun Rudrapal

Abstract

Vitamin E is a lipid-soluble substance that is the main component of the body's defense system against free radicals. It performs a range of important functions in the body as a result of its antioxidant action. Cancer, aging, and other diseases and ailments have all been related to oxidation. It has been shown that vitamin E protects against rheumatoid arthritis and cataracts. Additionally, vitamin E can help to prevent the production of prostaglandins like thromboxane, which encourage platelet clustering and hyper aggregation, which can lead to atherosclerosis. The present literature review examines the roles and functions of vitamin E in human health, different disorders, and the consequences of vitamin E deficiency. The tocopherol family of vitamers is the focus of the review's main points. In this review article, the part and actions of vitamin E are talked about, as well as the things that affect how well vitamin E treatment works. When given at the right time and for the right amount of time, Vitamin E should help people with oxidative stress caused by free radicals.

Keywords: vitamin E, health, antioxidants, tocopherols, free radicals

1. Introduction

A vitamin that dissolves in fat is vitamin E. Cereals, vegetable oils, meat, chicken, eggs, and fruits are just a few examples of the many meals that contain it. An essential vitamin, vitamin E is needed for the healthy operation of numerous organs in the body. As an antioxidant, it is also. RRR-alpha-tocopherol, a form of vitamin E that naturally occurs in foods, differs from the synthetic Supplemental vitamin E is available as all-rac-alpha tocopherol [1]. When patients have certain hereditary abnormalities or premature infants who were born very low in weight, vitamin E is used to treat vitamin E insufficiency, a condition that is uncommon but can occur. There is other more ailments for which vitamin E is utilized, however many of these other uses lack solid scientific backing. The antioxidant vitamin E has the potential to shield the body's cells from harm. Antioxidants may offer defense against major illnesses including cancer and heart disease. Additionally essential for the production of vitamin E and red blood cells also aids in the utilization of vitamin K [2].

The liver absorbs vitamin E once it is absorbed in the small intestine and stores it there until it is required. At that time, the liver only secretes alpha-tocopherol again, which the body can identify. Your immune system benefits from vitamin E, which also supports healthy skin, eyes, and brain function. Despite the rarity of vitamin E deficiency, maintaining health and preventing and treating disease depend on achieving daily vitamin E needs. Several circumstances can lead to vitamin E insufficiency. Premature infants with low birth weight are one example. It is also possible among those who have an illness that impairs the body's ability to effectively absorb dietary fat, such as Crohn's disease or cystic fibrosis. Both require supplementation to reduce the risk of complications [3].

Those who struggle to adequately absorb fats may become vitamin E deficient. Serious vitamin E deficiency symptoms include [4].

- Vision problems
- Muscle weakness
- muscle density loss
- Unsteady walking
- Abnormal eye movements

Continuing deficiencies may also result in issues with the liver and kidneys. Although the majority of Americans do not have severe vitamin E deficiencies, many may have slightly low amounts. Edible vegetable oils are the best sources of vitamin E in the diet because they have the highest concentrations of all the different homologs (**Table 1**) Red blood cells and serum contain alpha- and gamma-tocopherols, with alpha-tocopherol having the highest quantity [5]. Only trace amounts of beta- and delta-tocopherols can be found in plasma. The higher metabolism of the other forms of tocopherol and the α-tocopherol possess protein are the causes of

Oils	Alpha (α)-Tocopherol	G-Tocopherol	D-Tocopherol	A-Tocopherol
Coconut oil	1.6	—	1.0	1.1
Maize	10.5	59.3	0.9	—
Palm	24.8	30.9	6.9	13.5
Olive	5.3	Trace amounts	0	0
Peanut	12.8	20.8	2.5	0
Soybean	11.00	58.5	27.0	0
Wheat germ	130.9	25.7	26.8	2.4
Sunflower	46.9	5.5	1.0	0

Table 1.
Vegetable oils contain vitamin E.

alpha-preferential tocopherol distribution in humans (alpha-TTP). The reason why alpha-tocopherol is largely removed in the urine while the majority of the absorbed beta-, gamma, and delta-tocopherols are released into the bile and excreted in the feces, is because of the binding affinity of α-tocopherol with alpha-TTP. In non-hepatic tissues, such as the endoplasmic reticulum and mitochondrial membranes of the heart and lungs and other organs with high amounts of free radical production, the alpha-tocopherol form also accumulates [6].

Oxidative stress has been linked to the pathogenesis of a number of diseases, including atherosclerosis, fatty liver disease, cancer, and neurodegenerative disorders [7, 8]. Oxidation of biological molecules such as lipids, proteins, and DNA, mediated by reactive oxygen species (ROS) and free radicals, results in injury to biological membranes, modification of proteins, inactivation of enzymes, and modification of DNA, according to an abundance of experimental evidence. Therefore, the function of antioxidants against oxidative stress in disease prevention and treatment has received considerable attention from both scientists and the general public [9]. If oxidative stress plays a causal role, it is expected that antioxidants will reduce the risk or be beneficial in the prevention and treatment of these diseases.

Several 1980s epidemiological studies suggested the health benefits of antioxidants such as vitamin E and carotenoids. Cancer incidence is negatively associated with the ingestion of fresh fruits and green-yellow vegetables [10]. This is attributable, at least in part, to the numerous phytochemical compounds found in plant foods, the majority of which are powerful antioxidants [11]. The Multinational Monitoring of Trends and Determinants in Cardiovascular Disease (MONICA) study revealed an inverse association between plasma vitamin E concentration and mortality due to ischemic heart disease and malignancy. Several large cohort studies have supported the protective function of antioxidants with overwhelmingly positive findings. Antioxidant-containing dietary supplements are very popular among a sizable fraction of the population, as a result of the heightened enthusiasm.

Large-scale randomized clinical trials and meta-analyses, on the other hand, have shown disappointing and contradictory data about vitamin E's effects. Not only have there been a number of "null" results, but some studies have shown that taking a lot of vitamin E may be bad for you, which has caused a lot of debate [12–14].

This article talks about the roles and effects of vitamin E and uses the scientific data to answer the following questions. Are oxidative harm and diseases caused by free radicals? Does vitamin E stop damage caused by free radicals and diseases that come with it? Why do controlled, randomized human studies of vitamin E give mixed and disappointing results?

1.1 Chemistry of vitamin E

Vitamin E is not the same thing as tocopherol; rather, Tocopherol refers to tocol derivatives with a methyl substitution. The two homologous sets of natural tocochromanols are Unsaturated tocotrienols and tocopherols with saturated side chains are two types of tocopherols. (**Figure 1**) depicts the fundamental chemical structure of tocopherols and tocotrienols; it has two places on a 6-chromanol ring where a lengthy isoprenoid side chain is connected. In contrast to tocopherols, which have a saturated isoprenoid C16 side chain, tocotrienols have a farnesyl side chain. Natural tocopherols have the RRR-configuration, but all-racalpha-tocopherol, which is the synthesized version, has eight distinct stereoisomers. Natural tocotrienols only have the 2R, 3′E and 7′E structure, and tocotrienols only have the chiral stereo center at C-2 [7]. Chiral

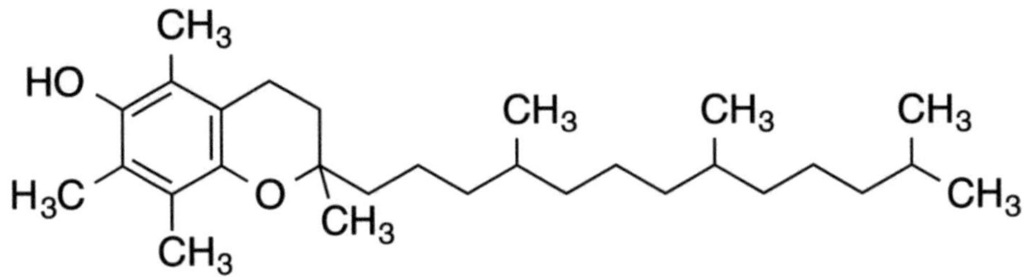

Figure 1.
Structure of tocopherol.

recognition is the process by which the enzymes and body's receptors solely communicate with the enantiomers of the one of chiral compounds. As a consequence, only one of the two enantiomers has the desired effect on the body, while the other two may either have no effect or a negative effect. E- Vitamin cannot be converted to one another within the human [15, 16].

1.2 Sources and recommended intakes

Various foods and oils contain vitamin E. Alpha-tocopherol is present in substantial quantities in vegetable oils, fortified cereals, green leafy vegetables, nuts, seeds.

food intake guidelines	amount of -tocopherol in each serving, in milligrams	Percent daily value
Spinach, raw, 1 cup	21.2	102
Almonds, dry roasted, 1 ounce	6.4	32.9
Kiwifruit, 1 medium	5.5	26.8
Hazelnuts, dry roasted, 1	4.1	23.1
Tomato, raw, 1 medium	2.72	14.2
Peanuts, dry roasted, 1 ounce	2.5	10
Spinach, boiled, ½ cup	1.6	9.8
Corn oil, 1 tablespoon	1.5	11
Broccoli, chopped, boiled, ½ cup	1.6	5.6
Sunflower oil,	1.0	6.1
Safflower oil,	3.8	24.5
Soybean oil,	0.9	5.4
Wheat germ oil, 1 tablespoon	0.9	4.1
Peanut butter, 2 tablespoons	1.0	3
Mango, sliced, ½ cup	0.5	2.5

Table 2.
Food that contain vitamin E found in some foods.

Age	RDA in mg (IU)	
	Females	**Males**
0 to 5 months	3.9 (2)	5 (7)
7 to 12 months	4 (9.5)	6. (6.2)
1 to 3 years	6 (10)	6 (7)
4 to 8 years		
9 to 13 years	11 (14.9)	7 (8.9)
>14 years	14 (45.8)	10 (15.2)
In pregnancy	16 (26.5)	16 (20.4)
If lacting	20 (26.7)	

Table 3.
Guidelines for intake of E vitamin.

(**Tables 1** and **2**) lists the top resources of vitamin E and their tocopherol quantity, and their percentage of daily values. The ideal supplementation number of mixed tocopherols is yet unknown, and no official recommendations has been established addressing the consumption of vitamin E. Vitamin E may not appear to be hazardous when only received from food sources. However, it has been discovered that supplements might cause pro-oxidant damage, but often only at high levels (for instance, >1000 mg/day) [17]. Listed below are the recommended daily amounts (RDAs) for vitamin E (alpha-tocopherol) (**Table 3**).

1.3 Dietary factors interaction

The breakdown of vitamin E is significantly influenced by selenium, glutathione, vitamin B3, and vitamin C. For a diet high in vitamin E to be successful, it must also contain foods that are high in these other nutrients. Beta-carotene and vitamin E have a potential cooperative connection, but vitamin C and beta-carotene have a very high likelihood of having one [3]. It has been discovered that the interactions between tocopherols, Thiols increase the potency of cellular antioxidant protection mechanisms [18]. Findings in the year of 2007 the study from the Women's Health manifestation that E vitamin and alpha-tocopherol supplementation lower the risk of dying from thromboembolism in healthy normal women [19]. Additionally, it was shown that vitamin E supplementation increased prothrombin's under-carboxylation in humans, which suggests that vitamin E lowers people's levels of vitamin K [20].

2. Vitamin E's functions

2.1 Controlling oxidative stress

The E Vitamin, a potent antioxidant that disrupts chemical bonds, stops the synthesis of reactive oxygen species molecules when fat is subjected to oxidation and during the spread of free radical reactions [21]. even though the phospholipid concentration level may only be one molecule for every 2000 of it, it is mostly found in the membranes of cells and organelles where it can exert its possessive impact.

To prevent lipid peroxidation, it serves as the first line of defense, shielding the cell membranes from oxidative damage. According to studies, alpha tocopherol alone is not as effective at inhibiting lipid peroxidation in human erythrocytes as a mixture of tocopherols is [22]. The polyunsaturated fatty acids found in membrane phospholipids and plasma lipoproteins are additionally safeguarded through its peroxyl radical scavenging action [23–25]. The resulting tocopheroxyl radicals can then go on to oxidize other lipids, go through extra oxidation to form tocopheryl quinones, combine with another radical to form non-reactive tocopherol dimers, or be reduced by additional antioxidants to tocopherol. It has been discovered that although gamma-tocopherol captures and neutralizes the free radicals already present, α-tocopherol primarily suppress the generation of anew free radicals. Numerous possible outcomes have been connected to oxidation, including illnesses and disorders, cataracts, cancer, aging, and cancer. As a result, E vitamin might aid in preventing or delaying the Reactive oxygen species molecules have been related to the emergence of chronic illnesses.

2.2 Evidence for in-vivo lipid peroxidation and its link to disease

Highly susceptible to oxidation are polyunsaturated fatty acids such as linoleic acid and arachidonic acid, as well as their esters. They are oxidized into numerous compounds, some of which are cytotoxic and reactive enough to modify proteins and DNA bases [26]. Enzymes, free radicals, and nonenzymatic, nonradical oxidants oxidize lipids. Frequently, lipoxygenase, cyclooxygenase, and cytochrome P450 induce oxidation in a regulated manner to produce specific physiologically essential products. In contrast, lipid peroxidation, which is oxidation mediated by free radicals, occurs arbitrarily and nonspecifically.

Free radicals attack proteins, DNA, and lipids without discrimination. Therefore, when free radicals are generated in vivo, as a result of high-energy irradiation or ischemia–reperfusion injury, the levels of oxidation products of proteins and DNA, as well as lipid peroxidation, are elevated. In addition, secondary lipid peroxidation products, such as unsaturated aldehydes, readily react with protein thiols, leading to the loss of protein function and cellular homeostasis. Among the lipid peroxidation products, hydroxy octadecadienoic acid (HODE) from linoleic acid and hydroxy eicosatetraenoic acid (HETE) and isoprostanes from arachidonic acid are frequently used as biomarkers of lipid peroxidation in vivo [27]. Vitamin E is a powerful radical-scavenging antioxidant that inhibits lipid peroxidation mediated by free radicals but not enzymatic oxidation by lipoxygenase and cyclooxygenase, as will be discussed in greater detail below. Both enzymatic and free radical oxidation produce HODE and HETE, but their isomer distribution depends on the type of oxidant. To evaluate the effects of vitamin E, it is crucial to understand the mechanisms and oxidants of lipid oxidation. The mechanisms of lipid oxidation have been thoroughly studied and are currently well understood [28, 29]. Trans, trans-forms of HODE and HETE have been shown to be specific products of lipid peroxidation mediated by free radicals.

Numerous studies have demonstrated that the levels of lipid peroxidation products, such as HODE, HETE, and isoprostanes, in biological fluids and tissues of diseased patients are generally, if not always, greater than those of healthy individuals. In addition, clinical research has established a connection between disease states and lipid peroxidation products.

2.2.1 Liver diseases

The importance of lipid peroxidation mediated by free radicals in liver injury induced by carbon tetrachloride and other halogenated alkanes has been investigated since the 1960s and documented in detail [30]. It was once thought that carbon tetrachloride affects the liver by the action of a simple solvent, but it is now understood that carbon tetrachloride must undergo metabolic activation to trichloromethyl radical by cytochrome P450, primarily by cytochrome P450 2E1, to exert its toxic effect [31]. The trichloromethyl radical reacts rapidly with oxygen to yield the trichloromethyl peroxyl radical, which attacks lipids and induces their peroxidation. Lipid peroxidation mediated by free radicals is involved in alcoholic liver disease caused by chronic alcohol ingestion [32, 33]. Plasma levels of several lipid oxidation products, including oxysterols and isoprostanes, have been shown to be elevated in alcoholic liver disease patients [34, 35].

Nonalcoholic fatty liver disease (NAFLD), a hepatic manifestation of metabolic syndrome, is now the most common liver disorder, affecting a high proportion of the global population. The incidence of NAFLD is increasing due to increases in the prevalence of two major risk factors, obesity, and type 2 diabetes, which are related to lifestyle and diet. The characteristic feature of NAFLD is an excessive accumulation of fat, notably triglyceride, in the liver, and it encompasses a wide spectrum from benign steatosis to nonalcoholic steatohepatitis (NASH), liver cirrhosis, liver failure, and hepatocellular carcinoma [36].

NAFLD and NASH are multifactorial diseases and oxidative stress has been implicated in their pathogenesis. Several human and animal studies have reported an association between NAFLD/NASH disease state and biomarkers of lipid peroxidation [37]. One such study reported that levels of 9- and 13-HODE, major products of linoleic acid peroxidation, were significantly elevated in patients with NASH compared to those with steatosis, and a strong correlation was observed between these oxidation products and liver histopathology such as inflammation, fibrosis, and steatosis [38]. These HODEs were racemic, suggesting them to be produced by free radical oxidation.

Numerous studies have reported a beneficial effect of vitamin E on NAFLD and NASH [39]. In one such study, the effects of vitamin E at a dose of 800 IU/day or placebo for 96 weeks were examined in adults with NASH and without diabetes; there was improvement in the histological features of NASH [40]. Another study reported measurable differences in the metabolic profile of subjects likely to respond to vitamin E treatment for NASH and those who experienced histological improvements following treatment [41]. In a recent retrospective study of the effects of 300 mg/day vitamin E for 2 or more years in patients with biopsy-proven NASH, vitamin E ameliorated NASH fibrosis, especially in those who showed improved transaminase activities and insulin resistance [42, 43].

2.2.2 Atherosclerosis

Atherosclerosis is a leading cause of cerebral infarction, myocardial infarction, coronary artery disease, and peripheral arterial disease. It is a chronic inflammatory disease characterized by excessive cholesterol deposition in the arterial wall and sluggish progression. It begins in childhood and remains asymptomatic for decades, but it is the primary cause of death in developed nations [44, 45]. It is generally recognized that oxidative modification of low-density lipoprotein (LDL) is a crucial initial event in the development of atherosclerosis. Incubation of macrophages with oxidized LDL, but not with native LDL, results in accumulation of cholesteryl esters within the cell. Oxidation of low-density lipoprotein increases its pro-atherogenic effect, whereas oxidation of high-density lipoprotein decreases its anti-atherogenic effect [46, 47].

In the 1990s, extensive research was conducted on the oxidative modification of LDL, and its mechanisms and products were elucidated; however, the epitope responsible for recognition by macrophage scavenger receptors has not yet been identified. However, it has been demonstrated that macrophages take up oxidatively modified LDL, which is the first step in the formation of foam cells. LDL oxidation generates diverse byproducts. Cholesteryl esters and phosphatidylcholine (PC) are the predominant lipids in LDL particles of humans. The oxidation of linoleic acid and arachidonic acid esters of cholesterol and PC produces the corresponding hydroperoxides, hydroxides, and degradation products [48]. Plasma levels of lipid peroxidation products are higher in atherosclerotic patients than in healthy individuals. LDL isolated from diabetic patients has higher levels of HODE and HETE than LDL isolated from healthy subjects. In addition, the molecular ratios of HODE and HETE to the primary lipids (linoleates and arachidonates, respectively) are greater in diabetic LDL than in control LDL [49]. In addition, the levels of oxysterols such as 7-b-hydroxycholesterol, 7-keto-cholesterol, and cholesterol-5,6-epoxide rise in the order normal artery fatty streak advanced lesion [50]. Moreover, hydroperoxide and hydroxide forms of cholesteryl linoleate are frequently found in human atherosclerotic plaque [51, 52].

The oxidative modification of LDL is mediated by multiple oxidants and distinct mechanisms, it should be noted. LDL is oxidized by both free radicals and nonradical oxidants, including lipoxygenases, cytochrome P450 enzymes, and hypochlorite. Singlet oxygen could also be a factor [53]. Importantly, various oxidants generate distinct oxidation products, necessitating the use of distinct antioxidants. No single antioxidant is capable of preventing all forms of oxidation.

2.3 Safeguarding the cell membranes

As a result of vitamin E's improved lipid packing orderliness, the membrane can be packed more tightly, which increases cell stability. Vitamin E was demonstrated to be essential for preserving the balance of skeletal muscle by Howard et al. in 2011. and that introducing alpha-tocopherol to cultured myocytes makes it easier to repair plasma membranes. This occurs because phospholipids in the membrane are frequently the target of oxidants, and vitamin E efficiently prevents lipid peroxidation. Contrarily, when cultivated cells are exposed to an oxidant assault without alpha-tocopherol supplementation, the repair is notably inhibited. Comparative measurements show that an antioxidant needs to join the membranes, like α-tocopherol does, or be able to regenerate α-tocopherol, to promote the repair [54]. As a result, vitamin E aids in membrane repair by reducing the oxidation of phospholipids, which might hypothetically obstruct membrane fusion processes.

2.4 Regulation of protein kinase c activation and platelet aggregation

It has been discovered that elevating alpha-tocopherol levels in endothelial cells prevents platelet aggregation and causes the endothelium to release prostacyclin. to reduce the components of blood cells' adherence to the endothelium, it was hypothesized that the vascular cell adhesion molecule and the intracellular cell adhesion molecule (ICAM-1) (VCAM-1) were downregulated. Additionally, vitamin E-induced up regulation of the arachidonic acid cascade enzymes cytosolic phospholipase [55] cyclooxygenase-1 and A2 [56] leads to the generation of prostacyclin, a strong vasodilator that prevents platelet clumping in people [57]. Earlier studies advise that tocopherols may prevent protein kinase C (PKC) [58] and increase the activity of nitric oxide synthase [59] in order tocopherols may inhibit platelet aggregation.

When it comes to lowering PKC activity, alpha-tocopherol in its natural RRR version has been shown to be twice as effective as other all-racemic (synthetic) alpha-tocopherols [60]. This is caused by how alpha-attenuating tocopherols affect the synthesis of diacylglycerol, a lipid that makes PKC more mobile and thus activates it. The activity of protein phosphatase type 2A is also increased by alpha-tocopherol, while PKC autophosphorylation and subsequently its activity is decreased. Compared to mixed tocopherols, alpha-tocopherol is less efficient at preventing platelet aggregation. Gamma-tocopherol-enriched vitamin E (100 mg of gamma-tocopherol, 40 mg of delta-tocopherol, and 20 mg of alpha-tocopherol per day) significantly reduced adenosine diphosphate-induced platelet aggregation in healthy people, but not in those taking pure alpha-tocopherol alone. (100 mg per day) [61].

3. Preventing disease with vitamin E

Due to its role as an antioxidant, its contribution to the reduction of inflammation, due to its immune-stimulating properties and suppression of platelet aggregation, many disease outcomes have been found to be prevented and reversed with the aid of vitamin E.

3.1 Cardiovascular diseases

Body's low-density lipoproteins are oxidized, which causes inflammation, which in turn leads to cardiovascular issues [62]. Gamma-tocopherol has been shown to enhance functions of cardiovascular by boosting the nitric oxide synthase activity, which generates nitric oxide that relaxes blood vessels [63]. This is accomplished by trapping molecules of peroxynitrite, a reactive nitrogen species, and improving endothelial function. Humans who take 100 milligrams of gamma-tocopherol daily have shown to have fewer arterial clotting risk factors, such as cholesterol, platelet aggregation [64].

Different research revealed that combined tocopherols were more successful than individual tocopherols at preventing lipid peroxidation and human platelet aggregation, indicating a cooperative platelet-inhibitory action. Tocotrienols were also discovered to reduce cholesterol production by inhibiting 3-hydroxy-3-methylglutaryl-CoA (HMG-CoA) reductase, in addition to tocopherols, which causes the liver cells to produce less cholesterol [65]. Contrary to what has been reported, the majority of recent large interventional clinical trials suggest that vitamin E use was related to a considerably elevated risk of hemorrhagic stroke in the participants [66]. It was therefore proposed that further extensive research including younger subjects could be necessary to fully comprehend the vitamin E's possibility for preventing coronary heart disease.

3.2 Cancer

Additionally, vitamin E has anti-cancer qualities. This may be due to vitamin E's several effects, which include activating heat shock proteins, down-regulating mutant p53 proteins, stimulating the p53 tumor suppressor gene, and having an anti-angiogenic action via the inhibition of transforming growth factor alpha [67]. As vitamin E compounds, alpha-, gamma-, and delta-tocopherols have developed, each with specific roles and anti-cancer properties. The development of PKC and collagenase [68], which promotes the proliferation of cancer cells, was found to be inhibited by alpha-tocopherol. Gamma-tocopherol was identified as superior to α-tocopherol in this situation for its capacity to suppress the proliferation of human prostate cancer cell lines, whereas delta-tocopherol has exhibited growth inhibitory efficacy against mouse mammary cancer cell lines [69]. In the culture, gamma-tocopherol suppress the cancer cells growth in a variety of methods.

The reactive nitrogen species of the compounds that modify the DNA strands and provide cells with cancerous changes are among the free radicals that are captured by it [70]. Additionally, it inhibits the activity of the cyclins, which stop the midst of the cancerous cell cycle and block the spread of the cancerous cells [71]. Additionally, it has been discovered that gamma-tocopherol performs better than alpha-tocopherol at inducing apoptosis, encouraging the activity of the peroxisome proliferators-activated receptor gamma, especially in colon cancer cells, and triggering several pathways that cause cell death [72, 73], inhibiting the growth of new blood vessels in tumors, and preventing tumors from receiving the supplements they require to grow. Tocotrienols was also found to have apoptotic and antiproliferative effects on both healthy and cancerous human cells in this setting [74]. This could be because apoptosis is induced by a mitochondria-mediated pathway or because cyclin D is suppressed, which would cause the cell cycle to be arrested [39]. Additionally, they impede vascularization and reduce 3-hydroxy-3-methyl coenzyme A (HMG-CoA) reductase activity, which stops the growth of cancer.

3.3 Cataracts

One of the most common causes of significant vision loss in older individuals is cataracts. They develop as a result of the buildup of proteins that have been harmed by free radicals. There may be a connection between vitamin E supplements and the likelihood of developing cataracts, according to several observational studies. *Leske et al.*, discovered that those who took vitamin E supplements and those with greater blood levels of vitamin [40]. Had clearer lenses. a sustained vitamin E treatment was linked in a different study [75] to a slower evolution of age-related lens opacification. Vitamin E, however, did not appear to have any impact on the onset or 6.3 years on average, the randomized Age-Related Eye Disease Study (AREDS) tracked the development of cataracts [76].

3.4 Alzheimer's disease

Through a mechanism involving oxidative stress and hydrogen peroxide, the beta-amyloid protein causes cytotoxicity, which eventually leads to the death of neuronal cells and AD. This process is what causes Alzheimer's disease (AD), which, through a free radical process, is brought on by protein oxidation and lipid peroxidation. Vitamin E can prevent hydrogen peroxide production and the cytotoxicity it produces. It lessens the cell mortality brought on by beta-amyloid in PC12 cells and rat hippocampus cell cultures [77, 78], as well as the neuroblastoma cell toxicity brought on by excitatory amino acids [79]. Vitamin E may decrease the progression of the disease in those with moderately severe Alzheimer's disease, according to the cooperative study on Alzheimer's from 1997.High vitamin E doses prevented the patient from losing the capacity to perform daily tasks and postponed their placement for several months in a private facility [80]. Another study revealed that people with AD had lower plasma levels of antioxidant supplements, indicating that the condition may be influenced by insufficient antioxidant activity. The neuroprotective impact of multiple vitamin E forms combined rather than just alpha-tocopherol alone is linked to older patients with high vitamin E plasma levels having a lowering risk of AD [81]. In a 2009 study, the effects of giving 847 individuals with and without taking an AD treatment 2000 IU of E vitamin were examined. It was discovered that combining an E vitamin with a cholinesterase inhibitor may be more advantageous than taking either medication by itself [82].

Using automated magnetic resonance imaging (MRI) measures and plasma levels of tocopherols and tocotrienols, Mangialasche et al. showed that it is possible to distinguish between Alzimers disease patients, people with moderate cognitive impairment, and healthy controls, as well as to predict the progression of with moderate cognitive impairment into Alzimers disease at the biomarker level. This demonstrates that plasma tocopherols and tocotrienols, which are nutritional indicators, may act as a surrogate for AD pathology [83]. However, due to its potential for harmful drug interactions in high doses, particularly those for lowering cholesterol, experts advise against patients taking vitamin E for AD treatment without a doctor's supervision.

3.5 Acute immunodeficiency syndrome and HIV

Although it is frequently discovered that Human immunodeficiency virus carriers (HIV) are low in E vitamin, unknown is whether E vitamin minerals is advantageous at any point during HIV infection. It has been shown that vitamin E to boost the growth of T helper cells (CD4 T-cells) and to restore delayed cutaneous hypersensitivity reactions at doses of 400 IU and higher [84]. Tang et al. examined the relationship between vitamin E and A levels and the development of HIV-1 illness in 1997. E vitamin levels in men above 24.2 m/l had a noticeably lower chance of illness development, according to this study. High amounts of vitamin E in the blood were shown to be strongly correlated with the consumption of vitamin E supplements at the time of study admission in this cohort [85].

The correction of immunological markers that are disrupted in HIV/AIDS was demonstrated in a study on the mouse acquired immunodeficiency syndrome utilizing a micronutrient intake rose by 15 times [86]. In addition, it has been demonstrated that increasing dietary vitamin E can guard against azidothymidine adverse effects such as bone marrow toxicity [87]. Similar findings were obtained from related research utilizing d-alpha-tocopherol supplementation on bone marrow cells from stage IV acquired immunodeficiency syndrome patients [88]. But it has also been observed that a higher fatality rate was associated with lower pre-infection vitamin E levels. As a result, more investigation is required to clarify the part of role of E vitamin E in the development of human immune deficiency virus-1 [89].

3.6 Immunity

It is now established that E vitamin boosts phagocytic activity, humoral immunological responses, and cell immunity in addition to stimulating the body's defenses. When immunological phagocytosis is engaged in infectious disorders, it has a noticeable impact, but it has less of an impact on cell-mediated immune defenses. Humans' cell-mediated and humoral immunological functions are markedly improved by its supplementation, especially in the adults. In healthy individuals' daily intake of 200 mg of E vitamin, who did not experience any adverse side effects increased the antibody response to several vaccines [90]. Higher plasma levels of E vitamin were linked with fewer infections throughout a three-year period [91], and additionally, vitamin E increased resilience to viral infections.in aged patients. According to *Kutty et al.* demonstrated that taking vitamin E supplements daily can improve the immunological reaction to a specific antigen [92]. When combined with vitamin CE has been proven to help treat a variety of illnesses, including photo dermatitis, pre-eclampsia/dysmenorrhea, and tardive dyskinesia, menstrual pain in addition to the conditions already listed [93].

3.7 Vitamin E deficiency

1. It is relatively uncommon for people to lack vitamin E. People who are unable to absorb dietary fat, suffer from uncommon disorders of the metabolism of fat, or acquire or acquired conditions that restrict their ability to absorb the vitamin are the main ones who are affected, as well as individuals who have these conditions (such as bowel syndrome, cystic fibrosis).

2. Recent studies have demonstrated that alpha-TTP controls the liver cells' production of alpha-tocopherol and that severe vitamin E shortage in humans can result from missense mutations of specific residues of arginine on the protein surface [94]. The target membrane's PIPs (phosphatidylinositol phosphates) made it easier for alpha-TTP to transfer alpha-tocopherol than it was for the arginine mutants. The arginine mutant a-alpha-TTPs failed to bind PIPs, in contrast to the wild-type alpha-TTP.

3. Muscle weakness, poor vision, immune system issues, tremors, trouble walking. Numbness and unsteadiness are all symptoms of vitamin E deficiency.

4. Dysarthria may accompany other deficit-related symptoms such as the lack of positive babinski reflexes, the loss of vibratory sensations, deep tendon reflexes and other symptoms, as well as neuromuscular conditions such spinocerebellar ataxia and myopathies.

5. Red blood cells are harmed by reactive stress, which leads to retinopathy [95–98], and the immune system is compromised;

6. A vitamin E deficit can also result in anemia. A E vitamin shortage will cause irreversible nerve damage, blindness, heart disease, and if neglected, cognitive impairment will occur.

7. According to certain research, a E-vitamin deficiency may be a factor in male infertility.

3.8 Complications of vitamin E

3.8.1 Safety and adverse effects

Oral vitamin E use is typically regarded as safe when administered in the proper dosages. Occasionally, taking vitamin E orally can result in:

- Gonadal dysfunction
- Rash
- Fatigue
- Weakness
- Headache
- Blurred Vision.
- Intestinal Cramps
- Increased Creatine Concentration in Urine (creatinuria).

Higher vitamin E dosages could make side effects more likely. Furthermore, there is a worry that those who consume high amounts of vitamin E but are in bad health-care more likely to pass away. Numerous illnesses can interact with vitamin E use. For instance, studies indicate that consuming vitamin E orally may raise the risk of prostate cancer. According to additional research, taking vitamin E may make it more likely that a person will pass away if they have a history of severe heart diseases, such as a stroke or heart attack. If any of the following apply to you before using vitamin E:

A lack of vitamin K, retinitis pigmentosa, a condition that damages the retina in the eyes, and bleeding issues.

Diabetes, a prior history of a heart attack or stroke, head and neck malignancy, liver illness, and diabetes.

Your risk of bleeding may go up if you take the supplement. You should stop taking vitamin E two weeks before surgery if you are planning on having it. If you are about to undergo or have recently undergone a technique to unblock arteries and restart your heart's regular blood flow, talk to your doctor about using vitamin E.

Several studies have documented adverse effects of vitamin E supplements in high doses. As stated previously, a meta-analysis of 19 clinical trials involving 135,968 participants indicated that high-dose vitamin E supplements (400 IU) may increase all-cause mortality. In contrast, a recent review article reported that a pooled analysis of 18 randomized controlled trials conducted in individuals who appeared to be healthy found no effect of vitamin E supplementation at doses ranging from 23 to 800 IU/day on all-cause mortality. In addition, meta-analyses of 33 and 57 trials found no correlation between vitamin E supplementation and mortality.

Unknown is the mechanism underlying the increased risk associated with high-dose vitamin E supplementation. The phenomenon may be attributable to the induction of cytochrome P450, an enzyme that speeds up the metabolism of other substances [99]. Vitamin E has eight isoforms, with -tocopherol being the most potent and abundant in humans, despite -tocopherol being ingested in equal amounts. High-dose -tocopherol supplementation accelerates the metabolism of non-tocopherol forms. It has not been established whether or not the different isoforms of vitamin E have specific functions in vivo that could be compromised by -tocopherol supplementation. In certain in vitro systems, −tocopherol has been found to act as a pro-oxidant. In the absence of reducing agents such as vitamin C, −tocopherol accelerates the oxidation of LDL because the a-tocopheroxyl radical induces LDL oxidation.

Nevertheless, vitamin C absolutely inhibits this pro-oxidant action by decreasing -tocopheroxyl radical levels [100]. Therefore, it is improbable that vitamin E functions as a pro-oxidant in vivo.

3.8.2 Interactions

Your levels of vitamin E may change if you use certain medications. Several interactions are possible:

- **Anti-tumor antibiotics and alkylating drugs**: High vitamin E dosages are feared to have an impact on how well this chemotherapy drug works.
- **Medicines and supplements that inhibit clotting and platelet function**: The risk of bleeding may increase if vitamin E is used along with these medications, supplements and herbs to reduce blood clotting.
- **Substrates for Cytochrome P450 3A4 (CYP3A4):** When using omeprazole, a medication impacted by these enzymes, as well as vitamin E, proceeds with caution (Prilosec, Zegerid).
- **Niacin and statins:** Niacin's effects may be diminished if taken with statins or vitamin E, which may help those with high cholesterol.
- **K-vitamin:** Combining vitamin K and vitamin E may lessen vitamin K's effects.

4. Conclusion

The doctor's Shute and Shute employed vitamin E as a supplement for the first time in Canada; as a result of the successful outcomes, they started utilizing it frequently in their offices. Since then, carefully planned clinical and experimental investigations have progressively advanced and advanced our understanding of E-vitamin. Vitamin E's antioxidant capabilities have been proven to be essential in the fight against a number of illnesses, including cancer, oxidative stress, atherosclerosis cataracts and AD, among others. Additionally, demonstrated to be efficient against diabetes, allergies, asthma, and other illnesses in addition to these. The emphasis of this review was on the vital roles that E-vitamin plays in a number of infections.

A side from the extensive advantages claimed, there has always been disagreement on the precise of E vitamin role and its connection to certain ailments. In the literature, there are several contradictory accounts of both favorable and unfavorable outcomes for the same biological processes. The absence of reliable indicators for monitoring E vitamin consumption and status, which would link intakes to potential clinical outcomes, is the main obstacle to understanding the functions in human health of E vitamin. In conclusion, despite the inconsistent facts surrounding vitamin E, the present body of research seems to be in favor of the idea that the advantages outweigh the drawbacks.

Vitamin E inhibits lipid peroxidation both in vivo and in vitro. These findings suggest that vitamin E may reduce the risk of diseases mediated by free radicals or be beneficial in their prevention and treatment. Numerous epidemiological studies have supported this theory, but the results of clinical intervention studies and meta-analyses have been controversial; some have reported positive findings, while others have reported null or negative findings.

To attain optimal results, numerous factors must be taken into account. Vitamin E should be beneficial for subjects enduring oxidative stress mediated by free radicals if administered at the appropriate dose, time, and duration.

Conflict of interest

The authors declare no conflict of interest.

Author details

Naresh Podila[1*], Sathish Kumar Konidala[1], Jithendra Chimakurthy[1], Srilatha Muddisetti[2], Suryaprabha Matangi[1], Natesh Gunturu[2], Yamarthi Venkateswara Rao[3] and Mithun Rudrapal[1]

1 Department Pharmaceutical Sciences, School of Biotechnology and Pharmaceutical Sciences, Vignan's Foundation for Science, Technology and Research, Guntur, A.P., India

2 Vikas College of Pharmaceutical Sciences, Suryapet, Telangana, India

3 Vignan Pharmacy College, Guntur, A.P., India

*Address all correspondence to: nareshtrcp10@gmail.com

References

[1] Abner EL, Schmitt FA, Mendiondo MS, Marcum JL, Kryscio RJ. Vitamin E and all-cause mortality: A meta-analysis. Current Aging Science. 2011;**4**(2):158-170

[2] Abraham GE. Nutritional factors in the etiology of the premenstrual tension syndromes. The Journal of Reproductive Medicine. 1983;**28**(7):446-464 View abstract

[3] National Institutes of Health Office of Dietary Supplements. Vitamin E, Niki E, Noguchi N, Tsuchihashi H, Gotoh N. Interaction among vitamin C, vitamin E, and beta-carotene. The American Journal of Clinical Nutrition. 1995;**62**:1322S-1326S

[4] Ames BN. Micronutrient deficiencies: A major cause of DNA damage. Annals of the New York Academy of Sciences. 2000;**889**:87-106

[5] Chow CK. Distribution of tocopherols in human plasma and red blood cells. The American Journal of Clinical Nutrition. 1975;**28**:756-760

[6] Drotleff AM, Ternes W. Determination of RS, E/Z-tocotrienols by HPLC. Journal of Chromatography A. 2001;**909**:215-223

[7] Halliwell B, Gutteridge JM. Free Radicals in Biology and Medicine. 4th. Oxford: Clarendon Press; 2007

[8] Sugamura K, Keaney JF. Jr reactive oxygen species in cardiovascular disease. Free Radical Biology & Medicine. 2011;**51**:978-999

[9] Niki E. Assessment of antioxidant capacity in vitro and in vivo. Free Radical Biology & Medicine. 2010;**49**:503-515

[10] Palmer S. Diet, nutrition, and cancer. Progress in Food & Nutrition Science. 1985;**9**:283-341

[11] Gey KF, Brubacher GB, Stahelin HB. Plasma levels of antioxidant vitamins in relation to ischemic heart disease and cancer. The American Journal of Clinical Nutrition. 1987;**45**(5 Suppl):1368-1377

[12] Bjelakovic G, Nikolova D, Gluud LL, Simonetti RG, Gluud C. Mortality in randomized trials of antioxidant supplements for primary and secondary prevention: Systematic review and meta-analysis. Journal of the American Medical Association. 2007;**297**:842-857

[13] Traber MG, Frei B, Beckman JS. Vitamin E revisited: Do new data validate benefits for chronic disease prevention? Current Opinion in Lipidology. 2008;**19**:30-38

[14] Niki E. Do free radicals play causal role in atherosclerosis? Low density lipoprotein oxidation and vitamin E revisited. Journal of Clinical Biochemistry and Nutrition. 2011;**48**:3-7

[15] Zingg JM. Molecular and cellular activities of vitamin E analogues. Mini Reviews in Medicinal Chemistry. 2007;**7**:543-558

[16] Ball GFM. Vitamins in Foods: Analysis, Bioavailability, and Stability. Boca Raton, Florida: CRC Press; 2006. pp. 119-136

[17] Brown KM, Morrice PC, Duthie GG. Erythrocyte vitamin E and plasma ascorbate concentrations in relation to erythrocyte peroxidation in smokers and non-smokers: Dose response to vitamin E supplementation. The American Journal of Clinical Nutrition. 1997;**65**:496-502

[18] Di Mascio P, Murphy ME, Sies H. Antioxidant defense systems: The role of carotenoids, tocopherols, and thiols. The American Journal of Clinical Nutrition. 1991;**53**:194S-200S

[19] Glynn RJ, Ridker PM, Goldhaber SZ, Zee RY, Buring JE. Effects of random allocation to vitamin E supplementation on the occurrence of venous thromboembolism: Report from the Women's health study. Circulation. 2007;**116**:1497-1503

[20] Booth SL, Golly I, Sacheck JM, Roubenoff R, Dallal GE, Hamada K, et al. Effect of vitamin E supplementation on vitamin K status in adults with normal coagulation status. The American Journal of Clinical Nutrition. 2004;**80**:143-148

[21] Burton GW, Ingold KU. Autoxidation of biological molecules: 1. The antioxidant activity of vitamin E and related chainbreaking phenolic antioxidants in vitro. Journal of the American Chemical Society. 1981;**103**:6472-6477

[22] Liu M, Wallin R, Wallmon A, Saldeen T. Mixed tocopherols have a stronger inhibitory effect on lipid peroxidation than alpha-tocopherol alone. Journal of Cardiovascular Pharmacology. 2002;**39**:714-721

[23] Burton GW, Joyce A, Ingold KU. Is vitamin E the only lipidsoluble, chain-breaking antioxidant in human blood plasma and erythrocyte membranes? Archives of Biochemistry and Biophysics. 1983;**221**:281-290

[24] Howard AC, Anna K, McNeil AK, McNeil PL. Promotion of plasma membrane repair by vitamin E. Nature Communications. 2011;**20**:597

[25] Tran K, Wong JT, Lee E, Chan AC, Choy PC. Vitamin E potentiates arachidonate release and phospholipase A2 activity in rat heart myoblastic cells. The Biochemical Journal. 1996;**319**:385-391

[26] Niki E. Lipid peroxidation: Physiological levels and dual biological effects. Free Radical Biology & Medicine. 2009;**47**:469-484

[27] Niki E. Biomarkers of lipid peroxidation in clinical material. Biochimica et Biophysica Acta. 2014;**1840**:809-817

[28] Yin H, Xu L, Porter NA. Free radical lipid peroxidation: Mechanisms and analysis. Chemical Reviews. 2011;**111**:5944-5972

[29] Cheeseman KH, Albano EF, Tomasi A, Slater TF. Biochemical studies on the metabolic activation of halogenated alkanes. Environmental Health Perspectives. 1985;**64**:85-101

[30] Recknagel RO. Carbon tetrachloride hepatotoxicity. Pharmacological Reviews. 1967;**19**:145-208

[31] Smathers RL, Galligan JJ, Stewart BJ, Petersen DR. Overview of lipid peroxidation products and hepatic protein modification in alcoholic liver disease. Chemico-Biological Interactions. 2011;**192**:107-112

[32] Sid B, Verrax J, Calderon PB. Role of oxidative stress in the pathogenesis of alcohol-induced liver disease. Free Radical Research. 2013;**47**:894-904

[33] Adachi J, Asano M, Ueno Y, Naito T. Identification of 7-hydroperoxycholesterol in human liver by liquid chromatography-mass spectrometry. Alcoholism, Clinical and Experimental Research. 2000;**24** (Suppl. 4):21S-25S

[34] Aleynik SI, Leo MA, Aleynik MK, Lieber CS. Increased circulating products

of lipid peroxidation in patients with alcoholic liver disease. Alcoholism, Clinical and Experimental Research. 1998;**22**:192-196

[35] Wierzbicki AS, Oben J. Nonalcoholic fatty liver disease and lipids. Current Opinion in Lipidology. 2012;**23**:345-352

[36] Sumida Y, Niki E, Naito Y, Yoshikawa T. Involvement of free radicals and oxidative stress in NAFLD/NASH. Free Radical Research. 2013;**47**:869-880

[37] Feldstein AE, Lopez R, Tamimi TA, et al. Mass spectrometric profiling of oxidized lipid products in human nonalcoholic fatty liver disease and nonalcoholic steatohepatitis. Journal of Lipid Research. 2010;**51**:3046-3054

[38] Terao K, Niki E. Damage to biological tissues induced by radical initiator 2,2′-azobis(2-amidinopropane) dihydrochloride and its inhibition by chain-breaking antioxidants. Journal of Free Radicals in Biology & Medicine. 1986;**2**:193-201

[39] Morita M, Ishida N, Uchiyama K, et al. Fatty liver induced by free radicals and lipid peroxidation. Free Radical Research. 2012;**46**:758-765

[40] Yoshida Y, Hayakawa M, Cynshi O, Jishage K, Niki E. Acceleration of lipid peroxidation in alpha-tocopherol transfer protein-knockout mice following the consumption of drinking water containing a radical initiator. Journal of Oleo Science. 2008;**57**:577-583

[41] Sanyal AJ, Chalasani N, Kowdley KV, et al. Pioglitazone, vitamin E, or placebo for nonalcoholic steatohepatitis. The New England Journal of Medicine. 2010;**362**:1675-1685

[42] Cheng J, Joyce A, Yates K, Aouizerat B, Sanyal AJ. Metabolomic profiling to identify predictors of response to vitamin E for non-alcoholic steatohepatitis (NASH). PLoS One. 2012;7(9):e44106

[43] Sumida Y, Naito Y, Tanaka S, et al. Long-term (>=2 yr) efficacy of vitamin E for non-alcoholic steatohepatitis. Hepatogastroenterology. 2013;**60**:1445-1450

[44] Steinberg D. The LDL modification hypothesis of atherogenesis: An update. Journal of Lipid Research. 2009;**50**(Suppl):S376-S381

[45] Barter PJ, Nicholls S, Rye KA, Anantharamaiah GM, Navab M, Fogelman AM. Antiinflammatory properties of HDL. Circulation Research. 2004;**95**:764-772

[46] Noguchi N, Numano R, Kaneda H, Niki E. Oxidation of lipids in low density lipoprotein particles. Free Radical Research. 1998;**29**:43-52

[47] Colas R, Pruneta-Deloche V, Guichardant M, et al. Increased lipid peroxidation in LDL from type-2 diabetic patients. Lipids. 2010;**45**:723-731

[48] Garcia-Cruset S, Carpenter KL, Guardiola F, Stein BK, Mitchinson MJ. Oxysterol profiles of normal human arteries, fatty streaks and advanced lesions. Free Radical Research. 2001;**35**:31-41

[49] Niu X, Zammit V, Upston JM, Dean RT, Stocker R. Coexistence of oxidized lipids and alpha-tocopherol in all lipoprotein density fractions isolated from advanced human atherosclerotic plaques. Arteriosclerosis, Thrombosis, and Vascular Biology. 1999;**19**:1708-1718

[50] Suarna C, Dean RT, May J, Stocker R. Human atherosclerotic plaque contains both oxidized lipids and relatively

large amounts of alpha-tocopherol and ascorbate. Arteriosclerosis, Thrombosis, and Vascular Biology. 1995;**15**:1616-1624

[51] Kuhn H, Romisch I, Belkner J. The role of lipoxygenase-isoforms in atherogenesis. Molecular Nutrition & Food Research. 2005;**49**:1014-1029

[52] Malle E, Marsche G, Arnhold J, Davies MJ. Modification of low-density lipoprotein by myeloperoxidase-derived oxidants and reagent hypochlorous acid. Biochimica et Biophysica Acta. 2006;**1761**:392-415

[53] Wagner JR, Motchnik PA, Stocker R, Sies H, Ames BN. The oxidation of blood plasma and low density lipoprotein components by chemically generated singlet oxygen. The Journal of Biological Chemistry. 1993;**268**:18502-18506

[54] Szczeklik A, Gryglewski RJ, Domagala B, Dworski R, Basista M. Dietary supplementation with vitamin E in hyperlipoproteinemias: Effects on plasma lipid peroxides, antioxidant activity, prostacyclin generation and platelet aggregability. Thrombosis and Haemostasis. 1985;**54**:425-430

[55] Freedman JE, Farhat JH, Loscalzo J, Keaney JF Jr. Alphatocopherol inhibits aggregation of human platelets by a protein kinase C-dependent mechanism. Circulation. 1996;**94**:2434-2440

[56] Li D, Saldeen T, Romeo F, Mehta JL. Different isoforms of tocopherols enhance nitric oxide synthase phosphorylation and inhibit human platelet aggregation and lipid peroxidation: Implications in therapy with vitamin E. Journal of Cardiovascular Pharmacology and Therapeutics. 2001;**6**:155-161

[57] Brigelius-Flohé R, Traber MG. Vitamin E: Function and metabolism. The FASEB Journal. 1999;**13**:1145-1155

[58] Liu M, Wallmon A, Olsson-Mortlock C, Wallin R, Saldeen T. Mixed tocopherols inhibit platelet aggregation in humans: Potential mechanisms. The American Journal of Clinical Nutrition. 2003;**77**:700-706

[59] Steinberg D, Parthasarathy S, Carew TE, Khoo JC, Witztum JL. Beyond cholesterol: Modifications of low-density lipoprotein that increase its artherogenicity. The New England Journal of Medicine. 1989;**320**:915-924

[60] Li D, Saldeen T, Romeo F, Mehta JL. Relative effects of alphaand gamma-tocopherol on low-density lipoprotein oxidation and superoxide dismutase and nitric oxide synthase activity and protein expression in rats. Journal of Cardiovascular Pharmacology and Therapeutics. 1999;**4**:219-226

[61] Singh I, Turner AH, Sinclair AJ, Li D, Hawley JA. Effects of gamma-tocopherol supplementation on thrombotic risk factors. Asia Pacific Journal of Clinical Nutrition. 2007;**16**:422-428

[62] McAnally JA, Gupta J, Sodhani S, Bravo L, Mo H. Tocotrienols potentiate lovastatin-mediated growth suppression in vitro and in vivo. Experimental Biology and Medicine (Maywood, N.J.). 2007;**232**:523-531

[63] Sesso HD, Buring JE, Christen WG, Kurth T, Belanger C, MacFadyen J, et al. Vitamins E and C in the prevention of cardiovascular disease in men: The physicians' health study II randomized controlled trial. JAMA. 2008;**300**:2123-2133

[64] Shklar G, Oh SK. Experimental basis for cancer prevention by vitamin E. Cancer Investigation. 2000;**18**:214-222

[65] Ricciarelli R, Maroni P, Ozer N, Zingg JM, Azzi A. Agedependent

increase of collagenase expression can be reduced by alpha-tocopherol via protein kinase C inhibition. Free Radical Biology & Medicine. 1999;**27**:729-737

[66] McIntyre BS, Briski KP, Gapor A, Sylvester PW. Antiproliferative and apoptotic effects of tocopherols and tocotrienols on preneoplastic and neoplastic mouse mammary epithelial cells. Proceedings of the Society for Experimental Biology and Medicine. 2000;**224**:292-301

[67] Christen S, Woodall AA, Shigenaga MK, Southwell-Keely PT, Duncan MW, Ames BN. Gamma-tocopherol traps mutagenic electrophiles such as NO(X) and complements alphatocopherol: Physiological implications. Proceedings of the National Academy of Sciences of the United States of America. 1997;**94**:3217-3222

[68] Gysin R, Azzi A, Visarius T. Gamma-tocopherol inhibits human cancer cell cycle progression and cell proliferation by down-regulation of cyclins. The FASEB Journal. 2002;**16**:1952-1954

[69] Takahashi S, Takeshita K, Seeni A, Sugiura S, Tang M, Sato SY, et al. Suppression of prostate cancer in a transgenic rat model via gamma-tocopherol activation of caspase signaling. Prostate. 2009;**69**:644-651

[70] Stone WL, Krishnan K, Campbell SE, Qui M, Whaley SG, Yang H. Tocopherols and the treatment of colon cancer. Annals of the New York Academy of Sciences. 2004;**1031**:223-233

[71] Wells SR, Jennings MH, Rome C, Hadjivassiliou V, Papas KA, Alexander JS. Alpha-, gamma- and delta-tocopherols reduce inflammatory angiogenesis in human microvascular endothelial cells. The Journal of Nutritional Biochemistry. 2010;**21**:589-597

[72] Wada S. Chemoprevention of tocotrienols: The mechanism of antiproliferative effects. Forum of Nutrition. 2009;**61**:204-216

[73] Jiang Q, Wong J, Fyrst H, Saba JD, Ames BN. Gamma-tocopherol or combinations of vitamin E forms induce cell death in human prostate cancer cells by interrupting sphingolipid synthesis. Proceedings of the National Academy of Sciences of the United States of America. 2004;**101**:17825-17830

[74] Chen CS. Study Shows How Vitamin E Can Help Prevent Cancer. Available from: www.researchnews.osu.edu/archive/silenceakt.htm [Accessed: Dec 2013]

[75] Lonn E, Bosch J, Yusuf S, Sheridan P, Pogue J, Arnold JM, et al. Effects of long-term vitamin E supplementation on cardiovascular events and cancer: A randomized controlled trial. JAMA. 2005;**293**:1338-1347

[76] Goodman Y, Mattson MP. Secreted forms of beta-amyloid precursor protein protect hippocampal neurons against amyloid beta-peptide-induced oxidative injury. Experimental Neurology. 1994;**128**:1-12

[77] Behl C, Davis J, Cole GM, Schubert D. Vitamin E protects nerve cells from amyloid beta protein toxicity. Biochemical and Biophysical Research Communications. 1992;**186**:944-950

[78] Murphy TH, Schnaar RL, Coyle JT. Immature cortical neurons are uniquely sensitive to glutamate toxicity by inhibition of cystine uptake. The FASEB Journal. 1990;**4**:1624-1633

[79] Sano M, Ernesto C, Thomas RG, Klauber MR, Schafer K, Grundman M, et al. A controlled trial of selegiline, alphatocopherol, or both as treatment

for Alzheimer's disease. The Alzheimer's disease cooperative study. The New England Journal of Medicine. 1997;**336**:1216-1222

[80] Mangialasche F, Kivipelto M, Mecocci P, Rizzuto D, Palmer K, Winblad B, et al. High plasma levels of vitamin E forms and reduced Alzheimer's disease risk in advanced age. Journal of Alzheimer's Disease. 2010;**20**:1029-1037

[81] Pavlik VN, Doody RS, Rountree SD, Darby EJ. Vitamin E use is associated with improved survival in an Alzheimer's disease cohort. Dementia and Geriatric Cognitive Disorders. 2009;**28**:536-540

[82] Mangialasche F, Westman E, Kivipelto M, Muehlboeck JS, Cecchetti R, Baglioni M, et al. Classification and prediction of clinical diagnosis of Alzheimer's disease based on MRI and plasma measures of α−/γ-tocotrienols and γ-tocopherol. Journal of Internal Medicine. 2013;**273**:602-621

[83] Meydani M, Vitamin E. Lancet. 1995;**345**:170-175

[84] Tang AM, Graham NM, Semba RD, Saah AJ. Association between serum vitamin a and E levels and HIV-1 disease progression. AIDS. 1997;**11**:613-620

[85] Wang Y, Huang DS, Watson RR. Vitamin E supplementation modulates cytokine production by thymocytes during murine AIDS. Immunologic Research. 1993;**12**:358-366

[86] Ganser A, Greher J, Volkers B, Staszewski A, Hoelzer D. Azidothymidine in the treatment of ATDS. The New England Journal of Medicine. 1988;**318**:250-251

[87] Geissler RG, Ganser A, Ottmann OG, Gute P, Morawetz A, Guba P, et al. In vitro improvement of bone marrow-derived hematopoietic colony formation in HIV-positive patients by alpha-D-tocopherol and erythropoietin. European Journal of Haematology. 1994;**53**:201-206

[88] Graham SM, Baeten JM, Richardson BA, Bankson DD, Lavreys L, Ndinya-Achola JO, et al. Higher pre-infection vitamin E levels are associated with higher mortality in HIV-1-infected Kenyan women: A prospective study. BMC Infectious Diseases. 2007;**7**:63

[89] Meydani SN, Meydani M, Blumberg JB, Leka LS, Siber G, Loszewski R, et al. Vitamin E supplementation and in vivo immune response in healthy elderly subjects: A randomized controlled trial. JAMA. 1997;**277**:1380-1386

[90] Chavance M, Herbeth B, Fournier C, Janot C, Vernhes G. Vitamin status, immunity and infections in an elderly population. European Journal of Clinical Nutrition. 1989;**43**:827-835

[91] Radhakrishnan AK, Mahalingam D, Selvaduray KR, Nesaretnam K. Supplementation with natural forms of vitamin E augments antigen-specific TH1-type immune response to tetanus toxoid. BioMed Research International. 2013;**2013**:782067

[92] University of Maryland Medical Center. Vitamin E. Available from: www.umm.edu/health/medical/altmed/supplement/vitamine#ixzz2c7gfq93R [Accessed: Dec 2013]

[93] Kono N, Ohto U, Hiramatsu T, Urabe M, Uchida Y, Satow Y, et al. Impaired a-TTP-PIPs interaction underlies familial vitamin E deficiency. Science. 2013;**340**:1106-1110

[94] Office of Dietary Supplements, National Institutes of Health. Dietary Supplement Fact Sheet: Vitamin E.

Available from: www.ods. od.nih.gov/factsheets/vitamine.asp [Accessed: Aug 2010]

[95] Food and Nutrition Board, Institute of Medicine. Dietary Reference Intakes for Vitamin C, Vitamin E, Selenium, and Carotenoids. Washington, DC: National Academy Press; 2000

[96] Kowdley KV, Mason JB, Meydani SN, Cornwall S, Grand RJ. Vitamin E deficiency and impaired cellular immunity related to intestinal fat malabsorption. Gastroenterology. 1992;**102**(6):2139-2142

[97] Slover HT. Tocopherols in foods and fats. Lipids. 1971;**6**:291-296

[98] United States Department of Agriculture (USDA), Agricultural Research Service. USDA National Nutrient Database for Standard Reference, Release 25. Available from: www.ars.usda.gov/SP2UserFiles/Place/12354500/Data/SR25/nutrlist/sr25a323.pdf [Accessed: Dec 2013]

[99] Colombo ML. An update on vitamin E, tocopherol and tocotrienol: Perspectives. Molecules. 2010;**15**:2103-2113

[100] Rathore GS, Suthar M, Pareek A, Gupta RN. Nutritional antioxidants: A battle for better health. Journal of Natural Pharmaceuticals. 2011;**2**:2-14

Chapter 6

Vitamin E Inhibits Osteoclastogenesis in Protecting Osteoporosis

Chen Yongjie, Yu Naichun, Zhou Daguo, Li Zongguang, Gong Fengqing, Yi Weijiang, Chen Botao and Ji Guangrong

Abstract

The most common orthopedic condition affecting senior adults is osteoporosis, which is defined by a decrease in bone mass and strength as well as microstructural degradation that leads to fragility fractures. Bone remodeling is a well-planned, ongoing process that replaces deteriorated, old bone with new, healthy bone. Bone resorption and bone creation work together during the cycle of bone remodeling to preserve the bone's volume and microarchitecture. The only bone-resorbing cells in the human body, mononuclear preosteoclasts fuse to form osteoclasts, are multinucleated cells. In numerous animal models or epidemiological studies, vitamin E's anti-osteoporotic characteristics have been extensively described. This review aims to summarize recent developments in vitamin E's molecular features as a bone-protective agent. In RANKL/RANK/OPG signaling pathway, vitamin E inhibits synthesis of RANKL, stimulation of c-Fos, and increase level of OPG. Vitamin E also inhibits inflammatory cytokines, such as TNF-α, IL-1, IL-6, IL-27, and MCP-1, negative regulating the JAK–STAT, NF-κB, MAPK signaling pathways. Additionally, vitamin E decreases malondialdehyde and increases superoxide dismutase, GPx and heme oxygenase-1, in suppressing osteoclasts. In this article, we aim to give readers the most recent information on the molecular pathways that vitamin E uses to enhance bone health.

Keywords: vitamin E, osteoclast, osteoporosis, RANKL, inflammatory, oxidative stress

1. Introduction

The most common orthopedic condition affecting elderly people is osteoporosis, which is characterized by a decline in microarchitecture and a loss of bone mass and strength. Approximately 25% of all women aged 65 or older are affected by osteoporosis, which results in the lifetime fracture risk of patients with osteoporosis being as high as 40% [1]. Osteoporosis fractures involving the hip, vertebrae, and wrist reduce life expectancy or quality of life. Some drugs classified as antiresorptive or anabolic have been used in the treatment of osteoporosis [2, 3].

Antiresorptive drugs, which include bisphosphonates, estrogens, denosumab, and selective estrogen receptor modulators (SERMs), increase bone mineral density

IntechOpen

(BMD) and reduce fragility fractures. Bisphosphonates are the most widely used antiresorptive drug, which bind avidly to bone mineral and have a continual effect in months to years. However, bisphosphonates have low bioavailability and induce gastrointestinal problems, myalgia, and flu-like symptoms. Osteoclast activity and development are significantly decreased by the monoclonal antibody denosumab, which specifically targets RANKL. Every six months, denosumab is given by subcutaneous injection. Denosumab's anti-fracture effects are comparable to those of bisphosphonates; however starting 7 months after the last injection, there is a noticeable reduction of anti-resorptive action, which can lead to clusters of rebound vertebral fractures. SERMs exhibit estrogen agonist or antagonist qualities and bind to the estrogen receptor with high affinity to prevent bone loss. However, SERMs can induce venous thromboemboembolism, stroke and hot flashes, which limit their use [4].

Anabolic agents, including teriparatide and abaloparatide, promote bone formation to reduce the risk of fractures. Teriparatide and abaloparatide are both limited to a maximum of 24 months of therapy due to elevated osteosarcoma incidence in rats treated with these medications for their entire lives. These drugs similarly induce some adverse effects, including dizziness, leg cramps, nausea, postural hypotension, and headache. Another anabolic agent is romosozumab, which blocks the actions of sclerostin to reduce bone resorption. However, the romosozumab group in the ARCH study had a higher rate of a composite endpoint that included cardiovascular death, nonfatal myocardial infarction, and nonfatal stroke. In general, different drugs have been used to treat osteoporosis (**Table 1**), but their different adverse reactions limit their long-term use against osteoporosis [2, 3, 5, 6].

Vitamin E has shown great potential in the treatment of osteoporosis. There are two subclasses of the lipid-soluble vitamin E: tocopherol (TF) and tocotrienol (T3). Each isoform can further be separated into four unique analogs, namely, alpha (α), beta (β), gamma (γ), and delta (δ). Some researchers have investigated the impact of vitamin E (TF) on osteoblast differentiation in a previous study. Early osteoblast development

Drugs	Administration	Dosage	Side effects
Anti-resorptive Bisphosphonates			
Alendronate	Oral	70 (or 35) mg/week; 10 (or 5) mg/day	Upper gastrointestinal symptoms/osteonecrosis of the jaw (ONJ)/atypical femoral fracture (AFF)
Risedronate	Oral	5 mg/day; 35 mg/week; 75 mg twice per month. 150 mg/month.	Upper gastrointestinal symptoms, ONJ, or AFF.
Ibandronate	Oral or intravenous	150 mg/month orally or 3 mg/month	Upper gastrointestinal Symptoms, flu-like illness ONJ or AFF.
Zoledronate or Zoledronic acid	Intravenous	Intravenously 5 mg/year	Flu-like illness ONJ or AFF.
SERMs			
Raloxifene	Oral	60 mg/day	Venous thromboembolism, stroke and flashes

Drugs	Administration	Dosage	Side effects
Monoclonal antibody against RANKL			
Denosumab	Subcutaneously	60 mg every 6 months	Hypocalcaemia, rebound vertebral fractures, and AFF
Anabolic agents			
Teriparatide	Subcutaneously	20 μg/day	Hypercalcemia
Abaloparatide	Subcutaneously	80 μg/day	Dizziness, leg cramp, nausea, headache, hypercalcemia
Romosozumab	Subcutaneously	210 μg/day	Serious cardiovascular events, AFF and ONJ

Table 1.
Principal medications for treatment of osteoporosis.

was demonstrated to be hindered by TF, as evidenced by a decrease in alkaline phosphatase (ALP) activity and osteocalcin expression (OCN) [7]. In another study, another researcher showed that incubating human mesenchymal stem cells (MSCs) with TF boosted their proliferation. ALP and runt-related transcription factor-2 (Runx-2) expression was markedly increased in cells treated with TF. Gamma-T3 has been shown to decrease apoptosis and increase osteoblastic cell proliferation, differentiation, and mineralization. When αTF and δT3 were compared to TF and T3, αTF and δT3 had superior effects on inhibiting osteoblast differentiation. In a biomechanical strength test, bone scaffolds treated with all types of vitamin E isomers, particularly αTF and δT3, displayed enhanced elasticity [8]. According to this study, αTF and δT3 may be the isomers that affect bone most actively. Many researchers have focused on the effect of vitamin E on osteoblasts. In contrast, few *in vitro* studies have examined how vitamin E affects osteoclasts. This chapter emphasizes recent developments in our knowledge of vitamin E's molecular actions in inhibiting osteoclasts, particularly through interactions between multiple signaling pathways and signal transduction molecules.

2. Osteoclasts in bone homeostasis

Bone is a dynamic tissue that changes regularly during the course of life. During the process of bone remodeling, osteoclasts remove old or damaged bone and osteoblasts replace it with fresh bone. Without appreciably altering the net bone mass and mechanical strength, the ratio of osteoblast-mediated bone formation to osteoclast-mediated bone resorption is tightly regulated in homeostatic situations [9]. However, when this equilibrium is dysregulated, aberrant bone remodeling occurs, leading to postmenopausal and secondary types of osteoporosis. Osteoclasts are the only bone-resorbing cells in the human body, which is significant since they are essential for reconstructing the skeletal system. Numerous cells originating from blood-circulating monocytes combine to become osteoclasts. They themselves come from the bone marrow. Although the majority have only 5 to 20 nuclei, osteoclasts can have up to 200 nuclei. The active area of osteoclasts is formed by numerous tiny projections (microvilli) that protrude into the surface of the bone on the side of the cell that is closest to the bone. One of the many enzymes produced by osteoclasts, acid phosphatase dissolves both the organic collagen and the inorganic calcium and phosphorus found in bone. The osteoclast then engulfs and digests the broken-down pieces of mineralized bone inside the cytoplasmic vacuoles. Communication between osteoblasts and osteoclasts is crucial for optimizing bone remodeling during bone homeostasis. In addition to M-CSF, RANKL, and WNT5A, osteoblasts also release OPG and WNT16, which limit osteoclast activity. In contrast, S1P, CTHRC1, and C3, which encourage osteoblast differentiation, as well as SEMA4D, which inhibits osteoblast differentiation, are secreted by osteoclasts.

A hematopoietic growth factor called M-CSF enables mononuclear phagocyte lineages, which include osteoclasts, to survive, proliferate, differentiate, and move around. M-CSF binds to its specific receptor, C-FMS, on the surface of osteoclasts and monocytes/macrophages after being produced by osteoblasts and bone marrow stromal cells. Significant amounts of RANKL are expressed in lymph nodes, activated T lymphocytes, osteoblasts, and osteocytes. The main regulatory transcription factors and enzymes are activated as a result of RANKL activation of RANK, which promotes osteoclast differentiation, fusion, and proliferation. For instance, RANKL binding to RANK activates TRAF6, which has been discovered to activate the protein kinase TGF-activated kinase (TAK1). The classical IB kinase (IKK) complex is then activated by TAK1. IKK activation results in the activation of NF-κB, protein kinase Tp12, and mitogen-activated protein (MAP) kinase (MKK) kinases. MKKs trigger the activity of p38 MAP kinases and c-Jun N-terminal kinases. Extracellular signal-regulated kinase 1 (ERK1) and ERK2 are stimulated by activated Tp12, which in turn activates MEK1 and MEK2. Eventually, the MAPK and NF-κB signaling pathways are activated, increasing NFATc1, a crucial regulator of osteoclastogenesis. OPG has been shown to be a secreted glycoprotein made by numerous cell types, including osteoblasts, cells living in the liver or lungs, and B lymphocytes in the bone marrow. Due to the suppression of the development of osteoclasts, overexpression of OPG causes severe osteopetrosis. OPG is thought to act as a decoy receptor for RANKL, blocking the RANKL-RANK interaction and impairing osteoclast development and activation. Because it controls osteoblastogenesis and osteoclastogenesis through both catenin-dependent (canonical) and catenin-independent (noncanonical) pathways, the WNT pathway is essential for preserving bone homeostasis. By increasing RANK expression in osteoclasts and activating the MAPK pathway, WNT5A promotes RANKL-induced osteoclastogenesis, which is also inhibited by WNT16 both directly and indirectly. A direct method of preventing osteoclastogenesis is provided by WNT16-induced phosphorylation of JUN, which increases OPG expression in osteoblasts in addition to directly inhibiting osteoclastogenesis *via* the noncanonical JNK/MAPK pathway (**Table 2**) [10].

	Osteoblast	**Osteoclast**
Transcription factors	RUNX 1/2, OSX, ATF 4, SATB 2, AP-1	NFATc 1, AP-1, CREB, MITF, NF-κB, c-FOS
Activating cytokines	IL-10, IL 11, IL 18, IFN-γ CT-1, OSM, S1P, CTHRC 1, C3 BMP-2, TGF-β, FGF, PTH	RANKL, TNF-α, IL-1, IL-6, IL-7, IL-8, IL-11, IL-15, IL-17, IL-23, IL-34
Activating signaling pathways	Wnt, Smad, MAPK, PI3K/AKT, ERK, cAMP/PKA/CREB, Hedgehog	Akt, ERK, NF-κB, MAPK, JNK, STAT3/Smad 7 DAP12/FCRγ-Syk-PLCγ

	Osteoblast	Osteoclast
Inhibiting factors	TNF-α, TNF-β, IL-1α, IL-4, IL-7, IL 12, IL-13, IL-23, IFN-α, IFN-β, SEMA4D	OPG, IFN-α, IFN-β, IFN-γ, IL-3, IL-4, IL-10, IL-12, IL-27, IL-33
Inhibiting signaling pathways	ERK/JNK/P2YR, JNK/P38 MAPK, STAT 1/3, NF-κB	JAK1/STAT3/c-Fos, STAT 6, Fas/FasL

RANX 1/2, runt-related transcription factor 1/2; CT-1, Cardiotrophin-1; OSM, oncostatin M; OSX, osterix; ATF 4, activating transcription factor 4; SATB 2, special AT-rich sequence-binding protein 2; AP-1, activator protein-1; IFN-γ, interferon-γ; TNF-α/β, tumor necrosis factor-α/β; MITF, microphthalmia transcription factor.

Table 2.
Key factors and signaling pathways of osteoblastogenesis or osteoclastogenesis.

3. The effects of vitamin E on bone cells

In the past, researchers have looked at the direct effects of vitamin E on bone cells. In a prior study, Soeta et al. looked at the effects of vitamin E (αTF and δT3) on osteoblast differentiation. It was demonstrated that whereas TF initially hindered osteoblast differentiation, at a later stage, osteoblast differentiation reverted to normal [7]. Ahn et al. discovered in another investigation that incubation with αTF increased the proliferation of MSCs. In cells treated with αTF, the expression of ALP and Runx-2 was dramatically upregulated. Gamma-T3 has been demonstrated to enhance osteoblastic cell proliferation, differentiation, and mineralization, as well as reduce apoptosis [11]. According to prior research, α T3's protective effects on osteoblastic cells were consistent, while TF's effects were varied. Osteoclastogenesis was inhibited by both the γT3 and αT3 isomers, with the latter showing a more significant suppression of osteoclast production and activity than the former [12].

Vitamins also affect osteoclastogenesis. Johnson showed that vitamin E administration inhibits osteoclastogenesis, potentially by preventing the development of monocytes and lymphocytes. In their study, the Ovx control group had a much higher number of osteoclasts, which were strongly inhibited by all three dosages of vitamin E, although more efficiently in the Ovx group that received 300 mg of vitamin E per kg of food. Vitamin E also reduced the increase in monocyte and lymphocyte production induced by Ovx [13]. Woon Kim discovered that tocotrienol reduced the amount of RANKL that IL-17 stimulated FLS generation. Tocotrienol reduced the activation of extracellular signal-regulated kinase, kappa B-alpha inhibitor, and mammalian target of rapamycin caused by IL-17. Osteoclasts were differentiated when monocytes were incubated with IL-17, RANKL, FLS that had been treated with IL-17, or Th17 cells, and tocotrienol inhibited this osteoclast differentiation. Tocotrienol decreased IL-17 and sRANKL synthesis as well as Th17 cell differentiation [14]. Polyethylene has been shown by Hazir et al. to increase the number of TRAP-positive cells and the expression of genes related to osteoclasts. TF therapy markedly decreased the quantity of TRAP-positive osteoclasts, bone resorption activity, antioxidant-related gene expression and mitochondrial function [15]. In Walker 256/B tumor osteolytic rats, Badraoui et al. investigated the effects of vitamin E supplementation on osteoclast resorbing activity and cytomorphometry. The results indicate that fewer osteolytic lesions were visible in W256VE. In addition, the W256VE group had less alteration of bone microarchitecture and OC activity [16]. Quercetin and vitamin E have been studied by Vakili et al. in relation to ovariectomy-induced osteoporosis.

The results showed that ovariectomy increased the total number of osteoclasts and serum osteocalcin and lowered the bone weight, bone volume, trabeculae volume, and total number of osteocytes and osteoblasts. The expression of LC3, beclin1, and caspase 3 was also upregulated in the tibia, while bcl2 expression was downregulated. Treatment with Q and vitamin E significantly improved osteoporosis by reversing these alterations [17]. In a double-blinded, randomized, placebo-controlled human trial study, vitamin E has significantly decreased the serum C-terminal telopeptide of type I collagen (CTX) [18]. These studies have indicated that vitamin E could inhibit osteoclastogenesis in the treatment of osteoporosis.

4. The molecular mechanism of vitamin E in osteoclastogenesis

4.1 The effect of vitamin E on the M-CSF and RANKL/RANK/OPG signaling pathways

The RANK/RANKL/OPG trimolecular complex system and M-CSF exert a considerable influence on osteoclast production and the regulation of bone resorption. The maturation of osteoclast precursor cells into mature osteoclasts in the presence of RANKL depends on the production of M-CSF by osteoblastic stromal cells. Other names include osteoclast differentiation factor (ODF), tumor necrosis factor-related activation-induced cytokine, and OPG ligand (TRANCE). When RANKL binds to its receptor, RANK, on myeloid cells, a series of intracellular signaling events are triggered, including interactions with TRAF6 adaptor molecules and the activation of NF-κB, nuclear factor of activated T cells cytoplasmic 1 (NFATc1), MAPK, and PI3K. It has been demonstrated that NF-κB regulates RANKL-induced osteoclast development, activating the Fos proto-oncogene (c-Fos) before NFATc1. The presence of NFATc1, which has been dubbed the "master regulator of osteoclast development," is critical for stem cells differentiate into osteoclasts. If RANKL and NF-κB p65 interact, it becomes transiently active in osteoclast precursors within an hour. Additionally, the c-Fos is also vital for osteoclastogenesis induced by M-CSF and RANKL. OPG is a protein that is largely secreted by osteoblast lineage cells, and it functions as a key endogenous regulator of the RANK/RANKL/OPG pathway. The same ligand, RANKL, has affinity for the receptors OPG and RANK as well. In order to stop osteoclastogenesis and the survival of existing osteoclasts, OPG acts as a decoy receptor by interacting with RANK. The ratio of RANKL/OPG may play a significant role in regulating bone resorption because OPG expression is normally downregulated and not as active as RANKL. Cortical and cancellous bone mass is considerably increased by OPG overexpression, while osteoclast levels are decreased.

Ambroszkiewicz et al. observed that children with cystic fibrosis (CF), the most prevalent deadly autosomal recessive genetic illness that results in a range of long-term health issues, including bone ailments, have lower average levels of fat-soluble vitamins (A, D, and E). Children with CF have significantly lower levels of the markers for bone production (osteocalcin) and bone resorption (CTX, TRACP5b) than children without the condition. In comparison with healthy individuals, CF patients had serum levels of OPG that were much lower and RANKL that were approximately two times greater. This finding suggested a link between vitamin E and RANKL or OPG in various bone diseases [19]. In another study, Kim et al. showed that RANKL synthesis was promoted by IL-17 in fibroblast-like synoviocytes (FLS) and was inhibited by tocotrienol. Tocotrienol reduced the activation of NF-kappaB, extracellular signal-regulated kinase, and mammalian target of rapamycin caused by IL-17. Osteoclasts were differentiated when monocytes were incubated with IL-17, RANKL, FLS that had been treated with IL-17, or Th17 cells, and tocotrienol inhibited this osteoclast differentiation [14]. Trolox is a water-soluble vitamin E analog that has been studied for its effects on osteoclastogenesis and RANKL signaling by Kim et al. By preventing RANKL production in osteoblasts, trolox effectively prevented interleukin-1-induced osteoclast development in bone marrow cell-osteoblast coculture. This decrease in RANKL was ascribed to a downregulation of cyclooxygenase-2 activity, which in turn resulted in less prostaglandin E [2] being produced. Trolox also reversibly reduced the production of osteoclasts by bone marrow macrophages stimulated by RANKL and M-CSF. Trolox targets early osteoclast precursors since its effectiveness depends on its presence early in the culture process. Trolox pretreatment had no effect on the early signaling pathways activated by RANKL, such as the MAPK, NF-κB, and Akt pathways. By inhibiting its translation, Trolox reduced the stimulation of the c-Fos protein by RANKL. Trolox's suppression of osteoclastogenesis in bone marrow macrophages was reversed by ectopic overexpression of c-Fos [20]. In cocultures of osteoblasts and bone marrow cells stimulated by either IL-1 or a combination of 1,25(OH)(2) vitamin D(3) and prostaglandin E, Ha et al. discovered that vitamin E inhibits osteoclastogenesis. This is supported by the finding that only tocotrienols prevented RANKL in osteoblasts. Additionally, c-Fos expression was suppressed by α-tocotrienols but not α-tocopherols, which may have been accomplished by preventing ERK and NF-κB activation. This may have prevented RANKL-induced osteoclast development from precursors. When c-Fos or an active version of NFATc1, a crucial downstream target of c-Fos during osteoclastogenesis, was overexpressed, this anti-osteoclastogenic impact was reversed [21]. Human osteoblastic SaOS2 cells were treated with wear particles from vitamin E-doped and regular UHMWPE, and the gene expression and protein synthesis of IL-6, RANKL, OPG, DKK-1, and Sclerostin were then evaluated. When compared to standard UHMWPE, vitamin E-blended UHMWPE lowered RANKL while increasing OPG [22]. According to research by Mohammed, annatto tocotrienol (AT) and self-emulsified annatto tocotrienol (SEAT) both increased the number of osteoblasts and the pace at which trabecular mineralization occurred. In OVX animals, AT also reduced the expression of skeletal sclerostin. Only SEAT significantly boosted the bone formation rate and decreased the RANKL/OPG ratio. SEAT increases bone growth, inhibits sclerostin expression, and lowers the RANKL/OPG ratio in rats with estrogen shortage, all of which may have skeletal anabolic effects [23].

Overall, the RANK/RANKL/OPG system is a crucial signaling channel involved in communication between bone cells, and there is substantial evidence that altering this signaling pathway has significant consequences for bone remodeling. While OPG inhibits osteoclast-induced bone resorption by negatively regulating RANKL binding to RANK and shortening the half-life of membranous RANKL, RANKL mediates osteoclastogenesis and activates mature osteoclasts. According to growing research, T3 therapy has the potential to protect bones by regulating this route. However, more research is required to support these conclusions.

4.2 Vitamin E suppresses inflammation in preventing bone loss

Inflammation, which is defined by the activation of immune cells and the consequent release of inflammatory cytokines, is the first line of defense against illnesses for both the innate and adaptive immune systems. The removal of damaged tissues and the beginning of tissue repair are both essential functions of inflammation. Continued inflammation, however, also promotes bone resorption and inhibits bone development. The main controllers of the inflammatory response seen in metabolic bone disorders, such as osteoporosis, are cytokines. Interferon, interleukin (IL), and tumor necrosis factor-alpha (TNF-α) are only a few of these cytokines. Previous research has shown that proinflammatory mediators can both positively and adversely affect osteoclast and osteoblast activities. Significant stimulators of bone resorption include TNF-α, IL-1, and IL-6. After TNF-α and IL-6 are recognized by their corresponding receptors on bone marrow stromal cells, the inhibition of MAPK, the activation of suppressor of mothers against decapentaplegic ubiquitylation regulatory factor 1 (SMURF1) and SMURF2, and the activation of signal transducers and activators of transcription (STAT) all work together to cause the downregulation of osteoblast gene products. Through the overexpression of Dickkopf-related protein 1 (DKK1) and sclerostin (SOST), proinflammatory cytokines block the transcription of osteogenic factors to block the Wnt/β-catenin pathway. Additionally, due of the link among inflammatory factors and their receptors, which promotes osteoclast development, proliferation, and activation, osteoblasts create more M-CSF and RANKL. TNF-α, IL-1, and IL-6 signal in osteoclasts and osteoclast precursors through the NF-κB, MAPK, and Janus kinase (JAK)-STAT pathways to upregulate osteoclast-related genes and intensify osteoclastogenesis. The downregulation of c-Fos, NFATc1, and TRAF6, as well as the subsequent impairment of the activation of downstream targets, including NF-κB and JNK, has been hypothesized as the underlying mechanisms of IFN-γ in suppressing osteoclast development. Additionally, IFN-γ increases the interaction between Fas and Fas ligands to promote osteoclast precursor death and induces osteoblasts to create nitric oxide. On the other hand, IFN-γ indirectly encourages osteoclast development by promoting T-cell activation and the release of osteoclastogenic substances, including TNF-α and RANKL, by T cells. Additionally, IFN-γ stimulates the production of DC-STAMP, which is in charge of fusing mononucleated osteoclasts into mature, functioning osteoclasts. It is speculated that inflammatory cytokines play an important role in bone homeostasis.

Several different laboratory experiments have shown how vitamin E can reduce systemic inflammation. In children with biopsy-proven NAFLD, hydroxytyrosol (HXT) and vitamin E may reduce oxidative stress, insulin resistance, and steatosis, according to Mosca. Children participating in the HXT + VitE experiment had their plasma levels of IL-6, IL-1, IL-10, TNF, 4-hydroxy-2-nonenal (4-HNE), and 8-hydroxy-2'deoxyguanosine (8-OHdG) analyzed. This study showed that both the placebo (Pla) group and the HXT + VitE group exhibited changes in indicators of systemic inflammation. Levels of IL-1 and TNF were decreased in both groups; however, IL-6 and IL-10 increased significantly only in the HXT + VitE group. Systemic inflammation in children caused by NAFLD was reduced by HXT and VitE therapy [24]. Hashem pretreated rats with vitamin E (60 mg/kg) and continued giving them vitamin E until the experiment was complete to test the hypothesis that acute pancreatitis (AP) and the regulation of the TNF-AMPK axis in the presence and absence of vitamin E are related. Infiltration of inflammatory cells and severe pancreatic tissue damage, which were significantly shielded by vitamin E, served as evidence that AP had occurred. Additionally, L-arginine injections markedly lowered phospho-AMPK, IL-10 mRNA, and protein expression, which were markedly protected by vitamin E [25]. Amevor et al. exposed 400 Tianfu breeders to 0.4 g/kg quercetin (Q) and 0.2 g/kg vitamin E, Q (0.4 g/kg) and vitamin E (0.2 g/kg) for 14 weeks to study the effects of dietary combinations of Q and vitamin E on the intestinal structure and barrier integrity in aged breeder chickens. The findings demonstrated that Q + vitamin E had synergistic effects on intestinal morphology by increasing villus height and crypt depth and by reducing inflammatory damage to the intestines of elderly chickens. Additionally, Q + vitamin E boosted the mRNA expression of intestinal tight junction proteins, such as occludin, ZO1, and claudin-1. Additionally, Q + vitamin E boosted the expression of anti-inflammatory genes (IL-10 and IL-4) while decreasing the mRNA expression of pro-inflammatory genes (TNF-α, IL-6, and IL-1) [26]. Cell-based, preclinical, and clinical intervention studies have investigated the mechanisms behind the impact of vitamin E on the immune system and inflammation. Vitamin E affects inflammatory mediators produced by other immune cells, which in turn affects the integrity of T-cell membranes, signal transduction, and cell division both directly and indirectly. Since it impacts the host's susceptibility to infectious diseases including respiratory infections as well as allergy and infectious diseases like asthma, vitamin E's control of immune function has therapeutic importance. In many responses to infection, inflammation, and immunological activation, IL-1 is a key player. In addition to osteoporosis and cancer-induced osteolysis, IL-1 has a role in the etiology of rheumatoid arthritis and osteolysis of orthopedic implants. It promotes osteoclast development and activity, which results in excessive bone resorption. Without the assistance of osteoblasts or stromal cells, IL-1 may directly encourage the development of osteoclasts. It works by activating NF-κB in osteoclasts and inhibiting apoptosis. It was shown that vitamin E has the capacity to prevent activated monocytes from producing IL-1. Both hemopoietic and nonhematopoietic cells produce IL-6 in response to diverse forms of stimulation. Under the control of parathyroid hormone, vitamin D3, growth factor, and other cytokines, stromal cells and osteoblasts release IL-6 during bone remodeling in nanomolar amounts. The production of IL-6 by osteoblasts in response to IL-1, TNF-α, and LPS stimulation has also been documented. Osteoclastogenesis would be encouraged, and NF-κB would be activated as a result of high cytokine levels.

Animals with ovariectomy-induced estrogen shortage had higher levels of IL-1 and IL-6. Treatment with palm-derived T3 at 60 mg/kg was successful in preventing cytokine levels from increasing. In addition to enhancing bone microstructure in MetS mice, palm-derived T3 and annatto-derived T3 decreased inflammatory marker levels (IL-1 and IL-6). Researchers found that annatto T3 improved trabecular and cortical microstructure, raised serum procollagen I intact N-terminal propeptide while decreasing serum carboxyl-terminal telopeptide of type I collagen, and inhibited the expression of inflammation markers (MCP-1, IL-2, IL-23, IFN-γ, and TNF-α) in a distinct investigation [27–29]. Animals treated with free radicals and nicotine were also used to clarify the unique ways that vitamin E affects inflammatory indicators. Free radicals produced by ferric nitrilotriacetate increased the levels of OCN, IL-1, and IL-6 and negatively affected bone histomorphometry. By supplementation with a palm oil T3 combination at a rate of 100 mg/kg, these unfavorable alterations were reversed. Inflammatory cytokines and markers of bone resorption were both elevated after two months of nicotine administration, whereas a marker of bone production was decreased. Vitamin E at a dose of 60 mg/kg demonstrated effectiveness in enhancing all these parameters when compared to the control group. In a different investigation, it was found that 60 mg/kg of a palm T3 combination might prevent the rise in IL-1 and IL-6 caused by nicotine administration [30]. In contrast, in experimental animal models, ultrahigh molecular weight polyethylene (VE-UHMWPE) combined with vitamin E has been shown to have higher mechanical performance and less unfavorable cellular responses than ordinary polyethylene.

	TT	TF
Inflammatory cytokines	TNF-α, IL-1β, IL-6, IL-8 PGE2, COX-2, MCP-1, 5-LOX	LTB4, exotoxin 3, TNF-α MCP-1, 5-LOX, MPO, MDA TNF-α, IL-1β, IL-6
Signaling pathway	NF-κB, IRE1α, PPAR caspase-4, Nrf2, MPK, PI3K/Akt/mTOR, Ras–Raf–MEK–ERK, JAK/STAT	JAK/STAT, NF-κB, JNK/p38, Nrf2, ERK1/2, PPAR

Table 3.
Inflammatory factor and related signaling pathways which vitamin E regulated.

Human macrophages stimulated by VE-UHMWPE particles showed a different transcriptional program from macrophages stimulated by UHMWPE particles. IL-27 was one of the upregulated genes, and it was discovered that in macrophages cultivated with VE-UHMWPE particles as opposed to those cultivated with UHMWPE particles, its levels were much higher. IL-27 decreased the inflammatory response induced by standard UHMWPE particles *in vitro* and prevented osteoclasts from differentiating [31].

When combined, the information from these previous articles showed that vitamin E had anti-inflammatory characteristics that can protect against osteoporosis (TNF-α, IL-1, IL-6, IL-27, MCP-1) (**Table 3**) [32]. Furthermore, other inflammatory cytokines, including IL-7, IL-8, IL-11, and IL-34, IL-3, IL-4, IL-10, IL-12, and IL-33, whether regulated by vitamin E in osteoclastogenesis need further research.

4.3 Vitamin E suppresses oxidative stress in preventing bone loss

Reactive oxygen species (ROS) and reactive nitrogen species are the two types of free radicals that are most frequently mentioned when discussing oxidative stress. An imbalance among these free radical oxidants and the cells' antioxidant defenses leads to cellular damage. Chain reactions affecting DNA, proteins, cell wall lipoproteins, and other biological components are triggered by damage from free radical molecules such as superoxide (O_2−). Additionally, NF-κB and other pathways are activated, which upregulate the TNF-α and IL-1 downstream cytokines. Through inhibition or activation of several cellular pathways, changes in the cell and its organelles modify cellular function. These alterations result in inflammation (through IL-4, IL-6, TNF-α, and other factors), changes to crucial cellular processes, and even mitochondria-initiated cell death and apoptosis, which, in severe cases, can result in organ damage. According to the mitochondrial theory of aging, chronic cumulative oxidative stress encourages disease conditions and early aging. Oxidative stress has been linked to a wide range of chronic diseases, including osteoporosis, in a significant body of literature ever since the mitochondrial theory of aging was first proposed [33]. The mitochondria of the cell are where ROS are typically present. An oxygen molecule (O_2) is combined with four electrons and four protons to create two water molecules during the electron transport process in the mitochondria of the cell. This transfer results in the formation of a superoxide ion (O_2^-) or, less frequently, a hydroxide (OH-) or peroxide ion (O_2^{2-}), each of which is a reactive-free radical in ROS form. Normally, the antioxidant enzymes superoxide dismutase (SOD) and peroxidase convert these ROS into the less reactive molecules hydrogen peroxide (H_2O_2) and water (H_2O). The nicotinamide adenine dinucleotide phosphate (NADP-NADPH) cycle, which neutrophils use to destroy viruses and bacteria, xanthine oxidase reactions during purine catabolism, nitric oxide synthase reactions during nitric oxide synthesis, and cyclooxygenases (COX) in the pathway for prostaglandin synthesis are some additional cellular pathways that produce ROS. Despite having the potential to cause damage, ROS and RNS are essential for healthy cell function because they play a role in cellular messaging that controls processes including gene transcription and transduction, differentiation, apoptosis, and repair. To minimize the negative effects of unchecked oxidative stress, natural cellular defenses against ROS and RNS work to preserve equilibrium rather than eliminate them. Oxidative stress is accelerated by anything that speeds up mitochondrial metabolism in the cell. Examples include fractures, surgery, and trauma injuries, infections, the detoxification of drugs or alcohol, high oxygen conditions, smoke inhalation, elevated salt concentrations, and heavy metal exposure. Oxidative stress is also heightened by radiation from radiographic, solar, cosmic, or electromagnetic sources. Even consuming foods devoid of naturally occurring antioxidants, such as processed sugars and some fatty acids, increases oxidative stress. Additional factors that may contribute to an increase in oxidative stress include dehydration, stress, anxiety, and possibly inadequate sleep. OPG and RANKL expression are both susceptible to the oxidative state, which lowers OPG expression and raises RANKL expression, tipping the scales in favor of bone loss. In response to excessive oxidative stress, sclerostin and DKK-1, two WNT pathway inhibitors, involved in the osteoimmunological regulation of bone remodeling, are increased, while OPG production is decreased. Aging, which is characterized by an increase in oxidative stress, and inflammatory mediators appear to be the two situations that most strongly promote the production of WNT pathway inhibitors. Increases in pro-oxidants have been associated with decreased osteoprogenitor development into the osteoblast cell lineage. Oxidative stress inhibits rabbit bone marrow stromal cells and calvarial osteoblasts differentiate into osteoblasts, according to research by Bai et al. Differentiation markers for this include decreased levels of ALP, type-I collagen, and nuclear phosphorylation of Runx2. OPG production is lowered as a result of decreased osteoblast activity. This decrease further modifies the overall RANKL/OPG ratio, which is essential for maintaining the balance of osteoblastic and osteoclastic activity. Oxidative stress causes an increase in apoptosis in osteoblasts and osteocytes. The cytokines for osteoblastic activity are reduced with osteocyte death, which further favors osteoclastogenesis. Additionally, osteocytes that are dead or dying induce osteoclastogenesis. Antioxidants, such as GSH, N-acetylcysteine, and alpha-lipoic acid, reduce these effects.

Given the rapidly changing potential role of oxidative stress in osteoporosis, reducing oxidative stress with an antioxidative agent may be a practical way to prevent osteoporosis. Strong free radical scavenger vitamin E has been shown to have bone-protective properties by reducing oxidative stress. Osteoblasts were protected against the negative effects of H_2O_2 by treatment with tocotrienol (1 μM) for 24 hours in a study using an H_2O_2-induced osteoblast model. In ovariectomized models, tocotrienol treatments both short-term and long-term successfully increased femur bone mineral density, bone formation, and strength. In addition, osteocalcin, bone morphogenetic protein 2 (BMP-2), and RUNX-2 mRNA expression in the tibia increased after tocotrienol treatments [34]. In an *in vitro* study that looked at the effects of tocotrienol on lipid peroxidation, antioxidant enzyme activities, and apoptosis of osteoblasts exposed to H_2O_2, tocotrienol was able to prevent malondialdehyde (MDA) elevation, decrease osteoblast apoptosis, and increase SOD, GPx, and CAT activities. It has also been shown that other natural compounds, including tocotrienols, rice protein, melatonin, and others, lower ROS by increasing heme oxygenase-1 (HO-1) activity, which may potentially counteract the negative effects of oxidative stress [35]. Nuclear respiratory factor 2(NRF2) significantly controls the transcription of these antioxidants. Under normal circumstances, NRF2 is recruited by Kelch-like ECH-associated protein 1 (KEAP1) into the Cul3-containing E3 ubiquitin ligase complex, where it is ubiquitylated and degraded. In contrast, oxidative stress inhibits the function of the E3 ubiquitin ligase, which causes NRF2 and KEAP1 to separate. The transcription of SOD, CAT, HO-1, GPxs is then activated once Nrf2 enters the nucleus and forms a heterodimer with tiny Maf proteins in antioxidant response elements (AREs), a conserved gene sequence in target gene promoter regions. By triggering the NRF2-KEAP1-ARE pathway, tocotrienols, rice protein, melatonin, and other naturally occurring substances that are NRF2 activators may reduce ROS. However, in Fujita's investigation, treatment with α-tocopherol had no effect on the proliferation of osteoclast precursors or the survival of mature osteoclasts. Instead, it increased the number of tartrate-resistant acid phosphatase (TRAP)-positive multinucleated osteoclasts. Additionally, they examined whether vitamin E's antioxidant characteristics were necessary for it to trigger the fusion of osteoclasts. None of the vitamin E isoforms, with the exception of α-tocopherol, induced osteoclast fusion [36]. This result indicated that vitamin E decreases bone mass by stimulating osteoclast fusion.

Vitamin E has shown tremendous promise in the treatment of postmenopausal osteoporosis because it is an efficient source for halting bone loss by suppressing oxidative stress.

4.4 Other mechanisms of vitamin E in preventing bone loss

Initially, a hormone mostly made by adipose tissue, leptin, regulates satiety and energy expenditure by acting on the hypothalamus. Leptin has a variety of effects, including the promotion of hemopoietic and osteoblastic differentiation, which it uses to control bone mass. Prior studies have shown that leptin has distinct effects on bone metabolism through both central and peripheral pathways. The brain's hypothalamic leptin receptors were activated in groundbreaking *in vivo* research, which showed that doing so decreased osteoblast formation and function and boosted osteoclast activity [37]. Two downstream molecular cascades are active: (a) an increase in RANKL through the PKA signaling pathway to encourage osteoclasts' effects on bone resorption; and (b) a decrease in c-myc expression, which encourages the production of cyclin D and prevents osteoblast development. An earlier study found that compared to healthy control animals fed a standard chow diet, rats on a high-carb, high-fat diet displayed symptoms of MetS, hyperleptinemia, and hypoadiponectinemia as well as damage to the trabecular microarchitecture. The increased leptin was decreased by palm and annatto T3 treatments [28, 29]. MicroRNA (miRNA) is a small, highly conserved endogenous noncoding RNA molecule with 20–22 bases that controls the expression of proteins. The differentiation of osteoclasts and miRNA expression are closely related processes. Several miRNAs inhibit osteoclast development and function by targeting the RANK/RANKL/OPG pathway. The expression of RANKL was suppressed by miR-17/20a and miR-26a. In bone marrow-derived macrophages, other miRNAs displayed suppression on osteoclast development through the TGF-β/SMAD signaling pathway. Insufficient vitamin E led to lower levels of miR-122a and miR-125b in the rat liver. These results suggest that vitamin E may regulate miRNA expression, which may be crucial for the regulation of bone homeostasis. The precise mechanism needs to be further researched [38, 39].

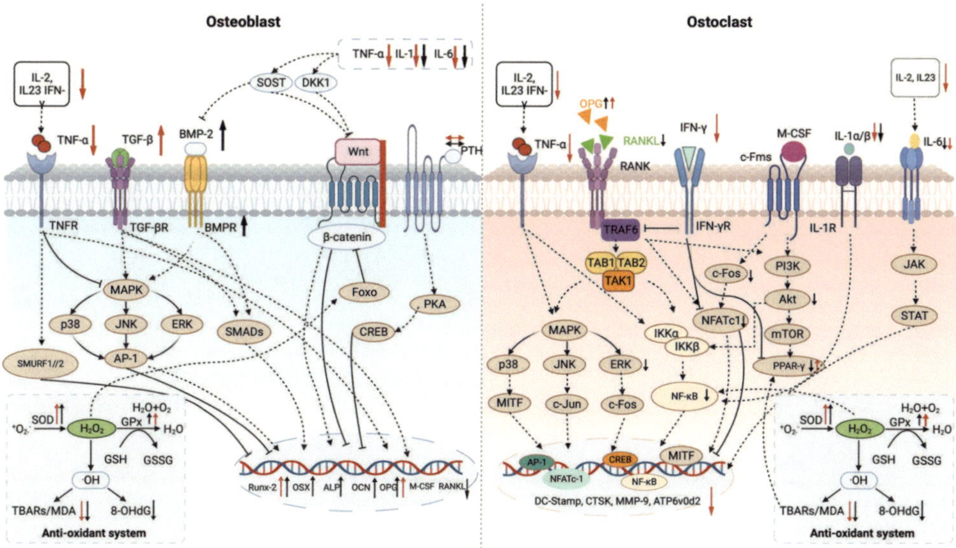

Figure 1.
The molecular mechanism of vitamin E on osteoblast or osteoclast. The effects of TF (red arrows) and TT (black arrows) on osteoblast and osteoclast are indicated.

5. Conclusion

Osteoporosis is the most prevalent orthopedic disease in elderly people. Vitamin E has shown great potential in treating osteoporosis. The close synchronization between bone resorption and formation is facilitated by the activation or inhibition of numerous downstream signaling pathways, which are controlled by endocrine and paracrine regulators. The molecular targets of vitamin E in regulating bone metabolism have been identified by numerous scientific studies (**Figure 1**). In order to inhibit osteoclasts, vitamin E altered the levels of inflammatory mediators, ROS, and hormones as well as the RANK/RANKL/OPG, NF-κB, MAPK, and oxidative stress signaling pathways. Vitamin E could be selected as a novel therapy against osteoporosis.

Acknowledgements

Chen YJ, Yu NC, and Zhou DG gathered literatures and drafted this manuscript. Li ZG, Gong FQ, Yi WJ, and Chen BT provided valuable opinions and assistance in the process of the research. GJ supervised this study and revised this manuscript. This study was supported by the National Natural Science Foundation of China (82074233), Starting Package of Xiang'an Hospital of Xiamen University (PM201809170009), National Natural Science Foundation of Xiamen, China (3502Z20227119), and Young Investigator Research Program of Xiang'an Hospital of Xiamen University (PM202103050009).

Conflict of interest

The authors declare no conflict of interest.

Author details

Chen Yongjie, Yu Naichun, Zhou Daguo, Li Zongguang, Gong Fengqing, Yi Weijiang, Chen Botao and Ji Guangrong*
Department of Orthopedics Surgery, Xiang'an Hospital of Xiamen University, School of Medicine, Xiamen University, Xiamen, China

*Address all correspondence to: jiguangrong@sina.cn

References

[1] Ensrud KE, Crandall CJ. Osteoporosis. Annals of Internal Medicine. 2017;**167**(3):Itc17-itc32

[2] Reid IR, Billington EO. Drug therapy for osteoporosis in older adults. Lancet. 2022;**399**(10329):1080-1092

[3] Song S, Guo Y, Yang Y, Fu D. Advances in pathogenesis and therapeutic strategies for osteoporosis. Pharmacology & Therapeutics. 2022;**237**:108168

[4] Anam AK, Insogna K. Update on osteoporosis screening and management. The Medical Clinics of North America. 2021;**105**(6):1117-1134

[5] Sánchez A, Blanco R. Osteonecrosis of the jaw (ONJ) and atypical femoral fracture (AFF) in an osteoporotic patient chronically treated with bisphosphonates. Osteoporosis International. 2017;**28**(3):1145-1147

[6] Khosla S, Hofbauer LC. Osteoporosis treatment: Recent developments and ongoing challenges. The Lancet Diabetes and Endocrinology. 2017;**5**(11):898-907

[7] Soeta S, Higuchi M, Yoshimura I, Itoh R, Kimura N, Aamsaki H. Effects of vitamin E on the osteoblast differentiation. The Journal of Veterinary Medical Science. 2010;**72**(7):951-957

[8] Wan Hasan WN, Abd Ghafar N, Chin KY, Ima-Nirwana S. Annatto-derived tocotrienol stimulates osteogenic activity in preosteoblastic MC3T3-E1 cells: A temporal sequential study. Drug Design, Development and Therapy. 2018;**12**:1715-1726

[9] Amarasekara DS, Kim S, Rho J. Regulation of osteoblast differentiation by cytokine networks. International Journal of Molecular Sciences. 2021;**22**(6):3288

[10] Amarasekara DS, Yun H, Kim S, Lee N, Kim H, Rho J. Regulation of osteoclast differentiation by cytokine networks. Immune Network. 2018;**18**(1):e8

[11] Abd Manan N, Mohamed N, Shuid AN. Effects of low-dose versus high-dose γ-Tocotrienol on the bone cells exposed to the hydrogen peroxide-induced oxidative stress and apoptosis. Evidence-based Complementary and Alternative Medicine. 2012;**2012**:680834

[12] Brooks R, Kalia P, Ireland DC, Beeton C, Rushton N. Direct inhibition of osteoclast formation and activity by the vitamin E isomer gamma-tocotrienol. International Journal for Vitamin and Nutrition Research. 2011;**81**(6):358-367

[13] Johnson SA, Feresin RG, Soung do Y, Elam ML, Arjmandi BH. Vitamin E suppresses ex vivo osteoclastogenesis in ovariectomized rats. Food & Function. 2016;**7**(3):1628-1633

[14] Kim KW, Kim BM, Won JY, Min HK, Lee SJ, Lee SH, et al. Tocotrienol regulates osteoclastogenesis in rheumatoid arthritis. The Korean Journal of Internal Medicine. 2021;**36**(Suppl. 1): S273-Ss82

[15] Mohamad Hazir NS, Yahaya NHM, Zawawi MSF, Damanhuri HA, Mohamed N, Alias E. Changes in metabolism and mitochondrial bioenergetics during polyethylene-induced osteoclastogenesis. International Journal of Molecular Sciences. 2022;**23**(15):8331

[16] Badraoui R, Ben-Nasr H, Amamou S, El-May MV, Rebai T. Walker

256/B malignant breast cancer cells disrupt osteoclast cytomorphometry and activity in rats: Modulation by α-tocopherol acetate. Pathology, Research and Practice. 2014;**210**(3):135-141

[17] Vakili S, Zal F, Mostafavi-Pour Z, Savardashtaki A, Koohpeyma F. Quercetin and vitamin E alleviate ovariectomy-induced osteoporosis by modulating autophagy and apoptosis in rat bone cells. Journal of Cellular Physiology. 2021;**236**(5):3495-3509

[18] Vallibhakara SA, Nakpalat K, Sophonsritsuk A, Tantitham C, Vallibhakara O. Effect of vitamin E supplement on bone turnover markers in postmenopausal osteopenic women: A double-blind, randomized, placebo-controlled trial. Nutrients. 2021;**13**(12):4266

[19] Ambroszkiewicz J, Sands D, Gajewska J, Chelchowska M, Laskowska-Klita T. Bone turnover markers, osteoprotegerin and RANKL cytokines in children with cystic fibrosis. Advances in Medical Sciences. 2013;**58**(2):338-343

[20] Lee JH, Kim HN, Yang D, Jung K, Kim HM, Kim HH, et al. Trolox prevents osteoclastogenesis by suppressing RANKL expression and signaling. The Journal of Biological Chemistry. 2009;**284**(20):13725-13734

[21] Ha H, Lee JH, Kim HN, Lee ZH. α-Tocotrienol inhibits osteoclastic bone resorption by suppressing RANKL expression and signaling and bone resorbing activity. Biochemical and Biophysical Research Communications. 2011;**406**(4):546-551

[22] Galliera E, Ragone V, Marazzi MG, Selmin F, Banci L, Corsi Romanelli MM. Vitamin E-stabilized UHMWPE: Biological response on human osteoblasts to wear debris. Clinica Chimica Acta. 2018;**486**:18-25

[23] Mohamad NV, Ima-Nirwana S, Chin KY. Self-emulsified annatto tocotrienol improves bone histomorphometric parameters in a rat model of oestrogen deficiency through suppression of skeletal sclerostin level and RANKL/OPG ratio. International Journal of Medical Sciences. 2021;**18**(16):3665-3673

[24] Mosca A, Crudele A, Smeriglio A, Braghini MR, Panera N, Comparcola D, et al. Antioxidant activity of Hydroxytyrosol and vitamin E reduces systemic inflammation in children with paediatric NAFLD. Digestive and Liver Disease. 2021;**53**(9):1154-1158

[25] Al-Hashem F, Abd Ellatif M, ShamsEldeen AM, Kamar SS, Al-Ani B, Haidara MA. Vitamin E protects against the modulation of TNF-α-AMPK axis and inhibits pancreas injury in a rat model of L-arginine-induced acute necrotising pancreatitis. Archives of Physiology and Biochemistry. 2023;**129**(1):148-156

[26] Amevor FK, Cui Z, Du X, Ning Z, Deng X, Xu D, et al. Supplementation of dietary quercetin and vitamin E promotes the intestinal structure and immune barrier integrity in aged breeder hens. Frontiers in Immunology. 2022;**13**:860889

[27] Muhammad N, Luke DA, Shuid AN, Mohamed N, Soelaiman IN. Tocotrienol supplementation in postmenopausal osteoporosis: Evidence from a laboratory study. Clinics (São Paulo, Brazil). 2013;**68**(10):1338-1343

[28] Wong SK, Chin KY, Suhaimi FH, Ahmad F, Ima-Nirwana S. The effects of vitamin E from Elaeis guineensis (oil palm) in a rat model of bone loss due

to metabolic syndrome. International Journal of Environmental Research and Public Health. 2018;**15**(9):1828

[29] Wong SK, Chin KY, Suhaimi FH, Ahmad F, Ima-Nirwana S. Exploring the potential of tocotrienol from Bixa orellana as a single agent targeting metabolic syndrome and bone loss. Bone. 2018;**116**:8-21

[30] Norazlina M, Lee PL, Lukman HI, Nazrun AS, Ima-Nirwana S. Effects of vitamin E supplementation on bone metabolism in nicotine-treated rats. Singapore Medical Journal. 2007;**48**(3):195-199

[31] Terkawi MA, Kadoya K, Takahashi D, Tian Y, Hamasaki M, Matsumae G, et al. Identification of IL-27 as potent regulator of inflammatory osteolysis associated with vitamin E-blended ultra-high molecular weight polyethylene debris of orthopedic implants. Acta Biomaterialia. 2019;**89**:242-251

[32] Ungurianu A, Zanfirescu A, Nițulescu G, Margină D. Vitamin E beyond its antioxidant label. Antioxidants (Basel). 2021;**10**(5):634

[33] Kimball JS, Johnson JP, Carlson DA. Oxidative stress and osteoporosis. The Journal of Bone and Joint Surgery. American Volume. 2021;**103**(15):1451-1461

[34] Shen CL, Klein A, Chin KY, Mo H, Tsai P, Yang RS, et al. Tocotrienols for bone health: A translational approach. Annals of the New York Academy of Sciences. 2017;**1401**(1):150-165

[35] Wang Z, Liang M, Li H, Cai L, Yang L. Rice protein exerts anti-inflammatory effect in growing and adult rats via suppressing NF-κB pathway. International Journal of Molecular Sciences. 2019;**20**(24):6164

[36] Fujita K, Iwasaki M, Ochi H, Fukuda T, Ma C, Miyamoto T, et al. Vitamin E decreases bone mass by stimulating osteoclast fusion. Nature Medicine. 2012;**18**(4):589-594

[37] Takeda S, Elefteriou F, Levasseur R, Liu X, Zhao L, Parker KL, et al. Leptin regulates bone formation via the sympathetic nervous system. Cell. 2002;**111**(3):305-317

[38] Kim K, Kim JH, Kim I, Lee J, Seong S, Park YW, et al. MicroRNA-26a regulates RANKL-induced osteoclast formation. Molecules and Cells. 2015;**38**(1):75-80

[39] Shi C, Qi J, Huang P, Jiang M, Zhou Q, Zhou H, et al. MicroRNA-17/20a inhibits glucocorticoid-induced osteoclast differentiation and function through targeting RANKL expression in osteoblast cells. Bone. 2014;**68**:67-75